Patrick Rauter

Optical transitions in silicon-based optoelectronic devices

Patrick Rauter

Optical transitions in silicon-based optoelectronic devices

Temporally and spectrally resolved studies in the mid-infrared and Terahertz region of the electromagnetic spectrum

Südwestdeutscher Verlag für Hochschulschriften

Impressum/Imprint (nur für Deutschland/ only for Germany)
Bibliografische Information der Deutschen Nationalbibliothek: Die Deutsche Nationalbibliothek verzeichnet diese Publikation in der Deutschen Nationalbibliografie; detaillierte bibliografische Daten sind im Internet über http://dnb.d-nb.de abrufbar.

Alle in diesem Buch genannten Marken und Produktnamen unterliegen warenzeichen-, marken- oder patentrechtlichem Schutz bzw. sind Warenzeichen oder eingetragene Warenzeichen der jeweiligen Inhaber. Die Wiedergabe von Marken, Produktnamen, Gebrauchsnamen, Handelsnamen, Warenbezeichnungen u.s.w. in diesem Werk berechtigt auch ohne besondere Kennzeichnung nicht zu der Annahme, dass solche Namen im Sinne der Warenzeichen- und Markenschutzgesetzgebung als frei zu betrachten wären und daher von jedermann benutzt werden dürften.

Verlag: Südwestdeutscher Verlag für Hochschulschriften Aktiengesellschaft & Co. KG
Dudweiler Landstr. 99, 66123 Saarbrücken, Deutschland
Telefon +49 681 37 20 271-1, Telefax +49 681 37 20 271-0
Email: info@svh-verlag.de
Zugl.: Linz, Johannes Kepler University, Diss., 2010

Herstellung in Deutschland:
Schaltungsdienst Lange o.H.G., Berlin
Books on Demand GmbH, Norderstedt
Reha GmbH, Saarbrücken
Amazon Distribution GmbH, Leipzig
ISBN: 978-3-8381-1800-0

Imprint (only for USA, GB)
Bibliographic information published by the Deutsche Nationalbibliothek: The Deutsche Nationalbibliothek lists this publication in the Deutsche Nationalbibliografie; detailed bibliographic data are available in the Internet at http://dnb.d-nb.de.

Any brand names and product names mentioned in this book are subject to trademark, brand or patent protection and are trademarks or registered trademarks of their respective holders. The use of brand names, product names, common names, trade names, product descriptions etc. even without a particular marking in this works is in no way to be construed to mean that such names may be regarded as unrestricted in respect of trademark and brand protection legislation and could thus be used by anyone.

Publisher: Südwestdeutscher Verlag für Hochschulschriften Aktiengesellschaft & Co. KG
Dudweiler Landstr. 99, 66123 Saarbrücken, Germany
Phone +49 681 37 20 271-1, Fax +49 681 37 20 271-0
Email: info@svh-verlag.de

Printed in the U.S.A.
Printed in the U.K. by (see last page)
ISBN: 978-3-8381-1800-0

Copyright © 2010 by the author and Südwestdeutscher Verlag für Hochschulschriften Aktiengesellschaft & Co. KG and licensors
All rights reserved. Saarbrücken 2010

Kurzfassung

Die spektralen Bereiche des mittleren Infrarot und der Terahertzstrahlung tragen eine Vielzahl von Informationen, weshalb sie für einen breiten Anwendungsbereich von Interesse sind. Des Weiteren wird die monolithische Integration von silizium-basierten optoelektronischen Komponenten als ein möglicher Ansatz zur Überwindung einer der Haupthürden bei der weiteren Verringerung der Strukturgrössen von integrierten Schaltungen gehandelt. Diese Hürde liegt in der limitierten Bandweite elektrischer Verbindungen, wobei ein möglicher Lösungsweg im Ersetzen der elektrischen Interchip-Verbindungen durch optische Pendants besteht. Es gibt demnach einen grosser Bedarf an auf Silizium basierenden Detektoren und Lasern. Eines der Konzepte zur Realisierung eines Laserbauteils im Silizium-Materialsystem ist das des Quanten-Kaskaden-Lasers, ein Bauteil das in den III-V Halbleitersystemen seit geraumer Zeit äusserst erfolgreich umgesetzt wird. Trotz der erfolgreichen Demonstration von Elektrolumineszenz in p-typ SiGe Quanten-Kaskaden-Strukturen bei einer Vielzahl von Wellenlängen ist die Umsetzung eines elektrisch gepumpten Gruppe-IV-Lasers bis jetzt ausgeblieben. Das breite wissenschaftliche Interesse an der Realisierung eines SiGe Quanten-Kaskaden-Lasers und das bisherige Unvermögen, in einer entsprechenden Struktur Populationsinversion zu erreichen, haben dazu geführt, dass umfassende Studien an Intersubbandübergängen in SiGe-Heterostrukturen und insbesondere zeitaufgelöste Experimente von ausschlaggebender Bedeutung bei der Entwicklung eines elektrisch gepumpten siliziumbasierten Lasers sind.

Der Hauptteil der voliegenden Arbeit behandelt zeitaufgelöste Photoleitungsexperimente an p-typ SiGe Quantentopfstrukturen, die auf die experimentelle Bestimmung von Intersubbandrelaxationszeiten abzielen, welche für die Umsetzung des Quanten-Kaskaden-Laser-Konzepts in diesem Materialsystem von Relevanz sind. Dabei wurden die Lochlebenszeiten mittels Photostrom-Pump-Pump-Experimenten gemessen, die mit Hilfe von ultrakurzen, von einem Freien-Elektronen-Laser generierten Laserpulsen durchgeführt wurden. Die Wellenlänge der Pikosekunden und kürzer dauernden Pulse lag im mittleren und fernen Infrarot. Mit Hilfe des ersten Laserpulses wurde eine Besetzung von angeregten Lochzuständen jenseits des Gleichgewichts erzeugt, die daraufhin von einem zweiten, zeitverzögert auf die Probe treffenden Puls ins Valenzbandkontinuum angeregt wurde. Der Photostrom, der sich aus dem Einfluss eines Pulspaares über einen Zwei-Photonen-Prozess auf die Halbleiterstruktur ergab, wurde in Abhängigkeit vom Ausmass der Zeitverzögerung zwischen den zwei Pulsen gemessen. Unter Zuhilfenahme eines Fit-Modells, dessen Relevanz durch Dichtematrixsimulationen untermauert wurde, wurden die Intersubbandlebenszeiten aus den gemessenen Pump-Pump-Kurven bestimmt.

Unter Anwendung von Photostrom-Pump-Pump-Experimenten konnte im Zuge dieser Arbeit die erste *direkte* Messung von Relaxationszeiten zwischen dem ersten angeregten schweren Lochzustand und dem Grundzustand des schweren Lochbandes mit einer Übergangsenergie über den longitudinal-optischen Phononenenergien in einem SiGe Quantentopf vorgenommen werden. Bei einer HH2-HH1 Übergangsenergie von 160 meV (HH steht für einen Zustand im schweren Lochband, Heavy-Hole), die deutlich über den Energien von longitudinal-optischen Phononen in einer SiGe Struktur liegt, wurden ultrakurze Intersubbandrelaxationszeiten um 550 fs gemessen. Demnach ist die Lochstreuung über die Wechselwirkung mit unpolaren optischen Phononen im SiGe-System ebenso effizient wie die Elektronenrelaxation mittels polar optischer Phononenstreuung in III-V Halbleitern.

Im Zuge der Bestrebung, die Auswirkung der örtlichen Diagonalisierung von Intersubbandübergängen im Terahertzbereich auf die charakterisischen Relaxationszeiten zu untersuchen, wurde eine Serie von zeitaufgelösten Messungen bei verschiedenen an die Probe gelegten Spannungen vorgenommen. Das Konzept diagonaler Laserübergänge in Quanten-Kaskaden-Strukturen stellt die vermutlich einzige Möglichkeit zur Manipulation von Intersubbandrelaxationszeiten in SiGe-Strukturen dar, um die individuellen Streuzeiten an die Voraussetzungen für Populationsinversion anzugleichen. Die Spannungsabhängigkeit von Relaxationszeiten zwischen angeregten Zuständen des leichten Lochbandes und dem Grundzustande im schweren Lochband des untersuchten Quantentopfsystems wurde ermittelt und ergab, dass die LH-HH-Relaxationszeiten (LH steht für leichtes Loch, Light-Hole) in der Grössenordnung von 10 ps mit Erhöhen der angelegten Spannung verdoppelt werden können. Diese Steigerung in den Intersubbandrelaxationszeiten wurde direkt mit einer spannungsinduzierten örtlichen Trennung zwischen dem angeregten Zustand und dem Grundzustand des untersuchten Übergangs in Verbindung gebracht. Durch Ändern der an die Struktur gelegten Spannung wurde folglich ein im Ortsraum direkter LH-HH-Übergang, dessen Energie 30 meV betrug, in einen diagonalen übergeführt. Die in dieser Arbeit vorliegenden Untersuchungen demonstrieren demnach eine Manipulation der nichtstrahlenden Relaxationszeiten einer SiGe-Heterostruktur über die angelegte Spannung, und stellen eine direkte Beobachtung des Unterschieds in der Intersubbandrelaxationszeit zwischen örtlich direkten und diagonalen Übergängen in ein und derselben Struktur dar.

Zusätzlich zu den zeitaufgelösten Experimenten an SiGe Heterostrukturen behandelt die vorliegende Arbeit den Entwurf und die umfassende Charakterisierung von spannungsdurchstimmbaren, in SiGe realisierten Zweifarb-Quantentopfdetektoren (QWIPs, für Quantum Well Infrared Photodetectors) mit konkurenzfähigen Leistungszahlen im mittleren Infrarot. Ein neuartiges Design ermöglichte das Umschalten zwischen zwei Responsivitätsmaxima eines SiGe QWIPs bei Photonenenergien von 115 meV und 260 meV durch Umpolen der angelegten Spannung. Die erzielten Responsivitäten betrugen bis zu 200 mA/W. Die Umsetzung eines weiteren Entwurfkonzepts, welches auf dem Wachstum des Detektors auf einem $Si_{0.5}Ge_{0.5}$-Pseudosubstrat beruht, ermöglichte die Einführung energetisch hoher Barrieren in die Quantentopfstruktur, die von Schichten reinen Siliziums gebildet werden. Diese Barrieren erhöhen die Wiedereinfangwahrscheinlichkeit von ins Valenzband angeregten Löchern und reduzieren damit die Verstärkung des Stromrauschens auf 0.11. Die Umsetzung dieses Konzepts zur Rauschminderung wurde ausschließlich durch die Verwendung des Pseudosubstrates mit hohem Germaniumgehalt ermöglicht, welches den Parameterbereich der möglichen Materialzusammensetzung beim Wachstum und damit die Designfreiheit stark vergrösserte. Die hier präsentierte Arbeit stellt die erste umfangreiche Charakterisierung eines SiGe Pseudosubstrat-QWIPs dar.

Zusätzlich zu der experimentellen Auseinandersetzung mit siliziumbasierten Detektoren im mittleren Infrarot behandelt die vorliegende Arbeit weiters einen neuartigen Zugang zur Herstellung von ultrasensitiven Terahertzdetektoren, so genannten BIBs (für Blocked Impurity Band Detectors). Ähnlich wie extrinsische Detektoren nutzen BIBs Übergänge zwischen Verunreinigungszuständen und dem energetisch nächsten Band des Halbleiters. Aufgrund von hohen Feldstärken im Bauteil und der resultierenden möglichen Lawinenmultiplikation von photogenerierten Ladungsträgern können BIBs als Festkörper-Photomultiplier eingesetzt werden. Die konventionelle Herstellung von QWIPs mittels Molekularstrahlepitaxie ist nun ein technologisch extrem aufwendiges Unterfangen und wegen der hohen Anforderungen an das Wachstum einer unumgänglichen ultrareinen Siliziumschicht zur Unterdrückung von Dunkelstrom nur in spezialisierten Anlagen möglich, weshalb ein Bedarf an alternativen Herstellungsmethoden besteht. Im Zuge dieser Arbeit wurde demonstriert, dass die Herstellung von konkurrenzfähigen Si:B BIBs unter völligem Verzicht auf epitxielle Verfahren möglich ist, nämlich durch die Ionenimplantation der benötigten Dotieratome in ein ultrareines, kommerzielles Siliziumsubstrat. Die optoelektronischen Eigenschaften der so erzeugten BIBs wurden in der vorliegenden Arbeit sorgfältig untersucht, wobei konkurrenzfähige Responsivitäten um 0.5 A/W bei niedrigem Dunkelstrom und einer Photonenenergie um 30 meV beobachtet wurden. Weiters wurde eine starke Lawinenverstärkung von photogenerierten Ladungsträgern mit einem Faktor bis zu 100 und demensprechende Responsivitäten bis 65 A/W beobachtet. In diesem Operationsmodus starker Photomultiplikation wies der Dunkelstrom jedoch untolerierbar hohe Werte auf. Dennoch, durch die Untersuchung des Einflusses der implantierten Dotierkonzentration auf die Leistung des BIBs in dieser Arbeit konnte die Basis für eine weiter Optimierung ionenimplantierter Si:B BIBs hin zur Einzelphotonendetektion im Terahertzbereich gelegt werden.

Abstract

The mid-infrared and terahertz region of the electromagnetic spectrum carry a wide variety of information, leading to a diverse field of applications for these spectral regimes. As further monolithic integrable silicon optoelectronics are one of the possible means of overcoming the challenge faced by integrated circuit scaling associated with the limited bandwidth of electric chip interconnects, there is a strong need for silicon-based detectors as well as laser sources. One of the concepts for implementing a lasing device in the silicon system is that of quantum cascade lasers, which has been successfully realized in III-V materials. Despite the successful demonstration of electroluminescence at various wavelengths in p-type SiGe quantum cascade devices, electrically pumped group-IV lasing has yet to be achieved. With an extensive research interest in the realization of a SiGe quantum cascade laser and due to the inability to achieve population inversion in such a device so far, the thorough study of intersubband transitions in SiGe heterostructures and in particular of intersubband relaxation times has become a crucial issue on the road to an electrically pumped group-IV laser.

The main part of this work presents time-resolved photocurrent experiments on p-type SiGe quantum well structures, aiming at the determination of intersubband relaxation times of relevance for pursuing the quantum cascade concept in this material system. The hole lifetimes were measured by photocurrent pump-pump experiments using a free-electron-laser for providing ultrashort pulses in the picosecond to sub-picosecond regime at wavelengths in the mid- to far-infrared. By establishing a non-equilibrium occupation of an excited hole state by one laser pulse and probing this occupation into the valence band continuum by a second pulse delayed from the first, the photocurrent through the individual samples originating from a two-photon process was measured as a function of the delay between the two pulses. By using a fit model, whose accuracy was confirmed by density matrix simulation, the intersubband relaxation times were extracted from the pump-pump characteristics.

By employing photocurrent pump-pump experiments in the course of this work, the first *direct* measurement of the relaxation time between an excited heavy-hole (HH) state and the HH ground state with a transition energy above longitudinal optical (LO) phonon energies in a SiGe heterostructure could be reported. For a HH2-HH1 transition energy of 160 meV, an ultrafast intersubband relaxation time of about 550 fs was determined. The hole scattering by non-polar optical phonons in SiGe structures was thus found to be of similar efficiency as the electron relaxation by polar optical phonons in III-V systems.

In an effort to study the effect of diagonal transitions on intersubband relaxation times for transition energies in the terahertz regime, time-resolved experiments were carried out for a series of voltages applied to p-type SiGe quantum well structures. The concept of diagonal lasing transitions in a quantum cascade device is probably the only means of manipulating intersubband relaxation times according to the requirements for population inversion in the SiGe system. The bias-dependence of the relaxation times between the excited light-hole (LH) state and the HH ground state of the investigated quantum well structures was determined, where the decay times of the order of ten picoseconds could be increased by a factor of two when increasing the voltage applied to the devices. This increase in intersubband relaxation times was directly associated with the voltage-induced spatial separation between the HH ground state and the excited LH state, whose occupation was monitored during the experiment. Thus, by altering the voltage applied to the structure, the nature of the LH-HH transition with a transition energy of 30 meV was changed from spatially direct to diagonal. The studies presented in this work therefore demonstrated a voltage-tuning of hole intersubband relaxation times in a SiGe heterostructure and a direct determination of the difference in the non-radiative relaxation time between a spatially direct and a diagonal transition within one and the same structure.

In addition to the time-resolved experiments, this thesis includes work on the design and thorough characterization of voltage-tunable two-color SiGe quantum well infrared photodetectors (QWIPs) featuring competitive figures of merit in the mid-infrared region. By employing a novel design, the peak responsivity of a SiGe QWIP could be switched between 115 meV and 260 meV while featuring responsivities up to 200 mA/W. By utilizing a second novel design concept based on the growth of a QWIP structure on a $Si_{0.5}Ge_{0.5}$ pseudosubstrate, layers composed of pure silicon could be introduced into the heterostructure. These Si layers form barriers in the valence band, leading to an increase in the re-trapping probability of holes excited into the continuum, and thus reduce the noise gain significantly to 0.11. The implementation of this concept for the reduction of noise was only made possible by employing a pseudosubstrate of high germanium concentration and by the resulting tremendous increase in the compositional parameter space available for growth. The work presented in this thesis constitutes the first thorough characterization of a pseudosubstrate SiGe QWIP.

Apart from the work on silicon-based detectors for the mid-infrared region, this thesis covers a novel approach to the fabrication of ultra-sensitive detectors operating in the terahertz regime, so-called blocked impurity band detectors (BIBs). Similar to extrinsic detectors, the detection process in BIBs involves transitions between impurity sites and the associated band. Due to the high fields present in BIB devices and the resulting possible avalanche multiplication of photogenerated charge carriers, BIBs can even be employed as solid state photomultipliers. However, the conventional fabrication of BIBs by molecular beam epitaxy is technologically extremely cumbersome due to the high requirements associated with the growth of an ultra-pure silicon layer, which is required in order to suppress dark-current, and alternative means of fabrication are needed. In the course of this work it could be demonstrated, that competitive Si:B BIBs can be fabricated by completely non-epitaxial means, namely by ion-implantation of the required impurities into an ultra-pure, commercially available silicon substrate. The optoelectronic performance of the ion-implanted Si:B BIBs was thoroughly characterized, where competitive responsivities of 0.5 A/W were observed at low dark-current and at a photon energy of 30 meV. In addition, strong avalanche gain of up to 100 and associated responsivies of up to 65 A/W were observed in a regime, where dark-current was unacceptably high for proper BIB operation. However, by concluding on the dependence of the sample performance on the featured impurity concentration, the basis for future optimization of the already competitive operation of ion-implanted Si:BIBs promising for reaching terahertz single photon detection is formed by this work.

Contents

1	**SiGe quantum well infrared photodetectors**		**1**
2	**SiGe QWIPs:**		
	Extending the tunability range		**5**
	2.1 Design and fabrication		5
		2.1.1 Structure design and growth	5
		2.1.2 Sample processing	8
	2.2 Photocurrent experiments		8
		2.2.1 Current-voltage characteristics	9
		2.2.2 Photocurrent spectra	14
	2.3 Summary and conclusion		26
3	**Pseudosubstrate QWIP**		**29**
	3.1 Design and fabrication		29
		3.1.1 Concept theory	29
		3.1.2 Structure design and growth	32
		3.1.3 Sample processing	34
	3.2 Photocurrent experiments		35
		3.2.1 Current-voltage characteristics	35
		3.2.2 Photocurrent spectra	38
	3.3 Noise measurements		46
	3.4 Figures of merit and discussion		48
	3.5 Summary and conclusion		49

4 BIB detectors: Introduction and theory — 51
4.1 Introduction — 51
4.2 Theoretical background — 53

5 Ion-implanted blocked impurity band detectors — 59
5.1 Alternative fabrication strategies — 59
5.2 Ion implantation — 62
5.3 Processing and backside etching — 64
5.4 Photocurrent experiments — 72
5.4.1 Dark-current characteristics — 72
5.4.2 Photocurrent spectra — 78
5.4.3 I-V characteristics under illumination — 79
5.4.4 Responsivity — 82
5.5 Simulation of figures of merit — 92
5.5.1 Analytical model — 92
5.5.2 Simulation results — 98
5.6 Summary and conclusion — 115

6 Intersubband relaxation times: An introduction — 119
6.1 Intersubband silicon optoelectronics — 119
6.1.1 Challenges and motivation for the Si system — 119
6.1.2 The quantum cascade laser: An intersubband device — 120
6.1.3 Potential advantages of SiGe over III-V materials — 129
6.1.4 Progress on the road to a SiGe QCL — 134
6.2 Intersubband relaxation processes in SiGe — 138
6.2.1 Non-polar optical phonon scattering — 138
6.2.2 Acoustic phonon scattering — 147
6.2.3 Alloy scattering — 151
6.2.4 Interface roughness scattering — 156
6.2.5 Lifetime limiting processes in SiGe structures — 160
6.2.6 Disadvantage of non-polar scattering for QCL development — 163

	6.2.7	Overlap integrals and selection rules	164
6.3		Measurement of intersubband relaxation times in SiGe	166
	6.3.1	Pump-probe technique	167
	6.3.2	Below the optical phonon energy: Experimental results so far	172
	6.3.3	Above the optical phonon energy: Experimental results so far	175
	6.3.4	Missing information on intersubband relaxation times	178

7 Monitoring the intersubband relaxation by LO phonons in SiGe — 181

7.1		Design and fabrication	181
	7.1.1	Structure design and growth	181
	7.1.2	Sample processing	184
7.2		Selection rules	185
7.3		Spectral characterization of the HH2-HH1 transition	188
	7.3.1	Voltage modulated waveguide transmission	188
	7.3.2	FEL photocurrent spectra based on two-photon-absorption	191
7.4		Principle of photocurrent pump-pump experiments	192
7.5		Experimental setup for pump-pump experiments	195
	7.5.1	The free electron laser source	195
	7.5.2	Pump-pump setup	198
7.6		FEL experiments	202
	7.6.1	Power dependence	202
	7.6.2	Role of the FEL intensity for the experimental time resolution	204
	7.6.3	Photocurrent pump-pump experiments: Results	207
	7.6.4	Photocurrent pump-pump experiments: Interpretation	210
7.7		Density matrix simulations	218
	7.7.1	Density matrix basics	218
	7.7.2	Model	220
	7.7.3	Simulation results for the FEL power dependence of the photocurrent	227
	7.7.4	Simulation results for the pump-pump characteristics	232
7.8		Summary and conclusion	238

8	Bias-tuning of intersubband relaxation times	241
	8.1 Design and fabrication	243
	8.1.1 Structure design and growth	243
	8.1.2 Sample processing	245
	8.2 Band structure simulations for T037 and T038	247
	8.2.1 Band structure results	247
	8.2.2 Selection rules	252
	8.3 Spectral characterization of the HH1-LH1 transition	254
	8.4 Pump-pump setup	258
	8.5 Time-resolved experiments	259
	8.5.1 Photocurrent pump-pump experiments: Results	260
	8.5.2 Photocurrent pump-pump experiments: Interpretation	264
	8.5.3 Relaxation time extraction	270
	8.5.4 Interpretation of the relaxation time tuning	288
	8.6 Summary and conclusions	293
9	Conclusions and outlook	297
	Bibliography	I
	List of figures	XI
	List of publications	XI
	Curriculum vitae	XVIII
	Awards	XX
	Acknowledgements	XXI

Chapter 1

SiGe quantum well infrared photodetectors

The mid-infrared region of the electromagnetic spectrum carries a huge variety of information. Black-body radiators of a temperature around 300 K emit in the mid-infrared, opening a diverse field of applications to detectors in this spectral region, ranging from breast cancer diagnostics to night vision. Vibrational modes of molecules result in absorption lines in the mid-infrared, which can be used as fingerprints to determine the composition of gases for purposes ranging from pollutant detection to astronomical research.

One possible means of realizing detectors in the infrared region are intersubband transitions in semiconductor heterostructures. Nanostructures built up by layers of compound semiconductors define potential wells in the valence and/or conduction band. At low temperatures charge carriers are trapped in this potential landscape, where the required energy to excite the carriers into the continuum and to produce a photocurrent depends on the band offset of the respective material combinations. Material systems, in which so-called quantum well infrared photodetectors (QWIPs) are conventionally realized, are GaAs/AlGaAs, InGaAs/InAlAs, InSb/InAsSb, InAs/GaInSb. The epitaxial control of the growth of QWIPs on a monolayer scale and within a large compositional range introduces an enormous design freedom and allows the optimization and tailoring of these devices by bandstructure engineering. For GaAs/AlGaAs, QWIP production is very well developed, large focal plane arrays can be produced at a high yield and low costs. Nevertheless, QWIPs have attracted a considerable amount of research interest, aiming at increasing the detectors' absorption and operation temperature [1], at understanding noise mechanisms [2] or at realizing QWIPs in alternative material systems [3].

Apart from the interest in QWIPs from a device-optimizing point of view, QWIPs continue to serve as model systems for understanding basic mechanisms of intersubband transitions and carrier transport in heterostructures, both coherent and incoherent. The results constitute a foundation for the detailed insight into the operation of more complex heterostructure devices such as quantum cascade lasers (QCLs). In recent years, special attention was drawn to the time-resolved study of the behavior of QWIPs regarding photocurrent transport [4, 5] and intersubband phase and energy relaxation times [6, 7]. Furthermore, due to the high electrical bandwidth of QWIPs, they are being employed in quadratic autocorrelation experiments based on two-photon absorption processes involving two bound quantum well states [8–10].

As already stated above, at present the mass production of QWIPs concentrates on the III-V composite semiconductors. However, the group-IV system promises the intrinsic advantages of CMOS compatibility, cheap raw materials and the most developed processing technology of all semiconductor systems. As a result, optical transitions in SiGe quantum well structures have recently attracted extensive research interest in the attempt to pave the way to integrable silicon optoelectronics and electrically pumped group-IV laser sources [11, 12]. While the indirect nature of the silicon band structure prevents its use for interband optics, intersubband transitions in the valence band hold the potential to form the base of group-IV optoelectronics [13, 14]. The structures used to study the properties of the SiGe system with respect to its suitability for intersubband detectors and emitters are composed of pseudomorphic $Si_{1-x}Ge_x$ layers commonly grown by molecular beam epitaxy (MBE). In [15] a competitive SiGe QWIP was demonstrated, while in [16] the use of SiGe heterostructures as voltage-tunable two-color mid-infrared detectors based on intersubband transitions was reported.

Chapter 2 of this thesis presents the work on extending the range of the voltage tunability of a two-color detector down to lower photon energies by means of bandstructure design. The design concept is introduced alongside with bandstructure calculations. Following a short description of the sample processing steps, a thorough characterization of the device by photocurrent spectroscopy is given.

SiGe intersubband structures are conventionally grown on Si wafers, where limitations regarding the Ge content are imposed on the pseudomorphic growth by the accumulation of strain. However, as recently demonstrated, the use of pseudosubstrates of a significant Ge content can overcome these limitations [17, 18]. For a correctly strain-balanced structure matched to the pseudosubstrate, the pseudomorphic growth of pure Si on a layer of ultra-high Ge concentration is feasible. Recent work on quantum well structures grown on $Si_{0.5}Ge_{0.5}$ pseudosubstrates demonstrated the excellent predictability of the absorption behavior of such structures by six-band $\mathbf{k}\cdot\mathbf{p}$ calculations [17, 19], but up to now no detailed characterization of any QWIP grown on a SiGe pseudosubstrate has been reported. SiGe pseudosubstrates offer an increase in design freedom of SiGe QWIP devices and thus the possibility to improve their figures of merits and extend their fields of application by tailoring their response spectra.

Chapter 3 demonstrates the fabrication and characterization of a SiGe QWIP grown on a $Si_{0.5}Ge_{0.5}$ pseudosubstrate, where the implementation of novel design concepts is made feasible by the extension of the compositional range introduced by the pseudosubstrate. A thorough analysis of the device's figures of merit is given.

Chapter 2

SiGe QWIPs: Extending the tunability range

The chapter at hand presents work on a voltage-tunable two-color SiGe quantum well infrared photodetector, whose design is based on that of a similar device H019 presented in [16]. The modifications applied to the design of H019 aim at extending the detection range of the device towards lower photon energies. The description of the structural design and fabrication of the presented SiGe heterostructure device is followed by a thorough characterization of its optoelectronic properties.

2.1 Design and fabrication

2.1.1 Structure design and growth

The design of sample K091 was based on the SiGe structure H019, on which competitive, voltage-tunable two-color detectivity was reported in [16]. H019 exhibits a strong absorption line around 160 meV, which originates from an optical transition between the heavy-hole (HH) ground state (HH1) and the first excited HH state (HH2) [20]. The transmission characteristic of H019 is shown in Fig. 2.1. It was measured in wave-guide geometry for incoming infrared radiation of two different polarizations. The lines in Fig. 2.1 represent the relative intensity of the transmitted radiation, that means the transmitted intensity in transversal magnetic (TM) polarization normalized to that in transversal electric (TE) polarization. The thin curve represents the simulated relative transmission. The strong dip in transmission seen at 160 meV lacks a counterpart in the photoresponse [16]. Put differently, even though the HH1-HH2 absorption is strong for TM polarized radiation, H019 does not show any responsivity at the respective 160 meV. This is due to the fact, that the HH2 state is

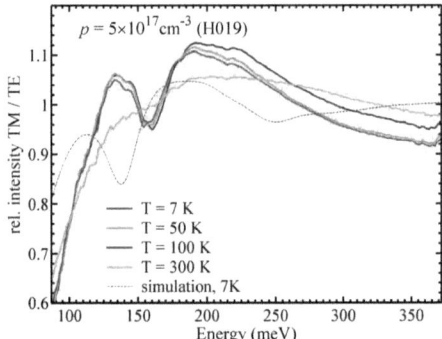

Figure 2.1: Relative transmitted intensity (transmitted intensity in TM polarization over transmitted intensity in TE polarization) as a function of photon energy for sample H019 at different temperatures, taken from [20], p. 43, and measured by T. Fromherz and S. Griesser. The thin dotted blue line represents the simulated transmission curve calculated as described in [20], pp. 17. H019 exhibits strong absorption for TM polarized radiation around 160 meV.

bound and therefore unsuitable for generating a photocurrent. The concept for the design of K091 was to utilize the excited HH2 state as a photoconductive level by introducing an additional, shallow well in order to enable the tunnelling of HH2 carriers into the continuum.

Structure K091 was grown pseudomorphically on a Si substrate by C. Falub and D. Grützmacher at the Paul Scherrer Institut in Villigen, Switzerland, using low temperature ($\sim$ 350°C) molecular beam epitaxy. Directly on the (100) substrate a highly p-type (boron) doped contact layer (doping concentration $2 \cdot 10^{18} \mathrm{cm}^{-3}$) was grown. This bottom contact layer was followed by ten periods of the sequence given in table 2.1. The values last in the table are those closest to the substrate. For comparison, the growth sequence of sample H019 is given along with that of K091. Finally, another highly doped (doping concentration again $2 \cdot 10^{18} \mathrm{cm}^{-3}$) Si layer was grown to act as top contact. The crucial structural difference between H019 and K091 is given by the introduction of two additional layers (3.5 nm of $Si_{0.93}Ge_{0.07}$ and 1 nm of Si), as seen from table 2.1. Further, the Ge concentration of the layer richest in Ge was increased to 43%. Additionally, both samples differ in the doping strategy. K091 is modulation-doped in the spacer layers separating subsequent well regions, while for H019 the well regions were constantly doped leaving the spacer layer undoped.

Figure 2.2 demonstrates the influence of the structural differences on the band offsets of

Table 2.1: MBE growth sequence and doping concentration of K091 and H019 [16]:

	K091			H019	
Layer thickness	Ge concentration	Doping concentration	Layer thickness	Ge concentration	Doping concentration
16 nm	0	0	16 nm	0	0
7 nm	0	$8 \cdot 10^{17} \text{cm}^{-3}$	7 nm	0	0
2 nm	0	0	2 nm	0	0
3.5 nm	7 %	0			
1 nm	0	0			
4.5 nm	28 %	0	3.5 nm	28 %	$5 \cdot 10^{17} \text{cm}^{-3}$
1.5 nm	0	0	2.5 nm	0	$5 \cdot 10^{17} \text{cm}^{-3}$
3 nm	34 %	0	2.3 nm	37 %	$5 \cdot 10^{17} \text{cm}^{-3}$
1.5 nm	0	0	2.5 nm	0	$5 \cdot 10^{17} \text{cm}^{-3}$
3 nm	36 %	0	2.4 nm	40 %	$5 \cdot 10^{17} \text{cm}^{-3}$
1.5 nm	0	0	2.5 nm	0	$5 \cdot 10^{17} \text{cm}^{-3}$
3 nm	38 %	0	2.6 nm	42 %	$5 \cdot 10^{17} \text{cm}^{-3}$
1.5 nm	0	0	3 nm	0	$5 \cdot 10^{17} \text{cm}^{-3}$
2.7 nm	43 %	0	3.9 nm	42 %	$5 \cdot 10^{17} \text{cm}^{-3}$
2 nm	0	0	2 nm	0	0
7 nm	0	$2 \cdot 10^{17} \text{cm}^{-3}$	7 nm	0	0
16 nm	0	0	16 nm	0	0

K091 and H019. The valence band offsets have been calculated according to [21], as described in detail in [20], pp. 13. In the envelope-function-approach the heterostructure of H019 forms five quantum wells of different depth, while the band edge profile of K091 features an additional, very shallow well (W6). This additional well shall, as mentioned above, enable the coupling of the HH2 state of W5 to continuum states and utilize this state as a base of high responsivity for TM polarized radiation at low wavenumbers. The modulation doping of the spacer layers of K091 causes strong band bending of the spacer adjacent to W6, pushing W6 up in energy (for negative charges) and therefore supporting tunnelling at lower voltages. Generally, the barriers between adjacent wells are thinner for K091 (1.5 nm) than for H019 (2.5 nm), resulting in more efficient charge carrier transfer between W1 and W5 when changing the applied voltage.

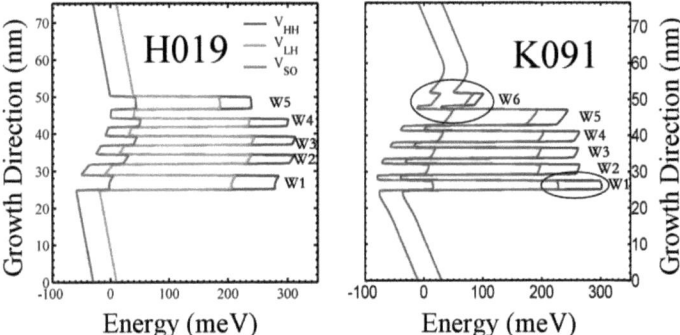

Figure 2.2: Valence band edges of the heavy- (blue), light- (green), and split-off(red)-bands for H019 and K091 at zero applied bias. The band bending is due to charge carrier redistribution. The low values on the growth direction scale are nearer to the substrate. The main differences between the two samples' respective band edges characteristics are highlighted.

2.1.2 Sample processing

In order to perform photocurrent experiments, which require to bias the structure and measure the vertical current through the sample caused by photoexcitation, mesas of different sizes were etched into the samples and contacted on bottom and top by an Al:Si metallization. The processing steps are given in table 2.1.2. The use of a sandwich top contact consisting of aluminum and silicon in its maximum solubility concentration prevents the spiking through the active quantum well region during the thermal processing step 23 and thus the formation of electric shortcuts. The mesa etching depth in step 6 was chosen in a way to penetrate both the top contact layer and the active structure (combined thickness of 867 nm), and to hit the highly doped bottom contact layer. The device fabrication sequence was followed by polishing steps, where a waveguide was created along with a side facette. This facette is used to couple radiation into the device and allows illumination of the active structure by both TM and TE polarized radiation.

2.2 Photocurrent experiments

The experimental results presented in this section were obtained for a mesa of 400×400 μm^2. The mesa was completely covered by its top contact metallization, preventing direct illumination from the front side. The active structure could be illuminated via a polished 30° side facette in waveguide geometry, or alternatively from the sample backside under normal incidence.

2.2.1 Current-voltage characteristics

Dark-current characteristics

Along with a detector's response to incoming radiation, its dark-current characteristics crucially determine the quality of its performance. This is due to the fact that the noise generated by the dark-current through the device ultimately limits the measurable photocurrent signal. For K091, the dark-current characteristics were measured by placing the sample in a liquid nitrogen vessel and performing a voltage sweep, where a Keithley 236 Source Measure Unit was employed. The result is shown in Fig. 2.3 as a black line. For positive voltages applied

Table 2.2: Sample processing steps for K091

Step No.	Process	Equipment	Parameters
1	cleaning	acetone, methanol	
2	resist deposition	resist 1818, spinner	40 s, 4000/s
3	softbake	oven	90°C, 15 min
4	photolithography mesa	mask aligner	9 s exposition
5	developing	developer	1 min
6	mesa etching	reactive-ion-etcher	100% SF_6, 50% O_2, 40 mT, 15% RF, 4 min, 930 nm
7	cleaning	acetone, methanol	
8	resist deposition	resist 1818, spinner	40 s, 4000/s
9	softbake	oven	90°C, 15 min
10	photolithography bottom contact	mask aligner	9 s exposition
11	developing	developer	1 min
12	native oxide removal	hydrofluoric acid	
13	vapor deposition	evaporation chamber, Al target	200 nm Al
14	lift-off	acetone	
15	cleaning	acetone, methanol	
16	resist deposition	resist 1818, spinner	40 s, 4000/s

Step No.	Process	Equipment	Parameters
17	softbake	oven	90°C, 15 min
18	photolithography top contact	mask aligner	9 s exposition
19	developing	developer	1 min
20	native oxide removal	hydrofluoric acid	
21	vapor deposition	evaporation chamber, Al, Si targets	20 nm Al, 2nm Si, 20 nm Al, 2nm Si, 20 nm Al, 2nm Si, 40 nm Al
22	lift-off	acetone	
23	contact alloying	rapid thermal annealing oven	380°C, 20 sec $N_2 + H_2$
24	cleaning	acetone, methanol	
25	resist deposition	resist 1818, spinner	40 s, 4000/s
26	softbake	oven	90°C, 15 min
27	photolithography gold contact	mask aligner	9 s exposition
28	developing	developer	1 min
29	vapor deposition	evaporation chamber, Ti, Au targets	10 nm Ti 100 nm Au
30	lift-off	acetone	
31	backside polishing	polishing wheel paper: 4000, 1000 diamond spray: 1 μm, 0.25 μm	
32	facette polishing	polishing wheel paper: 4000, 1000 diamond spray: 1 μm, 0.25 μm	30° to (100) surface
33	bonding	bonder, Au wire	70°C force: 3.5, 5 time: 7, 8 power: 9.1, 8.1

to the top contact, the dark-current rises slowly and exhibits a slight kink around 3 V. For negative bias, the dark-current rises strongly with the applied voltage. The reason for this asymmetric behavior lies in the asymmetric compositional structure of the sample.

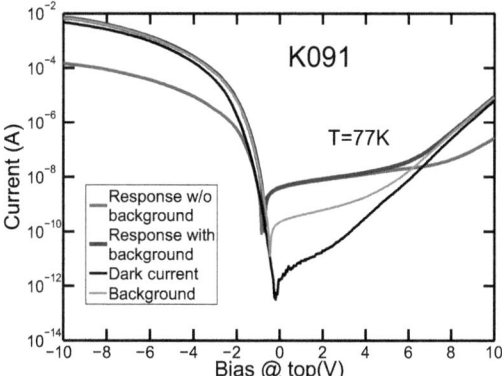

Figure 2.3: Current-voltage characteristics of K091. The black curve represents the sample's dark-current measured in a dark liquid nitrogen vessel under conditions of negligible background radiation. The green and blue curves show the total current through the sample in presence of the radiation background and an additional blackbody radiation at 550°C, respectively. The red curve presents the difference between the green and blue curves.

Band structure calculations

Band structure calculations enable a deep understanding of our samples' electro-optical properties. They were performed as described in references [21, 22]. Figure 2.4 shows the simulation results for the HH, light-hole (LH) and split-off (SO) band edges of one structural period as solid lines (blue, green and red, respectively) for two applied voltages (electric fields) of -3 V (-39 kV/cm) and +3 V (+39 kV/cm). The plots present the band offsets originating from the compositional differences between subsequent layers and their respective strain as a function of the structure's coordinate along the growth direction, where low values are closer to the substrate than high ones. The band edges were calculated self-consistently, including band bending due to the redistribution of charge carriers. The one-dimensional confinement generated by the band offsets results in quantized energy levels, which are shown as a contour plot. The calculation of the energetic states was performed in a $\boldsymbol{k.p}$ approach including six bands. At finite in-plane wave vectors, mixing of the different hole bands occurs. Accounting for the finite lifetimes of carriers in the individual levels, the absolute squared wave functions resulting from the six-band calculations were broadened by 7 meV, before their summed absolute square was plotted as a function of their energetic position relative to the band edges. The resulting plot reflects the spatially resolved density-of-states along the growth direction associated with the potential landscape induced by the heterostructure. The straight red line indicates the quasi-Fermi-level of one period of the structure, calculated at liquid nitrogen temperature.

For an applied voltage of + 3 V, the quasi-Fermi-level implies that all holes are confined in the HH1 ground state of the deepest well W1. Thus, for generating thermionic dark-current at a temperature of 77 K, a calculated activation energy of about 240 meV has to be overcome. The HH2 state of W1 is strongly confined. Charge carriers thermally excited into this state cannot contribute to a photocurrent.

At a bias of -3 V, the HH1 level of the shallow well W5 forms the ground state of each period. The HH2 state of W5 is weakly confined. As for increasing negative bias tunneling from the HH2 state of W5 into the continuum via W6 becomes increasingly efficient, the activation energy for dark-current generated by thermally assisted tunneling reduces to 100 meV. This partly accounts for the steep rise of the dark-current for negative voltages, which is thus a direct result of the design concept aiming at enabling HH2 carriers in W5 to tunnel into the continuum via the shallow additional well W6. However, even under consideration of the increasing tunnelling efficiency with rising bias, the strong dark-current rise for negative voltages is not fully accounted for. Possible explanations for the unexpectedly strong dark-current at negative bias are given in section 2.2.2.

Photocurrent characteristics

The photocurrent response of K091 to radiation from a black-body radiator at 550 °C was measured in a helium cryostat with ZnSe windows by illuminating the mesa under investigation from the polished sample backside. The black-body radiation was collimated by a CaF_2 lens of $f = 101$ mm focal length, where both the sample and the center of the radiator were placed at a distance of $2f$ from the lens. The Keithley 236 SMU was used to obtain the I-V characteristics. Further, the photocurrent generated by background radiation in absence of the black-body radiator was measured. The results of these measurements are given in Fig. 2.3.

The blue curve represents the experimental results as measured by the SMU under illumination by the black-body radiator, where the total current is built up by the contributions of the dark-current and the response to the black-body radiator. The sample's response to the background, including the dark-current, is given by the green curve. The red curve shows the difference between these two measurements, thus representing the current response to the black-body radiation without the dark- and background currents. This curve can be used

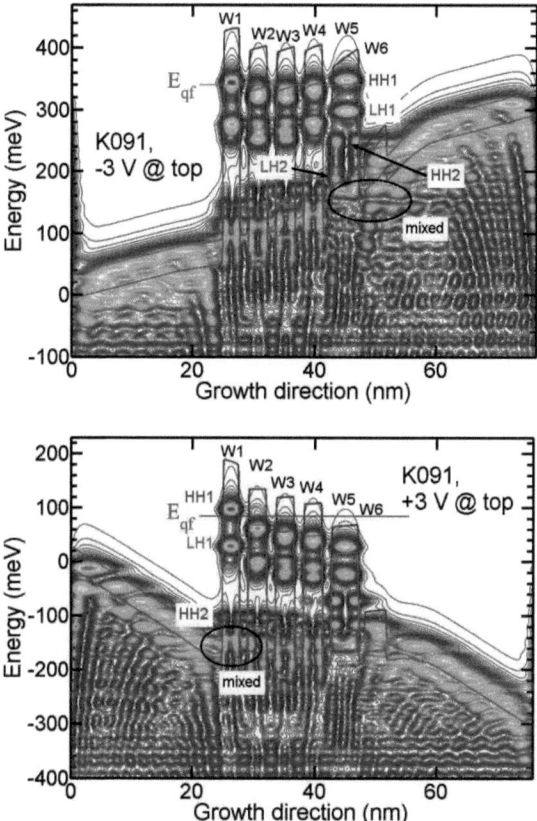

Figure 2.4: Dependence of the band structure of K091 on the applied voltage. The plots show the calculated heavy-, light- and split-off- (blue, green, red) hole band edges for voltages of ±3 V and a contour plot of the respective absolute squared wave functions. The spreading of the eigenstates along the energy-axis accounts for their homogenous broadening. For positive applied voltages, the charge carriers are located in W1. When changing to negative bias, holes are transferred to W5 via the intermediate wells W2-W4. The labels indicate the character of the excited states (HH, LH, mixed). The straight red line indicates the quasi-Fermi-level as predicted by the simulation.

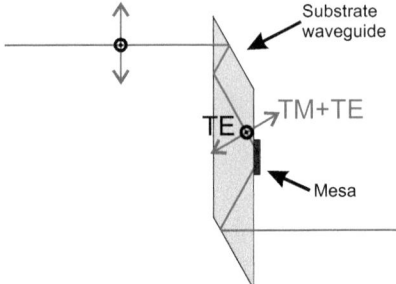

Figure 2.5: Scheme of the waveguide coupling geometry employed for illuminating the investigated mesa of K091. By adjusting a polarizer in front of the cryostat window, the mesa can either be illuminated by purely TE polarized radiation (black field vector out of the drawing plane), or by radiation of mixed TE+TM polarization (red field vector).

to calibrate the sample's photocurrent spectra, as presented in the next section, in order to determine the responsivity of K091. The red curve exhibits a behavior similar to that of the dark-current, with a slow rise with positive applied bias, and a steep increase for negative voltages. However, a detailed discussion of the I-V characteristics gained under illumination requires the information provided by spectroscopic experiments, and is therefore given in the following section.

2.2.2 Photocurrent spectra

Setup

The spectral shape of K091's response to mid-infrared radiation was determined using a Bruker IFS 66 Fourier transform interferometer (FTIR). The light of a globar source passed the interferometer, was condensed by an ellipsoid mirror and was coupled into the sample waveguide via a side facette. The waveguide setup is illustrated in Fig. 2.5. The sample was placed in a helium cryostat with ZnSe windows. Positioning a polarizer in front of the cryostat window, the active structure could be either illuminated by radiation of purely transverse electric (TE, electric field vector parallel to the sample surface, black vector out of the drawing plane in Fig. 2.5) polarization, or by transverse magnetically (TM, electric field parallel to the growth direction, red vector in Fig. 2.5) polarized radiation with a small TE component. Considering the side facette's angle of 30° to the sample surface, the TM:TE ratio of the radiation entering the waveguide was 3:1 for the polarizer set to TM mode.

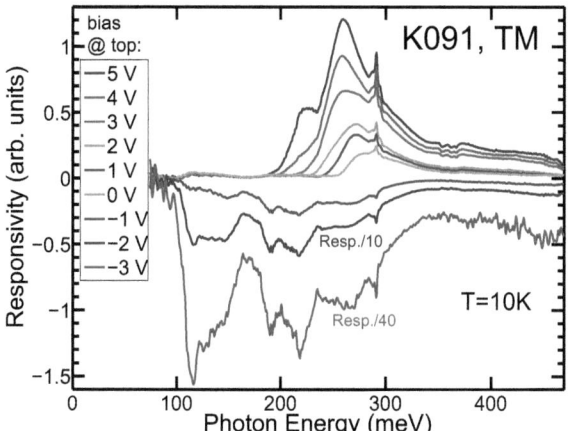

Figure 2.6: Voltage dependence of the responsivity of K091 for TM polarized radiation at a sample temperature of 10 K. Note the switching of the responsivity peak over a very wide range from 115 to 260 meV when reversing the sign of the applied voltage. The height of the peak at 115 meV exhibits an extremely strong dependence on the magnitude of the bias, for it is generated by a tunnelling process from W5 (see Fig. 2.4).

However, at the position of the active structure, the electric field of the TE component was suppressed by the gold metallization covering the mesa, further increasing the TM:TE ratio. Thus the structure was hit by highly TM or TE polarized light for the respective polarizer settings. The photocurrent through the biased sample generated by the polarized radiation was measured employing a Stanford Research Systems SR570 low-noise current preamplifier, where the voltage signal at the output was fed into the spectrometer. Further data processing involved the acquisition processor of the IFS 66 and a PC controlling the interferometer. The interferograms were recorded in step-scan mode. In order to suppress the photocurrent signal generated by excitation over the band gap (around 1.1 eV), the electronic band pass filter of the SR570 was tuned to a range from 300 Hz to 3 kHz. For a given mirror velocity of 7.5 kHz, this limits the acquired spectral information to a band between 80 and 780 meV. The spectra were measured for different temperature, controlled by a Lakeshore 331 Temperature controller, and for a series of voltages, where the bias was applied by the SR570 preamplifier.

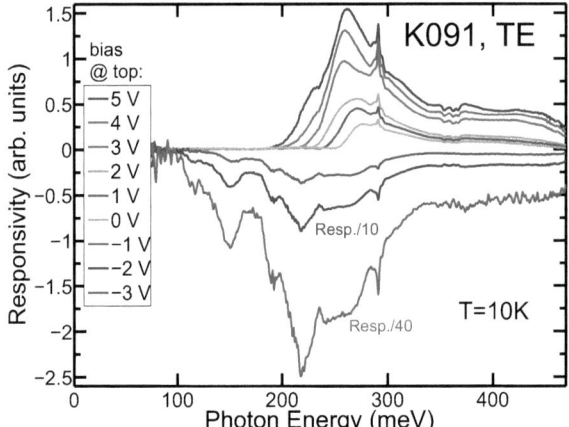

Figure 2.7: Voltage dependence of the responsivity of K091 for TE polarized radiation at a sample temperature of 10 K. The switching of the responsivity peak is far less pronounced for TE polarization than for TM in Fig. 2.6. This is due to the fact that selection rules influence the shape of the responsivity spectra crucially.

Results

Figures 2.6 and 2.7 show the results for TM and TE radiation, respectively, at a sample temperature of 10 K.

For TM polarized radiation, the responsivity of K091 exhibits a strong tunability with the applied voltage, as seen in Fig. 2.6. When changing the bias from -3 V to +3 V, the responsivity maximum switches from 115 meV to 260 meV, which is more than a factor of two in the photon energy. The sensitivity onset shifts from about 100 meV to 210 meV when performing this voltage change, again a factor of two in energy. Note that the responsivity onset can be continuously tuned, while the energetic position of the maxima exhibits switch-like behavior.

For TE polarized radiation, the voltage-tunability of the sample's responsivity is far less pronounced. A bias change from -3 V to +3 V induces a switching of the sensitivity maximum from 220 to 260 meV. Nevertheless, the continuous tunability of the responsivity onset is to be found in both polarizations. Note that for both TM and TE radiation the quantitative value of the photocurrent signal for negative applied voltages is two orders of magnitude higher than that for similar positive bias, a behavior analogous to that of the dark-current in Fig. 2.3.

Interpretation

The basis of the voltage and polarization dependence of the structure's photocurrent response to mid-infrared radiation lies in its band structure, as shown in Fig. 2.4. As already mentioned in the previous section, at low temperatures of 10 K and an applied bias of +3 V, the charge carriers of each period are confined in the HH1 state of W1. In order to generate a photocurrent, holes have to be excited into continuum states, which are not confined and enable a drift along the applied electric field. As seen in Fig. 2.4, both the LH1 and the HH2 states of W1 are well confined at a bias of +3V and therefore do not contribute to a photocurrent. To do so, carriers have to reach the mixed states built up by HH, LH and SO contributions close to the HH band edge.

Now, the band structure plot presents the density of hole states resulting from the heterostructure's potential landscape. However, for an efficient excitation of carriers from the HH1 state into the continuum not only a sufficient density of final states, but also a strong dipole matrix element between the two states is required. Thus Fig. 2.3 only provides part of the information needed to thoroughly understand the responsivity behavior at hand. The performed band structure calculations enable the determination of the dipole matrix element between initial state Ψ_i and final state Ψ_f, $<\Psi_f|\pi|\Psi_i>$, where π is the momentum operator including spin-orbit interaction following Eqn. 2.1 [20, 23, 24]. In this equation, p represents the momentum operator, σ the Pauli operator, $\hbar$ Planck's constant, c the velocity of light, and m_0 the electron mass.

$$\pi = p + \frac{\hbar}{4\,m_0\,c^2}(\sigma \times \Delta V) \qquad (2.1)$$

Applying Fermi's golden rule [24], the optical absorption coefficient for a photon of wave vector k and polarization λ is proportional to $|<\Psi_f|\pi|\Psi_i> \epsilon_{k,\lambda}|^2$. $\epsilon_{k,\lambda}$ represents the unit vector along the direction of the electric field of the electromagnetic wave of the respective polarization (TE or TM). In order to create a band structure plot indicating the relevance of the respective final state $<\Psi_f|$ for the absorption of light of wave vector k and polarization λ, the data seen in Fig. 2.4 were weighted by the factor $C_{relevant}^f$ given in Eqn. 2.2 ([20], p. 21).

$$C_{relevant}^f = \sum_i \int d\nu \, \frac{2\,\pi\,e^2}{m_{effav}^2\,\nu\,n_r\,c\,\epsilon_0} |<\Psi_f|\pi|\Psi_i> \cdot \epsilon_{k,\lambda}|^2 \cdot \frac{\frac{\Gamma}{\pi}}{(E_i + h\nu - E_f)^2 + \Gamma^2} \cdot f_i \qquad (2.2)$$

Here, m_{effav} is the is the averaged effective hole mass, ν the photon frequency, n_r the average refractive index, ϵ_0 the dielectric constant, and Γ the homogenous energy broadening, which is set to 7 meV, as stated above. The last term accounts for the finite lifetimes of the states and replaces the delta-function appearing in Fermi's golden rule. The summation is carried out over all initial states, which are occupied according to the calculated quasi-Fermi-level, where f_i is their occupation probability. Figures 2.8 and 2.9 show the results of these calculations for applied voltages of -3 V and +3 V, respectively.

Let us first consider the case of +3 V applied to the sample top contact. As already mentioned above and also seen in Fig. 2.4, the LH1 and HH2 states of W1 are strongly confined. Thus, even though absorption from the HH1 ground state into the LH1/HH2 states is strong for TE/TM polarized radiation, as seen in Fig. 2.9, charge transport to the sample contacts from these final states is not possible. Photocurrent is generated by the excitation of charge carriers into mixed states, which are built up by HH, LH and SO components. The energetic position of these conducting final states is in excellent agreement with the position of the responsivity peaks around 260 meV in Fig. 2.7 and 2.6.

For a negative bias of -3 V, charge carriers at low temperatures are confined to the HH1 state of W5. In contrast to that of W1, the HH2 state of W5 is only weakly confined. This is due to the introduced side well W6, which enables tunnelling of HH2 charge carriers. Thus, transport is possible by exciting holes into the HH2 level. As concluded from the peaks in Fig. 2.8, absorption into HH2 is strong in TM polarization only. This directly leads to the responsivity peak at low energies around 115 meV, as observed for negative biases in Fig. 2.6. The responsivity contributions at higher photon energies around 200 meV at this bias originate from the excitation of ground state carriers into mixed states, as seen in Fig. 2.8.

For a direct demonstration of the polarization dependence of K091's responsivity at negative voltages, Fig. 2.10 unites the responsivity spectra measured at -3 V in one plot. As seen in this figure, absorption into the mixed final states, which forms the base of the responsivity peak around 200 meV, is nearly independent of the polarization of the incoming radiation. This behavior is a direct consequence of the mixed nature of the final states. Contrary, the peak around 115 meV is only strong in TM polarization, and insignificant for TE polarized light, which is due to the optical transition rules between pure HH1 and HH2 states. However, even though the absorption of charge carriers into the HH2 state of W5 is significantly stronger than that into mixed states for TM polarized radiation, as concluded from Fig. 2.8, the responsivity peak around 115 meV is only brought to dominance for negative voltages stronger than -3 V. The reason for this behavior can be found in the fact that for lower voltages tunnelling of HH2 carriers into the continuum is still too inefficient to generate a high photocurrent. At a bias of -3 V, however, tunnelling of HH2 carriers results in a significant low-energetic responsivity peak, which is only present in TM polarization.

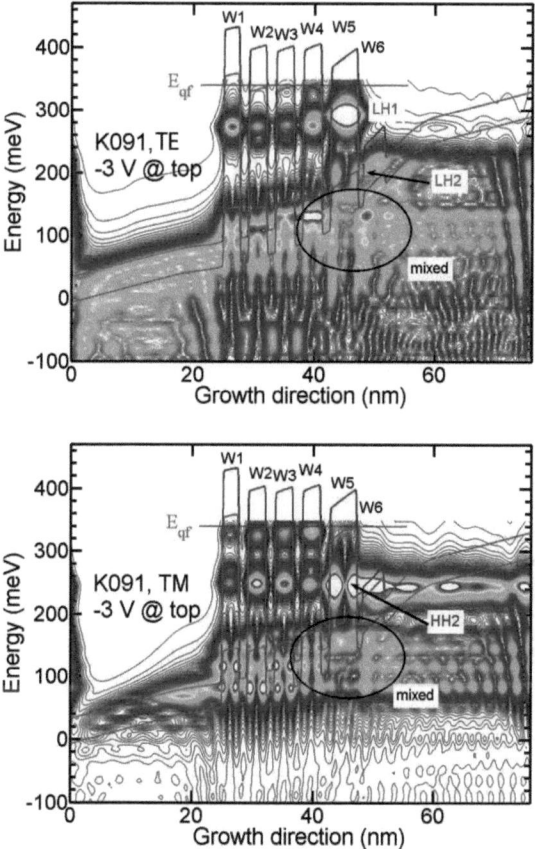

Figure 2.8: Final states relevant for the absorption of TE and TM polarized radiation from the ground state at -3 V bias. The plot is generated by weighting the band structure shown in Fig. 2.4 according to the states' relevance for the absorption of radiation as described in the text. Note that in TM polarization the transition from the ground state of W5 to its HH2 state is strong and results in a photocurrent via tunnelling. This transition is weak for TE polarized light, where the strong transition into the bound LH1 state does not lead to a photocurrent. Transitions into mixed states partly built up by SO states are strong in both polarizations.

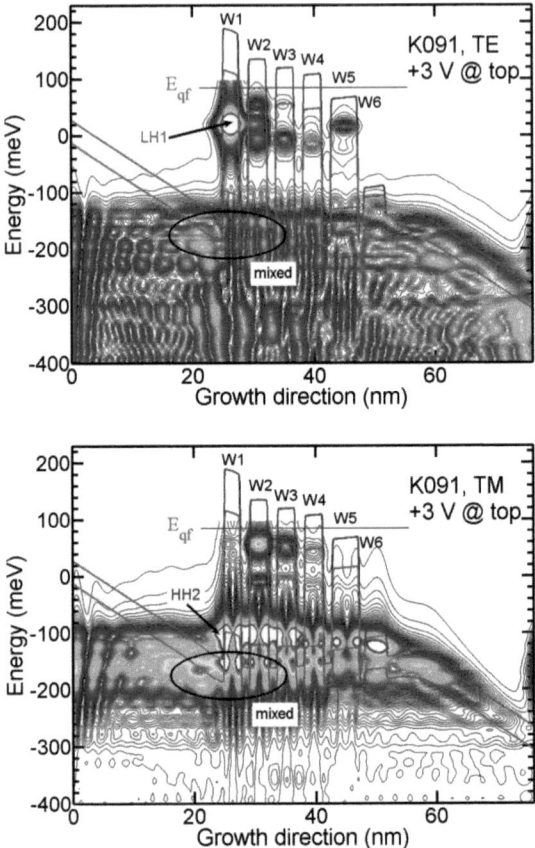

Figure 2.9: Final states relevant for the absorption of TE and TM polarized radiation from the ground state at +3 V bias. The plot is generated by weighting the band structure shown in Fig. 2.4 according to the states' relevance for the absorption of radiation as described in the text. Note that although the transition from the ground state of W1 into its HH2 state is strong in TM polarization, its contribution to the photocurrent is weak due to the strongly bound nature of HH2. The photocurrent is mostly generated by transitions into mixed states, which are partly built up by SO states. These transitions are strong in both polarizations.

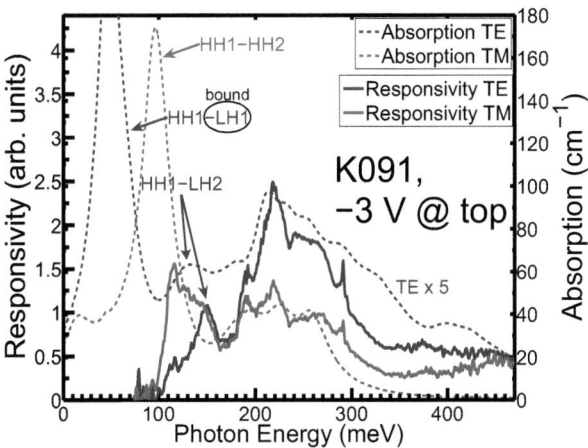

Figure 2.10: Simulated absorption spectra vs. measured photocurrent spectra for K091. The spectral dependence of the absorption coefficient calculated for different polarizations from the simulated band structure at a bias of -3 V is shown together with the measured responsivity. Experiment and simulation are consistent to an extremely high degree. The peaks are labeled according to their transitions of origin (see Fig. 2.8). Note the absence of the HH1-LH1 transition in the photocurrent spectrum due to the bound character of the LH1 state.

Slightly higher in energy, around 150 meV, a sensitivity peak emerges for negative voltages in TE polarization, as seen by the solid blue line in Fig. 2.10. This peak is generated by the absorption of ground state carriers into the LH2 state of W5, a transition which is strong for TE polarized radiation, as seen in Fig. 2.8. Again, the contribution of this transition to the photocurrent increases significantly with the applied bias due to an increase in tunnelling efficiency.

Even though the increasing tunnelling efficiency explains the increase in the responsivity for low photon energies and negative bias, it does not fully account for the general increase in sensitivity and dark-current shown in Fig. 2.3 by more than two orders of magnitude when changing the applied voltage from -2 V to -3 V. This drastic increase in photocurrent is also observed for the spectral measurements in Fig. 2.6 and 2.7. A possible explanation for the observed unexpectedly strong increase in both dark-current and photocurrent is the occurrence of impact ionization of charge carriers in the low side well W6. As the negative voltage applied to the sample is increased, W6 could be thermally occupied or filled with retrapped continuum carriers. These charge carriers in W6 could then be excited into the continuum via impact ionization by continuum carriers accelerated along the high-field region around W6. However, a study of the dependence of the photocurrent on the sample temperature and radiation intensity would be required to confirm this assumption.

Analogous to the considerations leading to the weighted band structure plots of Fig. 2.8, the data gained by the band structure simulations can be used to calculate the absorption spectra of K091. The expression given in Eqn. 2.2 is summed over all initial states, and weighted by the respective probability of occupation difference $f_i - f_f$, as shown in Eqn. 2.3.

$$\alpha(\nu) = \frac{e^2}{m_{effav}^2 \, \nu \, n_r \, c \, \epsilon_0 \, V} \sum_{i,f} |<\Psi_f|\boldsymbol{\pi}|\Psi_i>\cdot \boldsymbol{\epsilon}_{\boldsymbol{k},\lambda}|^2 \cdot \frac{\frac{\Gamma}{\pi}}{(E_i + h\nu - E_f)^2 + \Gamma^2} \cdot (f_i - f_f)$$

(2.3)

The results of these considerations are presented in Fig. 2.10 as broken lines for TE (blue) and TM (red) polarized radiation. As can be seen in this figure, the simulated absorption coefficient predicts the spectral shape of the photocurrent response with remarkable accuracy. The position and relative intensity of the responsivity curves' spectral features originating from absorption into higher-energetic, mixed continuum states are reflected in the calculated absorption spectra with a precision of 20 meV. This is especially remarkable, as transport is not included in the simulations. The influence of transport on the measured photocurrent spectra becomes crucial for absorption features with bound final states, such as that originating from the HH1-LH1 transition, which does not lead to a photocurrent at all.

As already mentioned above, side well W6 was included in the structure in order to enable tunnelling of HH2 carriers into the continuum. Even though the success of this design strategy is proven by the existence of the strong photocurrent peak around 115 meV, the differences both in the exact spectral position and in its height relative to features at higher energies imply that only the high energetic tail of the HH2 state can tunnel efficiently into the continuum, while its low energy component is bound to W5 and does not contribute to a current. This conclusion is supported by the spectral shape of the responsivity peak generated by HH2 carriers with its steep cut-off at low energies. Thus, even though W6 lives up to its purpose, careful redesign of the structure could enable more efficient utilization of the HH1-HH2 transition for the detection of low-energetic mid-infrared photons.

Calibration

Together with the I-V curves presented in Fig. 2.3, the photocurrent spectra measured at 77 K under illumination from the sample backside can be used to quantitatively determine the

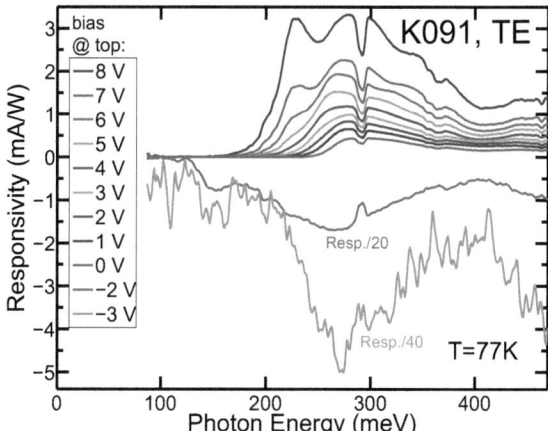

Figure 2.11: Quantitative spectrally resolved responsivity of K091 at 77 K. The photocurrent spectra measured under normal incidence at liquid nitrogen temperature were calibrated using the I-V characteristics presented in Fig. 2.3 as described in the text. The spectral shape of the responsivity to TE radiation at 77 K differs from that shown in Fig. 2.7 for 10 K due to a thermal redistribution of charge carriers.

structure's responsivity for TE radiation and varying applied voltages. In order to obtain the sample's responsivity spectrum $R(\nu)$ at 77 K and for TE polarized light, the photocurrent spectrum as measured $\widetilde{R}(\nu)$ is calibrated by the factor C_{cal} as given in Eqn. 2.4 following [20], p. 64.

$$C_{cal} = \frac{I_{blackbody}}{\zeta \cdot T_{windows} T_{lens} A_{mesa} \left(1 - \cos\left(2\arctan(\frac{d}{4f})\right)\right)}$$

$$\zeta = \int_0^\infty \widetilde{R}(\nu) \frac{\pi h \nu^3}{c^2} \left(\frac{1}{e^{h\nu/kT} - 1} - \frac{1}{e^{h\nu/kT_{bg}} - 1}\right) d\nu$$

$$R(\nu) = C_{cal} \cdot \widetilde{R}(\nu) \tag{2.4}$$

In this equation, ν is the frequency of the radiation, $I_{blackbody}$ is the current value of the red curve in Fig. 2.3 at the bias, for which the respective spectrum was measured. This current value represents the additional photocurrent through the sample under illumination by a black-body radiator at 550°C as compared to the current present under the influence of the radiation background only. The second term in the integral defining ζ accounts for the fact, that in absence of the black-body radiator the sample is hit by background radiation

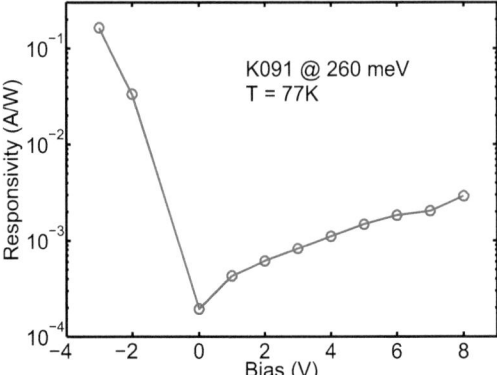

Figure 2.12: Voltage dependence of K091's responsivity to TE radiation at a photon energy of 260 meV and a temperature of 77 K. The values for positive applied voltages compare well with those reported in [16] for H019 at similar wavelength, while far exceeding the values reported for negative bias. The reason for this astonishingly high responsivity may be found in an avalanche multiplication of the photoconducting carriers.

within the dihedral angle attributed to the black-body radiator. T_{bg} thus represents the temperature of the background radiation, and is set to 300 K. The advantage of using the difference between the I-V characteristics measured under black-body illumination and that under the influence of the radiation background lies in the consequential consideration of any scattering light which might hit the sample in addition to the black-body radiation. Calibrating the photocurrent spectra by using the difference between the I-V curve under black-body illumination and the dark-current curve in Fig. 2.3 would not account for this potential additional sources of photocurrent, and could lead to an overestimation of the calibration factor C_{cal}. $T_{windows}$ and T_{lens} are the transmission coefficients associated with the pair of ZnSe cryostat windows and the CaF$_2$ lens used to collimate the black-body radiation and thus account for reflective losses. Using the refractive indices $n_{ZnSe} = 2.4$ and $n_{CaF2} = 1.3$ and including both surfaces of each optical element results in $T_{windows} = 0.5$ and $T_{lens} = 0.97$. A_{mesa} is the mesa area of 0.16 mm^2, d the lens diameter of 5 cm, f its focal length of 101 mm, and T the temperature of the black-body radiator. The result of this calibration is given in Fig. 2.11, where the responsivity of K091 for TE polarized radiation is shown for a temperature of 77 K.

Temperature dependence of the responsivity

Note the difference in spectral shape of the photocurrent curves in Fig. 2.11 recorded at 77 K compared to those at 10 K as seen in Fig. 2.7. For positive applied bias, the spectra broaden significantly, when increasing the temperature from 10 K to 77 K. This is due to the thermal occupation of the ground states of W2 and W3 at 77 K, when at 10 K only the HH1 state of W1 is significantly occupied. But not only the occupation of the initial states is energetically broader at 77 K. Also, carriers excited into at low temperatures bound LH2 states of W1 can reach the continuum by thermally assisted tunnelling at 77 K. The result of this thermal excitation into the continuum is an additional photocurrent peak around 230 meV at high applied voltages above 5 V.

At negative applied bias, a similar broadening of the responsivity spectra with increasing temperature can be observed. However, for negative voltages an increase in the sample temperature leads to an increasing occupation of the ground states of the deeper and narrower W4 to W3, which shifts the responsivity peak to higher photon energies. Put differently, as the sample temperature rises, the intermediate states W2 to W4 are increasingly occupied, and the photocurrent spectra for positive and negative applied biases become more and more similar. As a consequence, the voltage-tunability of the responsivity for TE polarized radiation is less pronounced at 77 K. Additionally, as temperature rises, the thermionic dark-current increases, as can be seen from the small signal-to-noise ratio of the spectrum for -3 V in Fig. 2.11.

Responsivity characteristics

Figure 2.12 highlights the voltage dependence of K091's responsivity at a photon energy of 260 meV. As already observed at a sample temperature of 10 K, the photocurrent induced by mid-infrared radiation increases gradually with positive applied voltage, and rises abruptly with increasing negative bias. The responsivity values for positive biasing compare well to those found for sample H019 in [16, 20], where the value for +4 V of K091 (1.1 mA/W @ 260 meV) is twice as large as that of H019 for -3.7 V (0.6 mA/W @ 220 meV). In contrast, for negative applied voltages, the photocurrent response of K091 exceeds that of H019 by orders of magnitude. Thus the quantitative responsivity of K091 is far above any expected value, and reaches impressive 200 mA/W for -3 V. The reason may not be found in the photoexcitation process itself, but in the presence of holes captured in the shallow well W6. Such carriers could be a cause of avalanche multiplication of photogenerated carriers and result in high responsivity figures. However, a sound confirmation of this explanation would require further investigation.

Detectivity

In addition to the responsivity, a second figure of merit is commonly used to give a measure for the quality of a detector device, namely the detectivity [20]. The definition of the detectivity is given in Eqn. 3.1, and its experimental determination requires the measurement of the dark-current noise through the detector device, as discussed in chapter 3.3 for sample K090. In case of sample K091, the experimental determination of the dark-current noise was prevented by experimental difficulties. However, an estimation for the detectivity of K091 can be given by assuming a noise gain equivalent to that of the similar structure H019, which was reported to be 0.48 at an applied bias of -3.7 V. Considering the measured dark-current value of $5 \cdot 10^{-10}$ A at a voltage of 4 V (indicated by the black line in Fig. 2.3), an estimated peak detectivity of $3.8 \cdot 10^9 \text{cm}\sqrt{\text{Hz}}/\text{W}$ is gained for K091 at 4 V. Sample K091 thus seems to feature detectivities for photon energies around 260 meV, which are four times as large as those observed for H019. This is partly due to the by a factor two stronger responsivity of K091, and partly owed to the weaker dark-current through K091 at a bias of 4 V as compared to H019 at -3.7 V. However, a thorough comparison between the detectivities of K091 and H019 would require an experimental determination of the dark-current noise present in K091.

2.3 Summary and conclusion

In this chapter the design, processing and thorough characterization of a voltage-tunable two-color mid-infrared quantum well detector in the SiGe system has been presented. The design concept was based on a similar device H019 demonstrated in [16] and enabled the extension of the detection range towards lower photon energies as compared to H019 by the introduction of an additional well. This well allowed the tunneling of HH2 carriers, which are strongly confined in case of H019, into the continuum and enabled the generation of a photocurrent. The performance of the QWIP device was thoroughly studied and characterized by FTIR measurements, dark-current experiments and by employing a black-body radiator for calibration purposes. Further, the optoelectronic behavior was discussed in detail utilizing band structure calculations, where a remarkable consistency between simulation and experiment was observed.

The novel QWIP design enabled a switching of the responsivity peak for TM polarized radiation over a remarkable range from 115 to 260 meV (factor 2.3 in photon energy) when reversing the sign of the applied bias. The onset of the responsivity could be continuously tuned between 100 and 250 meV. For comparison, H019's responsivity peak could be switched between 220 and 370 meV (factor 1.6 in photon energy). For TE polarized radiation, K091 features responsivity values of up to 3 mA/W for positive and of up to 200 mA/W for negative biasing. Thus, the responsivity performance of the novel K091 is slightly superior to that of H019 for positive applied biases, and exceeds it by two orders of magnitude for negative voltages. The cause of this surprisingly superior responsivity is not fully understood and requires further investigation.

To conclude, K091 succeeded in extending the range of voltage-tunable two-color SiGe QWIPs to lower photon energies and thus in opening further wavelength regions, in which integrable SiGe devices provide an alternative to QWIPs realized in conventional material systems for low-sensitivity applications. Our device exhibits highly competitive figures of merit, both in comparison to tunable multiwavelength QWIPs [16] and single well devices [15] in the SiGe system, and demonstrates the suitability of the SiGe system for flexible QWIP design and for spectral responsivity tailoring as a basis for specialized sensing applications.

Chapter 3

Pseudosubstrate QWIP

3.1 Design and fabrication

3.1.1 Concept theory

The most basic requirement for the measurement of any signal is the ability to distinguish it from the noise present in the sensing device. For detectors, the figure of merit quantifying the ability to distinguish the signal produced by incoming radiation from the detector noise is the detectivity, defined according to Eqn. 3.1 [25].

$$D^*(\nu) = \frac{R(\nu)\sqrt{\Delta f}\sqrt{A_{mesa}}}{\sqrt{<i^2>(f)}} \qquad (3.1)$$

In this equation, $R(\nu)$ is the spectrally resolved responsivity, A_{mesa} the detector area, and $\frac{<i^2>(f)}{\Delta f}$ the variance of the current through the device per frequency interval, whose square-root is equivalent to the spectral noise density in the device. The noise affecting the detection process in a device is generated by various processes. It is composed of the noise originating from the statistical process of the absorption of the photons to detect, the respective noise for the absorption of background photons, the Nyquist or Johnson noise present for any resistor [26], the $\frac{1}{f}$-noise, and, last but not least, the noise caused by the dark-current through the system, the contributions of which are discussed in more detail below. The noise fundamentally correlated with the detection process is also referred to as generation-recombination noise, which is "[...]associated with random thermal excitation and decay of mobile carriers in the sample, leading to fluctuations in the *number* of mobile carriers in the sample [in contrast to Johnson (thermal) noise which is associated with fluctuations in the *velocity* of the carriers]", as Beck puts it in [27]. However, the Nyquist noise is negligible for highly resistive devices, and the $\frac{1}{f}$-noise insignificant for relevant detection frequencies. The background noise is inevitably coupled to the detection process itself, and can be reduced by careful background radiation control. Thus the only noise contribution, which fundamentally limits the performance of the detector structure itself is the dark-current noise. Thus, for the common definition of the detectivity, the ratio of the responsivity and the spectral dark-current noise density is taken, and Eqn. 3.1 becomes Eqn. 3.2, where $\frac{<i^2_{dark}>}{\Delta f}$ is independent of the frequency f (white noise).

$$D^*(\nu) = \frac{R(\nu)\sqrt{\Delta f}\sqrt{A_{mesa}}}{\sqrt{<i^2_{dark}>}} \tag{3.2}$$

In the following section, different models for quantum well photoconduction will be presented. The model of Beck [27] is used as a foundation for the QWIP concepts discussed in this chapter. In order to give a detailed picture of this concept, a comparison of Becks model to a simpler original approach by Liu [25] will follow. Note that the labels **Liu** and **Beck** in Eqns. 3.3 to 3.12 indicate, on which of the two models (presented in [25] and [27], respectively) they are based on. However, more recent publications by Liu's group are in line with Beck's model [31]. Nevertheless, in this chapter the label 'Liu's model' refers to the conventional, simple approach by Liu in [25]. In this conventional model of Liu the dark-current noise is given by Eqn. 3.3, where $<i_{dark}>$ is the average dark-current, g_{noise} the noise gain, N the number of quantum wells, and p the carrier trapping probability per well.

$$\frac{<i^2_{dark}>}{\Delta f} = 4eg_{noise}<i_{dark}> \qquad \textbf{Liu, Beck} \tag{3.3}$$

$$g_{noise} = \frac{1}{Np} \qquad \textbf{Liu} \tag{3.4}$$

In Liu's model, the dark current is generated by carriers thermally excited from quantum wells into the continuum, their average emission current being $<i_{em}>$, and experiencing gain, as stated in Eqn.3.5.

$$<i_{dark}> = g_{noise}<i_{em}> \qquad \textbf{Liu, Beck} \tag{3.5}$$

The spectral responsivity is given by Eqn. 3.6, where e is the elementary charge, $\eta(\nu)$ the quantum efficiency (number of excited charge carriers per incident photon over all periods), $g(\nu)$ the photoconductive gain (number of charge carriers injected through the contacts per excited carrier), h Planck's constant, and c the velocity of light.

$$R(\nu) = \frac{e\eta(\nu)g_{photo}(\nu)}{h\nu} \qquad \textbf{Liu, Beck} \tag{3.6}$$

$$g_{photo}(\nu) = \frac{1-p}{Np} \qquad \textbf{Liu} \tag{3.7}$$

Combining Eqns. 3.2 to 3.7 leads to the proportionality in Eqn. 3.8.

$$D^*(\nu) \sim 1-p \qquad \textbf{Liu} \tag{3.8}$$

Thus, according to Liu's model, for achieving high detectivity values in a QWIP, a low carrier trapping probability would be favorable. However, the model of Liu assumes a homogenous trapping probability over the well region and an absence of direct tunneling between the wells.

The model of Beck in [27] takes a different approach by distinguishing between continuum charge carriers recaptured in the same quantum well period as they were generated and those captured by a well in a subsequent period. This is done by defining the emission rate as the number of charge carriers per second emitted *out of the well region of a period*, not counting those which are recaptured in the same period, and by treating the capture probability likewise. Additionally, Beck included direct tunneling between subsequent periods into his rate equations, and neglected them in his final statement. This treatment of the trapping probability allows the consideration of a case, in which each quantum well period is terminated on the extraction side by a barrier ensuring the trapping of all charge carriers reaching it by drifting in the continuum along the applied field. In such a case the generation-recombination noise would reduce to mere shot noise associated with the carrier emission. Therefore, the detectivity would increase due to the elimination of the statistical process of re-trapping. This conclusion is not reflected in the model of Liu, while it can be drawn from the expression for the dark-current noise and photocurrent gain found by Beck and given in Eqn. 3.9

$$\frac{<i_{dark}^2>}{\Delta f} = 4eg_{noise}<i_{dark}> \qquad \textbf{Beck} \qquad (3.9)$$

$$g_{noise} = \frac{1}{Np}(1-\frac{p}{2}) \qquad \textbf{Beck} \qquad (3.10)$$

$$g_{photo} = \frac{1}{Np} \qquad \textbf{Beck} \qquad (3.11)$$

In the frame of Beck's model, combining Eqns. 3.2, 3.5, 3.6 and 3.9 to 3.11 gives the proportionality in Eqn. 3.12.

$$D^*(\nu) \sim \frac{1}{\sqrt{1-\frac{p}{2}}} \qquad \textbf{Beck} \qquad (3.12)$$

The expression in Eqn. 3.12 reflects the case of $p=1$ accurately by giving a squared darkcurrent noise reduced by a factor of two as compared to Eqn. 3.3. When using Eqn. 3.12 for optimizing a QWIP's detectivity figure, it implies to push the capturing probability p as high as possible. This was already concluded by Schönbein, who states in [29]: "Our results indicate that a photoconductive QWIP with $p_c \approx 1$ should have a higher detectivity than a conventional GaAs/AlGaAs QWIP."

The following sections present the conception, design, growth and characterization of a SiGe QWIP based on the high-p strategy implied by Eqn. 3.12.

Table 3.1: MBE growth sequence and doping concentration of K090:

	K090		
Thickness	Ge Concentration	Doping Concentration	Purpose
16 nm	45 %	0	spacer
7 nm	45 %	$5 \cdot 10^{17} \mathrm{cm}^{-3}$	spacer & mod. doping
2 nm	45 %	0	spacer
3 nm	50 %	0	W6
1 nm	45 %	0	barrier
3 nm	90 %	0	W5
1.5 nm	0 %	0	barrier
2.3 nm	95 %	0	W4
1.5 nm	0 %	0	barrier
4.1 nm	83 %	0	W3
1.5 nm	0 %	0	barrier
4.1 nm	83 %	0	W2
1.5 nm	0 %	0	barrier
2.1 nm	100 %	0	W1
1 nm	15 %	0	barrier
2 nm	45 %	0	spacer
7 nm	45 %	$5 \cdot 10^{17} \mathrm{cm}^{-3}$	spacer & mod. doping
16 nm	45 %	0	spacer

3.1.2 Structure design and growth

The concept introduced in the previous section requires to form quantum wells deep enough for the targeted detection wavelength in the mid-infrared next to barriers high into the continuum in order to increase the re-capturing probability. The required well depth for transition energies from bound to continuum states in the mid-infrared range of interest is about 400 meV. To further introduce continuum barriers of sufficient energetic height and spatial width, an additional band offset of 300 meV is required. Thus, for our design concept we end up with a targeted HH band offset of about 700 meV between a well and a barrier region. This tremendous band offset cannot be realized with conventional plain Si substrates, as can be concluded from the following considerations. When using a pure Si substrate, the highest band offset for a $Si_{1-x}Ge_x$ quantum well layer can be intrinsically achieved by an interface to pure silicon. As can be seen from the band structure calculations performed according to [22], in order to gain a band offset of about 700 meV for the HH band, the pseudomorphic growth of a layer of a Ge concentration around 90% is necessary. As reported in [30], the critical

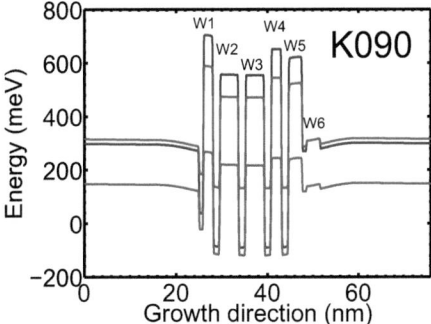

Figure 3.1: Valence band edges of the heavy- (blue), light- (green), and split-off(red)-bands for K090 at zero applied bias. The band bending is due to charge carrier redistribution. The low values on the growth direction scale are nearer to the substrate.

thickness for such a layer grown on a pure Si substrate is about 15 Å, which is insufficient for the design of a QWIP in the wavelength range of interest. In other words, the growth of the heterostructure according to the QWIP design considerations discussed in the last section is ultimately hindered by the limitations of growth on a plain Si substrate. Therefore, if the concept introduced in the last section is to be realized, a different approach in terms of the used substrate has to be taken.

As already mentioned in chapter 1, the use of pseudosubstrates of a significant Ge content can overcome the limitations imposed by a pure Si substrate [17, 18]. It enabled the growth of a heterostructure following the implications for noise reduction given in the previous section, thus featuring both layers of pure Si and pure Ge of significant thicknesses within one and the same layer sequence. Structure K090 was grown pseudomorphically on a commercial $Si_{0.5}Ge_{0.5}$ (100) pseudosubstrate by C. Falub and D. Grützmacher at the Paul Scherrer Institut in Villigen, Switzerland, using low temperature ($\sim 300°C$) molecular beam epitaxy. Directly on top of the pseudosubstrate, a highly p-type (boron) doped contact layer (doping concentration $2 \cdot 10^{18} cm^{-3}$) of $Si_{0.5}Ge_{0.5}$ was grown. This bottom contact layer was followed by ten periods of the sequence given in table 3.1. The values last in the table are those closest to the substrate. Finally, another highly doped (doping concentration again $2 \cdot 10^{18} cm^{-3}$) layer was grown to act as top contact. The basic design of K090 exhibits similarities to that of K091 in chapter 2. Figure 3.1 shows the band edge profile induced by the structure's layer sequence, calculated as described in chapter 2.1.1. Similar to K091, the sequence of $Si_{1-x}Ge_x$ layers of K090 results in the formation of six quantum wells per period, where the main wells W1 and W5 are of similar energetic and spatial dimensions as those of K091. This is the case even though their absolute Ge concentration values are completely different. Again, the intermediate wells W2 to W4 serve the purpose of transfer states for shifting charge carriers between W1 and W5.

The main difference in design between K091 and K090 are the high barriers between subsequent quantum wells, whose inclusion was exclusively made possible by the use of the pseudosubstrate. These barriers serve two purposes. Their main contribution to the detector performance should be a significant increase of the re-trapping probability of continuum carriers by simply blocking their path when drifting between two subsequent well periods along the applied electric field. As stated in chapter 3.1.1, this should decrease the sample's noise gain and lead to an increased detectivity. However, this measure should also decrease the photoconductive gain significantly. In order to compensate the consequential drop in responsivity, the quantum efficiency of the structure, labelled $\eta(\nu)$ in Eqn. 3.6, shall be increased for the photon energies of interest. Again, this task is served by the Si barriers, as they should form continuum resonances narrowing the spectral range of final states for photoconduction and thus increasing the sample's peak quantum efficiencies. For the detection of low-energetic photons, K090 should operate analogous to K091 by employing a quasi-bound state in W5 into which holes can be excited by TM radiation and from which tunnelling into the continuum via W6 is possible, as described in detail in chapter 2. The average Ge content of sample K090 is matched to the Ge content of the pseudosubstrate (50%), the structure is thus quasi-relaxed.

3.1.3 Sample processing

The processing of sample K090 into mesas was performed analogous to that for K091, as given in chapter 2.1.2. One exception is the duration of the etching process in step 6, which was 7.2 minutes for K090 and resulted in an etching depth of 980 nm.

Several runs of sample processing resulted in samples featuring exclusively mesas, which exhibited short-cut I-V characteristics. During one experimental session on a mesa, which operated well, a bond broke loose. After a new, seemingly successful bonding process, the same mesa behaved like a short-cut. This incident led to the suspicion, that the ultrasonic-based bonding process induced dislocations in the heterostructure, thus degrading its performance. Following a thorough re-examination of any involved step in sample processing, an error in the growth of the sample was discovered: Due to a design error, the growth of the QWIP structure was determined by highly doped *pure Si*, instead of a highly doped $Si_{0.5}Ge_{0.5}$ layer, which would match the average composition of the active structure and thus induce no strain. The actual pure Si layer induces strain due to its compositional mismatch, and leads to a mechanically unstable structure. As a consequence, defects might have been created by subjecting the sample to ultrasonic bonding. In order to overcome this technological difficulty, the bonding process was performed employing ultra-low powers (setting 2). However, this difficulties could be easily avoided by the proper design of the structure, which should lead to a quasi-relaxed, mechanically insensitive QWIP device.

3.2 Photocurrent experiments

The experimental results presented in this section were obtained for a mesa of 100×100 μm^2. The mesa was completely covered by its top contact metallization, thus direct illumination from the front side was not possible. The active structure could be illuminated by a cleaved side facette, or from the sample backside under normal incidence. All experiments were carried out completely analogous to those on sample K091 in chapter 2.1.1, except where explicitly mentioned.

3.2.1 Current-voltage characteristics

Experiment and results

While the dark-current curve presented in Fig. 3.2 was measured under the same condition as that of K091, the structure's response to black-body radiation was recorded differently. The radiation emitted by a black-body radiator at 550°C was collimated by a spherical mirror instead of a lens, and was used to illuminate the sample backside. The mirror had a focal length of 150 mm and a diameter of 60 mm. Further, the response of the sample to the black-body radiator was determined using lock-in technique. Therefore a chopper was placed in front of the cryostat window, operated at a frequency of 364 Hz. The investigated mesa's top contact was connected to output-high-pin of a Keithley 236 Source Measure Unit, while the sample bottom contact was connected to the input of a Femto DLPCA-200 low-noise current amplifier. The DLPCA-200 thus measured the current through the sample for an applied bias given by the Keithley 236. The output of the amplifier was fed into the voltage-input of a Stanford Research System DSP SR830 lock-in amplifier (LIA), which was triggered by the chopper signal. Both the Keithley 236 and the SR830 were connected to a PC via IEEE interfaces. The Keithley SMU was used to perform a voltage sweep of the mesa's top contact, where for each bias value the output of the LIA was recorded. Further, the measurement was split into six parts, where the appropriate amplification of the DLPCA-200 was chosen separately between 10^6 V/A (L) and 10^9 V/A (L) for each section of the voltage sweep.

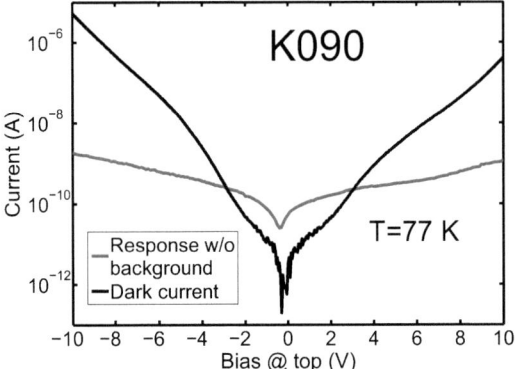

Figure 3.2: Current-voltage characteristics of K090. The black curve represents the sample's dark-current measured under low background conditions. The red curve presents the sample response to black-body radiation at 550°C without the radiation background and the dark-current, measured in lock-in technique.

Before each recorded value, the system was allowed to stabilize for 20 s. The resulting curves were scaled according to the preamplification factor, and multiplied by a factor $2 \cdot \sqrt{2}$. The latter accounts for the LIA-specific scaling factor between the LIA output voltage and the peak-to-peak value of a sinus-shaped input signal at the chopper frequency. The value of interest for the data evaluation is the peak-to-peak voltage of the LIA input equivalent to the signal difference between the sample response under black-body illumination and that under background conditions. The scaled curves were stitched together, and the result is presented as a red curve in Fig. 3.2. Note that this curve constitutes the complete analog to the red curve for K091 in Fig. 2.3.

As expected from the asymmetry of the structure, both the dark-current and the photocurrent characteristics in Fig. 3.2 exhibit an asymmetry in respect to the applied voltage, even though the asymmetry is far less pronounced than that for K091 in Fig. 2.3. The asymmetric doping of K091 is the reason for its steeper dark-current rise for negative applied bias, where the built-in field is in favor of the generation of dark-current by tunneling, and for its flatter rise for positive voltages, where the built-in field compensates the applied field and thus suppresses the dark-current generation. The modulation doping of K090 is distributed symmetrically over each period. The effect of the difference in the doping of the two samples becomes evident when comparing Figs. 3.1 and 2.2.

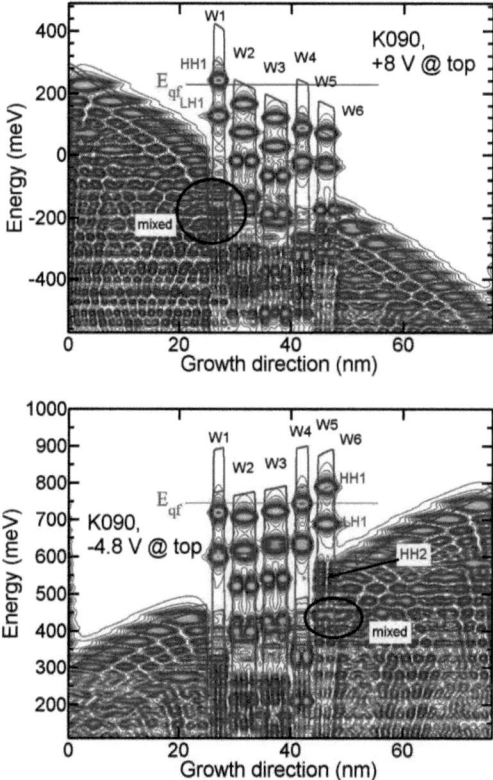

Figure 3.3: Dependence of the band structure of K090 on the applied voltage. The plots show the calculated heavy-, light- and split-off- (blue, green, red) hole band edges for voltages of +8 V and -4.8 V, and a contour plot of the absolute squared wave functions. The spreading of the eigenstates along the energy axis accounts for their homogenous broadening. For positive applied voltages, the charge carriers are located in W1. When changing to negative bias, holes are transferred to W5 via the intermediate wells W2-W4. The labels indicate the character of the excited states (HH, LH, mixed). The straight red line indicates the quasi-Fermi-level as predicted by the simulation.

Interpretation

The dark-current characteristics can be understood on the basis of band structure calculations in an analogous way to K091. The calculations were performed as described in refer-

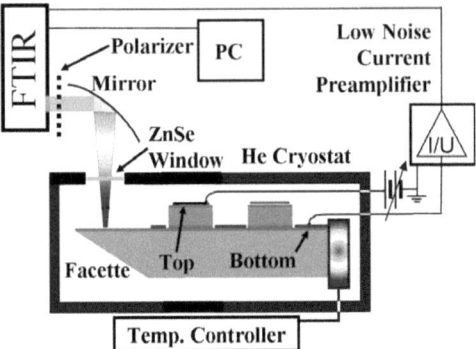

Figure 3.4: Setup for the spectroscopic measurements on K090.

ences [21, 22]. Figure 3.3 shows the simulation results for the HH, LH and SO band edges of one structural period as solid lines (blue, green and red, respectively) for two applied voltages (electric fields) of -4.8 V (-62.6 kV/cm) and +8 V (+104.4 kV/cm).

According to the quasi-Fermi-level calculated for a temperature of 77 K and indicated by the red line in Fig. 3.3, for positive applied bias the charge carriers occupy the HH ground state of W1. For the generation of thermionic dark-current, an activation barrier of about 340 meV has to be overcome. In case of a negative voltage applied to the mesa top contact, the holes are located in the ground state of W5, and the activation energy for thermionic dark-current reduces to 230 meV. Thus, the generation of thermionic dark-current is more efficient for negative applied voltages, where the same holds true for dark-current caused by the tunneling of carriers from occupied states into the continuum at high bias. Thus, the asymmetry in the sample's band structure is directly reflected in its dark-current characteristics.

3.2.2 Photocurrent spectra

Experiment and results

The spectral shape of K090's response to mid-infrared radiation was determined using a Bruker Vertex 80 FTIR. The interferometer radiation was coupled out of the spectrometer using one of its parallel exits, and focused to the position of the sample's side facet by a parabolic 90°off-axis gold mirror with a diameter of 58 mm and a parent focal length of 76.2 mm. The top contact of the mesa under investigation was biased using a battery-driven high

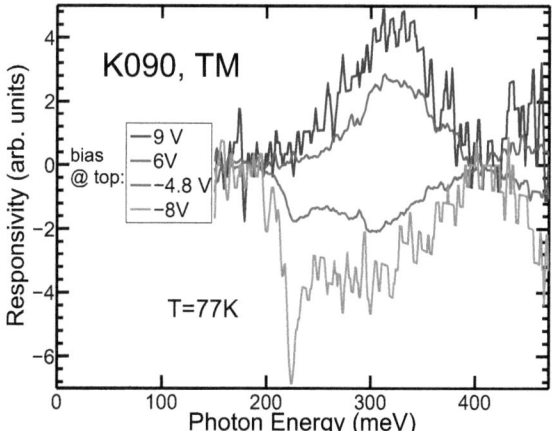

Figure 3.5: Voltage dependence of the responsivity of K090 for TM polarized radiation at a sample temperature of 77 K. Note the switching of the responsivity peak from 225 to 310 meV when reversing the sign of the applied voltage. The height of the peak at 225 meV exhibits an extremely strong dependence on the magnitude of the bias, for it is generated by a tunnelling process from W5.

precision voltage source. A DLPCA-200 current amplifier was employed to determine the vertical current through the sample. The input of the DLPCA-200 was connected to the bottom contact of the sample, its output to the external input of the spectrometer. The spectrometer was controlled by a PC and operated in fast-scan mode. The sample temperature was controlled by a Lakeshore 331 temperature controller. Figures 3.5 and 3.6 show the results for TM and TE radiation, respectively, and for a sample temperature of 77 K.

Analogous to sample K091 in chapter 2.2.2, for TM polarized radiation the responsivity of K090 exhibits a strong tunability with the applied voltage. When changing the bias from -8 V to +6 V, the responsivity maximum switches from 225 meV to 310 meV. As seen from Figs. 2.6 and 3.5, the peak responsivities of K090 are shifted to higher photon energies as compared to those of K091. The reason for this shift in the sensitivity range is discussed below.

For TE polarized radiation, the voltage-tunability of the K090's responsivity is far less pronounced. A bias change from -4.8 V to +8 V induces a switching of the sensitivity maximum

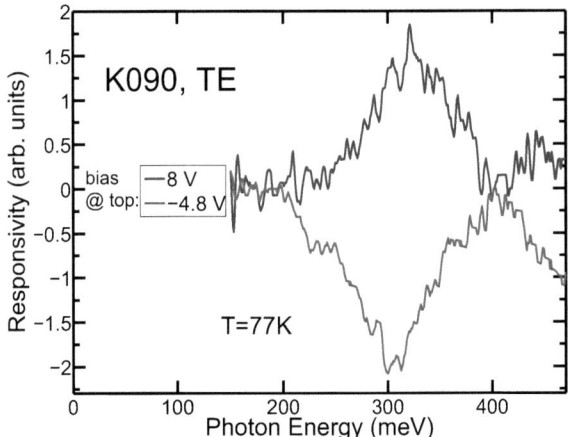

Figure 3.6: Voltage dependence of the responsivity of K090 for TE polarized radiation at a sample temperature of 77 K. The switching of the responsivity peak is far less pronounced for TE polarization than for TM in Fig. 3.5. This is due to the fact that selection rules influence the shape of the responsivity spectra crucially.

from 300 to 320 meV. This nearly complete absence of responsivity tunability is, again, a direct result of the design of K090.

Interpretation

As already discussed in detail for sample K091 in Chap.2.2.2, the consideration of K090's band structure leads to a detailed understanding of the dependence of its sensitivity on the applied bias and the polarization of the incident radiation. Figures 3.7 and 3.8 show the absolute squared wavefunctions weighted by their relevance for absorption processes in TM and TE polarization for an applied bias of -4.8 V and +8 V, respectively.

For negative applied bias, the charge carriers of each period are confined to the HH1 state of W5, as indicated by the quasi-Fermi-level in Fig.3.3. For negative voltages and TM polarized light, the mid-infrared sensitivity of the sample is based on a high dipole matrix element between the HH1 and HH2 states of W5, and the subsequent tunneling of HH2 carriers into the continuum via W6. This becomes evident when considering the strong polarization dependence of the responsivity peak at 225 meV in Fig. 3.5, which is associated with the

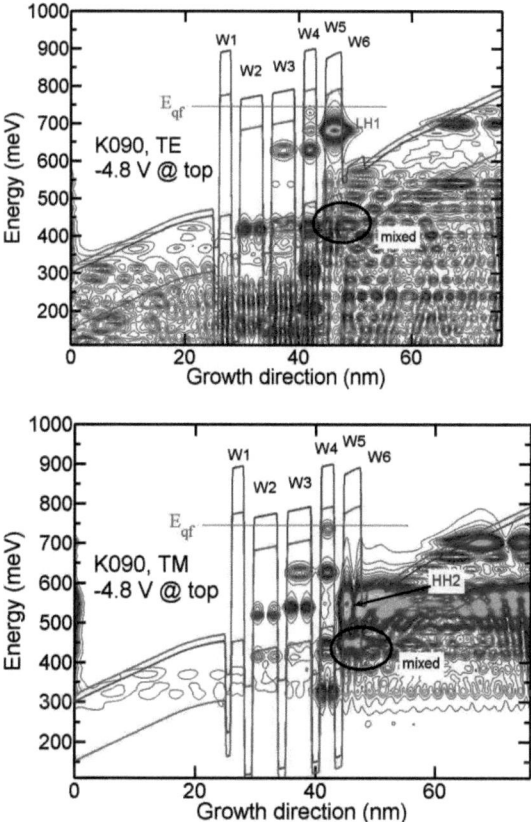

Figure 3.7: Final states relevant for the absorption of TE and TM polarized radiation by ground state carriers at -4.8 V bias. The plot is realized by weighting the band structure shown in Fig. 3.3 according to the respective states' relevance for the absorption of radiation as described in the text. Note that in TM polarization the transition from the ground state of W5 to its HH2 state is strong and results in a photocurrent via tunnelling. This transition is weak for TE polarized radiation, where the strong transition into the bound LH1 does not lead to photocurrent. Transitions into mixed states partly built up by SO states are strong in both polarizations.

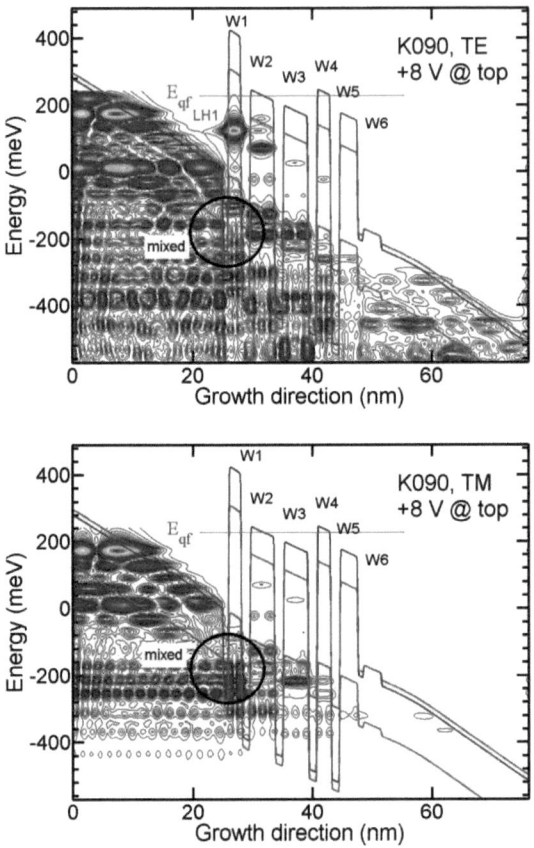

Figure 3.8: Final states relevant for the absorption of TE and TM polarized radiation from the ground state at +8 V bias. The plot is realized by weighting the band structure shown in Fig. 3.3 according to the respective states' relevance for the absorption of radiation as described in the text. The photocurrent is mostly generated by optical transitions into mixed states partly built up by SO states, which are strong in both polarizations.

HH1-HH2 transition in W5. The HH1-HH2 transition is only strong in TM polarization, which is illustrated in Fig. 3.7. Analogous to the responsivity peak of K091 around 115

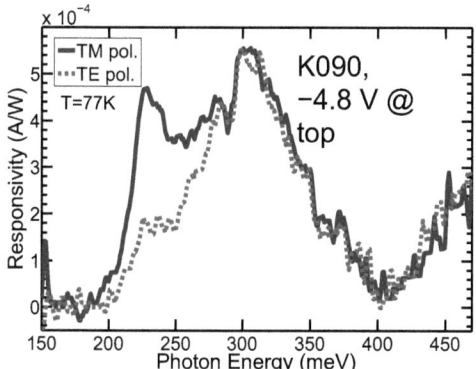

Figure 3.9: Polarization dependence of the responsivity spectra. The curves highlight the strong polarization dependence of the responsivity spectra at negative applied bias. The sharp peak at 225 meV originates from a transition between the HH1 ground state of W5 and an excited quasi-bound level mostly composed of HH states, which is only strong in TM polarization.

meV, the relative height of the responsivity peak of K090 around 225 meV depends strongly on the applied voltage. This is due to the fact, that carriers occupying the HH2 state of W5 have to tunnel into the continuum via W6 in order to contribute to the photocurrent, a process whose efficiency increases significantly with rising voltage. The shift of the HH1-HH2 transition of K090 towards higher energies of 225 meV, when compared to the 115 meV of K091, originates from the increased energetic depth of K090's W5 as compared to that of K091. For TE polarized radiation and negative biasing, the sensitivity peak around 300 meV is based on a transition of HH1 carriers in W5 into mixed states high up in the continuum. This transition is strong in both polarizations due to its mixed nature, and creates a high-energetic shoulder in the responsivity for TM polarization at negative bias, as seen in Fig. 3.5. This polarization-insensitive shoulder dominates the sample response for low bias. Figure 3.9 highlights the polarization dependence of the responsivity at -4.8 V, featuring both the TM-sensitive peak around 225 meV and the polarization-insensitive shoulder around 300 meV. With increasing applied bias, the TM-sensitive peak at 225 meV can be brought to dominance over the responsivity spectrum, as seen in Fig. 3.5. Now, as concluded from the plots in Fig. 3.7, the transition from the ground state into the LH1 state is very strong in TE polarization, but does not contribute to the photocurrent, as this final state is strongly confined.

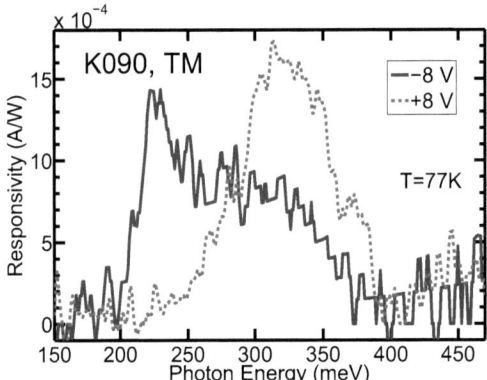

Figure 3.10: Voltage dependence of the responsivity spectra. The solid and broken lines show K090's responsivity for TM polarized radiation as as a function of the photon energy for an applied bias of -8 V and +8 V, respectively. The responsivity peak can be switched from 220 meV to 320 meV by changing the sign of the applied voltage. Note the in comparison to the structure in [16] very narrow linewidth of the responsivity peaks, which is due to the quasi-bound nature of the excited states.

For positive applied voltage, the holes are localized in W1 of K090, where they mainly occupy the HH1 ground state. Photocurrent is generated by the excitation of these HH1 carriers into quasi-bound continuum resonance states, as can be seen in Fig. 3.8. The final states with a strong dipole matrix element to the ground state are high up in the continuum, where the cause of these high-energetic continuum resonances can be found in the barriers introduced by the pure Si layers. As already mentioned, the introduction of barriers into the continuum was only made possible by the use of a $Si_{0.5}Ge_{0.5}$ pseudosubstrate. The Si barriers not only result in a shift of the responsivity peak towards higher photon energies, but also increase the dipole matrix element with the ground state and therefore the quantum efficiency η in Eqn. 3.6. When comparing Figs. 3.6 and 3.5, it becomes evident that the spectral shape of the sensitivity at positive applied bias is nearly independent of the polarization of the incident radiation. The reason for this behavior can be found, again, in the mixed nature of the final states contributing to the responsivity peak.

To sum up the spectral results for the sensitivity to TM polarized radiation, it can be stated that both the high and low energetic peaks originate from quasi-bound final states. While the responsivity peak around 225 meV for negative applied bias is based on a bound well state, which is enabled to tunnel into the continuum by the side well W6, the peak around 320 meV involves continuum resonances introduced by the Si barriers. Both types of final states should feature a strong quantum efficiency η due to their increased dipole matrix element to the ground state in comparison to spread, unstructured continuum states.

For the realization of a photoconducting bound-to-quasi-bound transition at high photon energies, the use of a pseudosubstrate is indispensable. Barriers of high Si content (in our case pure Si) are required to localize the excited continuum state above W1 and therefore introduce a bound nature along with a high dipole matrix element. Furthermore, the use of quasi-bound excited states narrows the responsivity peaks significantly. This becomes evident by comparing the polarization-insensitive peak at 320 meV exhibiting a full-width-at-half-maximum (FWHM) of 80 meV with the results for the tunable SiGe device in reference [16]. This device was grown on a Si substrate and exhibits a responsivity peak of equal strength at 220 meV with a significantly wider FWHM of 120 meV. For K090, the purely TM-sensitive transition at negative bias results in an even narrower responsivity peak. However, the absorption into mixed states around 300 meV creates a shoulder at the high-energy edge of this peak around 225 meV and thus degrades the sharpness of the spectral characteristics.

Responsivity

Analogous to chapter 2.2.2, the I-V characteristics in Fig. 3.2 were used to quantitatively determine the structure's responsivity for TE polarized radiation and varying applied voltages. For TM polarization, the sensitivity curves in Fig. 3.5 were scaled according to the calibration of the respective TE curve. In other words, C_{cal} was calculated from experimental data gained under illumination by TE polarized radiation, and was also applied to the spectra for TM polarization. As the I-V characteristics were obtained under backside illumination and normal incidence, Eqn. 2.4 has to be corrected into Eqn. 3.13, where $\widetilde{I}_{blackbody}$ is the actually measured current under backside illumination, and $L_{waveguide}$ the fractional loss of globar source intensity due to coupling into the waveguide during the spectral measurements. Therefore, $I_{blackbody}$ is the current that would be measured when coupling the black-body radiation into the sample waveguide, instead of illuminating from the backside. The results given in Figs. 3.9 and 3.10 thus represent the *internal* responsivity of the structure, neglecting coupling losses.

$$C_{cal} = \frac{\widetilde{I}_{blackbody}}{\int_0^\infty (1 - R_{window})\widetilde{R}(\nu) A_{mesa}\left(1 - \cos\left(2\arctan(\frac{d}{4f})\right)\right)\frac{\pi h \nu^3}{c^2}\frac{1}{e^{h\nu/kT}-1}\,d\nu}$$

$$\widetilde{I}_{blackbody} = \frac{I_{blackbody}}{(1 - L_{waveguide})} \quad (3.13)$$

At a sample temperature of 77 K and for TM polarized radiation, K090 exhibits peak responsivities of 1.5 mA/W at 225 meV and 1.7 mA/W at 320 meV. These responsivity values

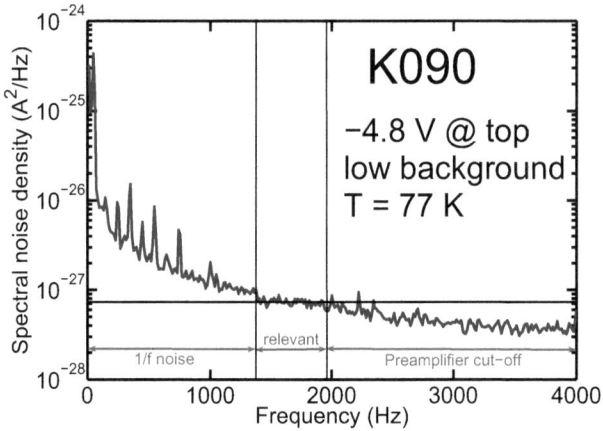

Figure 3.11: Spectral noise density of K090. The low-frequency section of the spectrum is dominated by the $\frac{1}{f}$-noise of the system. The high-frequency section is suppressed by the amplifier cut-off. The fraction of the noise spectrum which lies in between these two sections delivers a value of $7 \cdot 10^{-28} \mathrm{A}^2/\mathrm{Hz}$ for the white spectral dark-current noise density of K090 at -4.8 V.

are in the same range as those of K091 (3.3 mA/W for TE at 230 meV and 300 meV, +8V, see Fig. 2.11) and those of the device demonstrated in [16] (1.4 mA/W at 230 meV for TE, -7 V). Note that these competitive responsivity figures can be reported for K090 despite the fact that the transport of continuum carriers is hindered by the pure Si barriers, which have been introduced in order to reduce the noise gain.

3.3 Noise measurements

The structural design of K090 includes Si layers, which form energy barriers high up into the valence band continuum. These barriers were introduced in order to realize a concept for the reduction of the noise gain presented in chapter 3.1.1. For investigating the efficiency of this concept, the quantitative determination of the dark-current noise through the device is required. The noise measurements on K090 were performed under low-background conditions, by placing the sample in a liquid nitrogen vessel. The bottom contact of the mesa under investigation (the same mesa as for the experiments in the previous sections) is grounded. A voltage is applied to its top contact by a Femto DLPCA-200 low-noise current amplifier, which is also used to measure the current flowing through the biased sample. The voltage output signal of the amplifier is traced using a Tektronix TDS 3032B, and read out by a PC. The result of this measurement therefore is a current trace $i(t)$. A fast-Fourier transformation is performed in order to obtain $i(f)$, which is related to the spectral noise density function $S(f)$ via Eqn. 3.14, where Δt represents the time interval covered by the scope trace.

$$S(f) = \frac{2}{\Delta t} |i(f)|^2 \qquad (3.14)$$

The current noise $<i^2>(f)$ influencing measurements performed at a frequency f can be calculated from the spectral noise density following Eqn. 3.15, where Δf represents the experimental bandwidth.

$$<i^2>(f) = \int_{f-\frac{\Delta f}{2}}^{f+\frac{\Delta f}{2}} S(f') \mathrm{d}f' \qquad (3.15)$$

If the noise current density is constant within Δf, Eqn. 3.15 simplifies to Eqn. 3.16.

$$\frac{<i^2>(f)}{\Delta f} = S(f) \qquad (3.16)$$

Under low-background conditions, the current noise is dominated by the dark-current contribution, which is frequency independent or white above a threshold frequency, at which $\frac{1}{f}$-noise becomes negligible. Under these conditions, Eqn. 3.16 is equivalent to Eqn. 3.17.

$$\frac{<i^2_{dark}>}{\Delta f} = S(f) \qquad (3.17)$$

In this equation, $<i^2_{dark}>$ is equivalent to the expression used in chapter 3.1.1, and can thus directly be used to calculate the noise gain of K090 by using Eqn. 3.9. The experimentally gained $S(f)$ was averaged 100 times, and the result of the measurement performed at a bias of -4.8 V is presented in Fig. 3.11. In order to experimentally access the low dark-current noise of K090, a current amplification of 10^9 V/A at a cut-off frequency of 1 kHz had to be chosen. At this amplification, the bandwidth of the amplifier reaches up to 1 kHz. As a consequence, the constant region in the measured noise density spectrum is rather narrow. The relevant region of white dark-current noise is marked by the vertical lines in Fig. 3.11. It lies between the low-frequency-section, in which the $\frac{1}{f}$-noise is dominant, and the amplifier-cut-off-section at high frequencies. Despite these experimental difficulties, a value of $7 \cdot 10^{-28}$ A^2/Hz could be determined for the dark-current noise density of K090 at -4.8 V.

During the noise experiments on K090, the control and reduction of the noise of the measurement system itself proofed to be a demanding and tiresome procedure. An especially crucial aspect of the noise measurement setup was the proper grounding of any part of the setup, which included the detailed optimizing of the position of any cables. Additionally, the noise of the experimental setup itself depended on the operational status of other devices in the

Table 3.2: Comparison between the figures of merit of K090 and H019 [16, 20] at a temperature of 77 K:

	K090 @ -4.8 V, 225 meV	H019 @ -3.7 V, 250 meV [20]
R_{peak}	0.58 mA/W	0.62 mA/W
$\frac{<i_{dark}>}{A_{mesa}}$	$9.3 \cdot 10^{-5}$ A/cm^2	$7.3 \cdot 10^{-7}$ A/cm^2
$\frac{<i_{dark}^2>}{A_{mesa}}$	$7 \cdot 10^{-24}$ A^2 Hz^{-1} cm^{-2}	$2.2 \cdot 10^{-25}$ A^2 Hz^{-1} cm^{-2}
D_{peak}^*	$2.2 \cdot 10^8$ cm$\sqrt{\text{Hz}}$/W	$1 \cdot 10^9$ cm$\sqrt{\text{Hz}}$/W
g_{noise}	0.11	0.48
g_{photo}	0.16	0.53
p	0.62	0.21
η	$8 \cdot 10^{-4}$	$3.3 \cdot 10^{-4}$

room, if not in the institute. As a result of these ultra-sensitive experimental conditions, the determination of the noise actually originating from the sample was successful only for one particular voltage. A tremendous improvement in the noise measurement conditions could be achieved by replacing the external current amplifier by a preamplifier mounted next to the sample, which could be dipped into the liquid nitrogen and would enable the use of very short cables within the shielding provided by the nitrogen vessel itself.

3.4 Figures of merit and discussion

Employing Eqn. 3.9, the measured dark-current noise density can be used together with the dark-current value extracted from Fig. 3.2 to calculate the noise gain g_{noise}. Further, the re-trapping probability p of one structural period of K090 can be gained via Eqn. 3.10. Table 3.2 summarizes the figures of merit for K090 at an applied bias of -4.8 V and a temperature of 77 K. The sample's noise gain is remarkably low, confirming an efficient re-trapping of continuum carriers as intended by introducing the Si barriers. In fact, it differs by a factor of about four from the value of 0.48 reported in [20] for sample H019. Even though the noise gain of K090 is low, its responsivity exhibits highly competitive values. The reason for this can be partly found in the value of the photoconductive gain, which is according to Eqn. 3.11 by a factor 1.45 higher than the noise gain. An additional contribution to the competitive responsivity originates from the enhanced dipole matrix element between the HH ground state and conducting continuum states, and the therefore high quantum efficiency η. The strong dipole matrix element is induced by the quasi-bound nature of the final states of the transition, on which the mid-infrared sensitivity is based. For the quantum efficiency η, a high value of $8 \cdot 10^{-4}$ is gained from Eqn. 3.6. The peak quantum efficiency of sample K090 is therefore more than a factor of two enhanced as compared to sample H019 at the same wavelength and experimental conditions.

The factor preventing K090 from exhibiting superior detectivity in comparison to H019 is its unexpectedly high dark-current density of $9.3 \cdot 10^{-5} \text{A/cm}^2$, which is two orders of magnitude above that of sample H019. To put it differently, the detectivity of K090 ($2.2 \cdot 10^8$ cm$\sqrt{\text{Hz}}$/W) is *only* a factor five lower than that of H019 ($1 \cdot 10^9$ cm$\sqrt{\text{Hz}}$/W), even though the dark-current densities of the two samples differ by two orders of magnitude. As the peak responsivities of both samples are nearly equal for similar photon energies, K090 *would* feature a detectivity superior by a factor 2 due to its low noise gain, *if* its dark-current could be reduced to the values featured by H019. However, the reason for the high dark-current of K090 is not fully understood, and an explanation requires further investigation.

The realization of remarkable changes in some figures of merit, namely the reduction of the noise gain along with a simultaneous increase in quantum efficiency, was exclusively made feasible by the use of a $Si_{0.5}Ge_{0.5}$ pseudosubstrate. By introducing barriers into the valence band continuum, which are formed by pure Si layers and therefore feature a lower Ge content than the spacers between subsequent quantum well regions, the transport of continuum carriers was efficiently blocked, thus reducing the noise contribution originating from carrier trapping. This manifests itself in a high re-trapping probability of 0.62 and a consequently low noise gain of 0.11. At the same time, the barriers form quasi-bound resonances in the continuum, which result in an increase in peak quantum efficiencies to $8 \cdot 10^{-4}$ as compared to sample H019. Further considering the extraction of quasi-bound carriers from well W5 via W6, a principle which has been discussed in chapter 2, both mid-infrared transitions from the ground states of W1 and W5 resulting in a strong photocurrent have quasi-bound final states and thus feature a strong dipole matrix element. Thus, the use of a pseudosubstrate enabled an improvement of the concept of two-color detection demonstrated in [16,20]. Energy barriers into the continuum enabled the suppression of noise gain by a factor of nearly five and an increase in quantum efficiency by a factor of more than 2. Along with the increase in the spatial carrier density overlap between ground state and excited state, the utilization of quasi-bound states results in a spectral narrowing of the responsivity. The responsivity peak of K090 around 320 meV features a FWHM of 80 meV at 77 K. In comparison, that of H019 exhibits a FWHM of about 120 meV even at lower photon energies around 220 meV.

3.5 Summary and conclusion

In conclusion, in this chapter the fabrication and characterization of a tunable two-color mid-infrared SiGe QWIP grown on a $Si_{0.5}Ge_{0.5}$ pseudosubstrate was presented. Enabled by the use of the pseudosubstrate, design concepts have been realized, whose implementation has up to now been hindered by the limitations of pseudomorphic growth on pure Si wafers. Energetically high Si barriers stretching far into the valence band continuum have been employed to create quasi-bound states, which directly result in the narrowing of the peaks in the structure's responsivity spectra and in an increase in wavefunction overlap between the ground state and the photoconducting excited state. Further, the introduced Si barriers lead to a high re-trapping probability of continuum carriers of 0.62 per period. As a consequence of both the increase in the re-trapping probability and the increase in the spectral peak quantum efficiency ($8 \cdot 10^{-4}$) due to the energetic narrowing of the transition, the device features competitive responsivities of around 1.5 mA/W at 220 meV (-8 V bias) and 320 meV (+8V bias), while exhibiting a low noise gain of 0.11. Therefore, this work demonstrates the possibility of further improving and modifying voltage-tunable two-color mid-infrared QWIPs by employing quasi-bound excited states and by reducing the recombination noise while preserving the feature of tunability.

The work presented in this chapter constitutes the first thorough characterization of a SiGe QWIP grown on a pseudosubstrate. It demonstrates, that the use of SiGe pseudosubstrates for SiGe QWIP growth allows the fabrication of QWIP devices combining both pure Si and pure Ge within one structure. It thus opens the full compositional parameter space for advanced SiGe QWIP design with the prospect of figures of merit in excess of what has been demonstrated on plain Si substrates. The design concepts realized in sample K090 only serve as an example of what can be achieved in the SiGe system by growing a sophisticatedly tailored QWIP on a pseudosubstrate, and thus constitute a step towards employing SiGe QWIPs as a CMOS compatible alternative to QWIPs grown in conventional material systems.

However, the findings presented in this work not only hold relevance for the improvement of SiGe QWIPs, but give insight into the properties of structures grown on a type of substrate, which shows high promise as the base for advanced electro-optical structures on the road to a group-IV quantum cascade laser.

Chapter 4

BIB detectors: Introduction and theory

4.1 Introduction

The expression terahertz radiation is commonly used for electromagnetic waves within the frequency range between 10 THz and 300 GHz. As visualized in Fig. 4.1, the terahertz frequency range can be found between the microwave region and the infrared section of the spectrum, where it overlaps with the far-infrared. Due to difficulties in generating and exploiting electromagnetic waves in the Terahertz band, this region of the electromagnetic spectrum is addressed as terahertz gap.

The terahertz frequency range carries information about interstellar dust, which exhibits temperatures between 10 and 20 K and thus emits in this spectral region. Additionally, a variety of molecules exhibits characteristic lines in this spectral range. However, it was the transparency of organic material for terahertz radiation, which moved this part of the spectrum into the spotlight of general interest, and reliable sources were developed in order to explore the application of active terahertz optics. Some terahertz sources have been developed by adapting the technology used for microwave generation, like backward wave oscillators. Others were realized by implementing sources commonly used for the generation of infrared radiation, like quantum cascade lasers [32] and difference-frequency-mixers of near-infrared radiation. Since the availability of these reliable sources, a huge variety of applications for the sensing of terahertz radiation has been pioneered. The spectrum of applications includes DNA spectroscopy [33], pharmaceutical applications [34], gas sensing, as well as astronomic imaging. As organic material, alongside plastic and clothes, is highly transparent in the terahertz frequency band, while water and metals give some contrast, terahertz replacements for

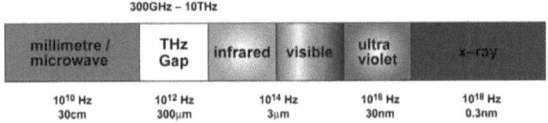

Figure 4.1: The terahertz region of the electromagnetic spectrum is found between the infrared and the microwave regions.

x-ray security systems are under development. In addition, medical imaging systems based on terahertz detection for e.g. tumor identification have been emerging recently. One possible means of realizing sensors in the terahertz region are blocked impurity band (BIB) detectors. BIBs are state-of-the-art devices for highly sensitive detection of mid- to far-infrared radiation. They are employed for astronomical imaging e.g. in the orbital Spitzer Space Telescope.

The BIB concept as introduced in [35] is based on that of extrinsic detectors. However, as this sort of devices conventionally features relatively low concentrations of impurity sites, huge layer thicknesses in the mm range are required in order to achieve sufficient internal quantum efficiencies and thus responsivities. Consequently, conventional intrinsic detectors cannot be processed into arrays required for imaging purposes. The basic idea behind BIBs is a tremendous increase in the dopant concentration in order to push the internal quantum efficiency into regions, where the use of thin devices in the micrometer range and thus the fabrication of arrays becomes possible. Naturally, this increase in the impurity concentration would lead to an intolerably high dark-current, whose suppression demands new strategies. The purpose of blocking the strong dark-current in a highly doped device is served by a so-called blocking layer (BL), which is composed of ultra-pure material and is located between the highly doped impurity layer (IL) and one of the electric contacts. The efficient blocking of the dark-current in a BIB is crucially dependent on an extremely low concentration of impurities in the blocking layer. Therefore, the realization of a sufficiently pure blocking layer is one of the major technological requirements for the fabrication of BIB devices of high detectivities.

In order to induce a well-controlled electric field profile in the biased BIB device, a thoroughly defined low counter-doping is included in the impurity layer along with the majority doping. In the vicinity to the blocking layer, a depletion zone develops in the impurity layer. In an optimized BIB device, the high fields present in this depletion region enable the operation of the device as a solid state photomultiplier [36].

Up to now the BIB concept has been applied to Si-based structures grown by molecular beam epitaxy, where the growth of a sufficiently pure BL on top of a highly doped IL is a technologically extremely delicate process. As a consequence, the growth of BIB devices is only possible in highly specialized MBE systems. Si BIBs are sensitive to infrared radiation down to 200 cm^{-1} [37], where the goal to push their sensitivity further into the far-infrared has motivated considerable endeavor to implement the concept in the Ge [38, 39] and GaAs [40] material systems. In the GaAs system, the formation of an impurity band in epitaxial layers has been demonstrated. However, the realization of an operational BIB device in alternative material systems was prevented by technological difficulties in growing a BL of sufficient quality. Theoretical considerations aiming at finding alternate operation modes for circumventing these difficulties [41] have not been able to overcome the technological barriers so far.

The proposal of abandoning the conventional, but troublesome and Si-confined, BIB fabrication technology of MBE growth by using ion-implantation as a means of introducing the required doping concentrations into an ultra-pure bulk substrate is highly promising. It holds the possibility of finally extending the BIB concept to other material systems as well providing technologically simpler means of fabrication for Si-based BIBs [42]. However, the device presented in [42] is a planar one, burdened by intrinsic disadvantages concerning array fabrication and exhibiting a non-competitive performance.

In the following chapter the fabrication and characterisation of the first ion-implanted vertical Si:B BIB is demonstrated. The next section of this chapter gives a theoretical background about the general operation of a BIB. Further, the fabrication of BIB structures will be described, where the fabrication process is based on two key technologies, namely ion-implantation and deep back-side etching through the Si substrate. Further, a characterization of the fabricated devices by Fourier transform infrared spectroscopy and by measuring the photocurrent response to black-body radiation will be given, revealing highly competitive figures of merit.

4.2 Theoretical background

The section at hand follows partly [43].

In a detector based on photoconduction, the generation of a measurable current signal by the absorption of an incident photon is fundamentally dependent on two processes: The excitation of a bound, immobile charge carrier into a state, which allows drifting along a field, and the subsequent transport of this charge carrier to the connections to external instruments. In most photoconducting structures, the photocurrent and the dark-current are caused in an analogous way, merely differing in the detailed mechanism of excitation. In contrast, in an optimally operating BIB, the generation of dark-current and photocurrent are spatially sep-

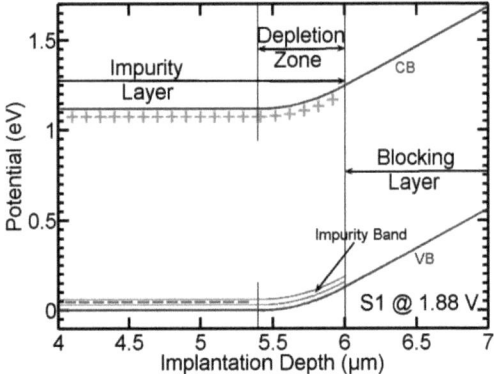

Figure 4.2: Band structure of a Si:B blocked impurity band detector. Due to a high majority doping concentration, an impurity band of significant width forms, originating from the negatively charged acceptor levels. Additionally, all counter doping sites (donors, phosphorus) are ionized. The bias applied to the sample drives impurity band electrons out of the impurity layer region next to the blocking layer, thus creating a space charge region, or depletion zone. In this depletion zone, high fields are present. The device operation is discussed in the text.

arated, thus resulting in fundamentally different generation mechanisms for both processes.

The absorption layer of a BIB consists of a highly doped semiconductor, conventionally Si, where the main doping concentration N_m is high enough for the formation of an impurity band, within which efficient hopping transport is possible. At the same time, a low counter-doping concentration N_c is present. At low temperatures, the charge carriers freeze out in the impurity levels. The detection of radiation is based on the excitation of charge carriers frozen out in the impurity band into the continuum of the associated band. Due to the high concentration of impurities, the quantum efficiency of this process is sufficient for achieving high responsivities even for device layer thicknesses in the micrometer range. This, in turn, allows the fabrication of BIB arrays. A blocking layer next to the impurity layer prevents the generation of dark-current via hopping transport through the device, where the purity of this blocking layer is the most critical issue for BIB operation and fabrication. This is due to the fact that the ability of the blocking layer to suppress dark-current essentially depends on the absence of impurities. The sample is contacted on both the impurity layer and the blocking layer in order to apply a voltage and extract a current.

Figure 4.2 shows a schematic of the band structure of a device with p-type main doping and an n-type counter doping. Due to the high ratio between main and counter-doping concentration, all counter-doping sites are ionized at operating temperature. Charge neutrality gives the same amount of ionized main doping atoms. When a positive external bias is applied to the impurity layer in respect to the blocking layer, the excess electrons of these ionized main doping sites are driven out of a region close to the blocking layer via hopping transport, and a depletion zone is formed. Thus, this depletion zone becomes charged with a charge concentration given by the counter doping concentration N_c (donors). The width of the depletion zone w depends on the amount of the applied bias V_b and is obtained by solving Poisson's equation. The result is given in Eqn. 4.1, where t_B is the width of the blocking layer.

$$w = \sqrt{\frac{2\epsilon\epsilon_0}{eN_c}|V_b| + t_B^2} - t_B \qquad (4.1)$$

If incoming radiation of sufficient photon energy hits the biased structure, electrons are excited from the valence band into the impurity band, and a hole in the valence band continuum is generated. The required photon energy depends on the energetic depth of the doping impurities, but also on the main doping concentration. This is due to the fact that for the concentrations employed in a BIB, the formation of a significantly broad impurity band results in a decrease of the minimal ionization energy of the impurities. Therefore, responsivity onsets in the terahertz frequency range can be achieved with BIBs. If the generation of a valence band hole occurs in the depletion zone, this charge carrier is driven by the present electric field into the blocking layer. At the same time, the electron excited into the impurity band is driven away from the depletion region.

Now, in the absence of dark-current, the total of the bias applied to the device drops across the depletion zone, leaving the neutral region of the impurity layer field-free. Due to the lack in a separating force, holes generated in this neutral region recombine with the associated impurity band electron, before they can reach the depletion region. In such a case, exclusively carriers generated in the depletion zone of the IL contribute to a current through the sample contacts. The internal quantum efficiency η for the detection of radiation is therefore given by Eqn. 4.2, where $\alpha(\nu)$ is the absorption coefficient induced by the impurities.

$$\eta = 1 - e^{-\alpha_{abs}(\nu)w} \qquad (4.2)$$

As can be concluded, the quantum efficiency can be driven to values above 90%, and rises with the applied bias up to a voltage, for which the depletion width is equivalent to the impurity layer thickness. Now, by carefully adjusting the counter-doping concentration and the applied field, a controlled dark-current can be drawn from the resulting high field region by tunnelling of valence band electrons into the impurity band. According to Ohm's law, this dark-current induces a field in the impurity layer even outside of the depletion zone, causing photoexcited holes generated in *any* part of the impurity layer to drift into the depletion zone. The details of this mode of operation are discussed below.

An important aspect of BIBs are the high fields present in the depletion zone in combination with the low ionization energy of the shallow main dopants. As holes generated by the absorption of photons drift along a sufficiently high field in the depletion zone, they gain enough energy to ionize additional impurity band sites. These secondary holes in turn gain energy while drifting along the strong field in the depletion zone and generate another pair of holes and ionized impurity sites by impact ionization. In case of a correctly designed and biased BIB, each absorbed photon generates an avalanche of holes in the valence band of the depletion zone, which further drifts across the blocking layer to the sample contact. Note that this avalanche multiplication process is restricted to valence band holes only. The electrons in the impurity band generated during the detection process drift to the opposite sample contact via hopping transport. This kind of transport is based on the sequential scattering of electrons between the individual impurity sites. Thus, the electrons are not continuously accelerated by the electric field like the valence band holes, but rather hop from impurity site to impurity site, thus unable to acquire sufficient kinetic energy for the impact ionization of additional carriers. Consequentially, the train of impurity band electrons generated along with the avalanche of valence band holes drifts across the impurity layer to the sample contact without inducing any further impact ionization. This restriction of the avalanche multiplication process to holes is highly favorable, as it reduces noise due to a limitation of the possibilities for impact ionization. The inability of impurity band electrons drifting across the impurity layer to experience avalanche gain is further highly valuable for the suppression of dark-current, as discussed later on in this section.

The avalanche gain g of photoexcited holes allows BIBs to feature responsivities, where one single photon can cause a charge burst easily distinguishable from the dark-current noise. Thus, properly designed BIBs can be employed as solid state photomultipliers. The impact ionization coefficient α_i depends on the electric field E as expressed in Eqn. 4.3, where E_c is the critical field. Thus the impact ionization gain rises quickly with the voltage applied to the BIB device.

$$\alpha_i = \sigma_i N_c e^{\frac{E_c}{E}} \qquad (4.3)$$

Due to the fact that the process of avalanche multiplication involves a series of individual events, this gain mechanism is quite noisy. But, even though the noise generated by the detection process itself is quite high, this does not necessarily hold true for the noise associated with the dark-current, which requires separate consideration.

As seen from Fig. 4.2, the electric field in a BIB reaches its maximum at the blocking layer/impurity layer interface. At low temperature, at which the thermionic dark-current is negligible, dark-current is merely generated by tunnelling of carriers from the valence band continuum into the impurity band. This tunnelling process is only efficient in regions, where the bands are highly bent due to the presence of a strong field. For conventional BIB operation, the fields present in the device are not high enough to generate a significant dark-current. However, for devices which qualify for single photon detection, a controlled tunnelling current has to be drawn in order to cause a drift of photoexcited holes into the gain region, that were generated outside of the depletion zone. Such devices feature counter doping concentrations high enough to generate fields sufficient for the generation of dark-current via tunnelling. Due to the localization of the field maximum, tunnelling current is generated preferentially near the blocking layer/impurity layer interface. This dark-current originating from tunnelling is driven across the blocking layer and reaches the sample contact, but due to the fact that the involved holes do not pass the depletion zone of the impurity band, *they do not experience gain*. As discussed above, impurity band electrons contributing to the dark-current cannot experience avalanche gain in principle, even though they drift across the depletion zone. Thus, for a properly designed BIB the dark-current does not experience avalanche gain, and the noise associated with it is expected to be negligibly low up to high voltages, for which avalanche multiplication of photoexcited charge carriers is already feasible. Nevertheless, the dark current in such devices is sufficiently high to induce a field in the neutral region of the impurity layer and thus a drift of *any* excited charge carrier into the gain region.

An additional feature of BIB devices originates from their special mechanism of charge carrier transport. Electrons excited into the impurity band can be removed via hopping transport, which is fast and resembles carrier drift along the electric field without any statistical processes like re-trapping or tunnelling. Consequentially, BIBs are not only highly sensitive, but also ultra-fast. While a single photon produces pulses of a few nanoseconds in temporal width, the read-out electronics usually broadens the response to some microseconds. However, in reference [35] electronic responses in the sub-microsecond regime were reported for a BIB device. According to [43], "the SSPM [solid state photomultiplier, BIB] provides time resolution some 10^8 times better than can be achieved with conventional photoconductors with similar NEPs [noise equivalent powers]."

Chapter 5

Ion-implanted blocked impurity band detectors

5.1 Alternative fabrication strategies

Conventionally, BIBs are fabricated by epitaxial methods, usually molecular beam epitaxy. As already stated in the introductory chapter 4, the proper operation of a BIB fundamentally depends on the purity of the blocking layer. The impurity band forming in the impurity layer exhibits efficient hopping-transport characteristics, and fields in a BIB have to be typically very high, especially when operating in the solid state photomultiplier mode. It is therefore essential, to keep the doping concentration of the blocking layer very low. An experimental value between $1 \cdot 10^{13} \mathrm{cm}^{-3}$ and $5 \cdot 10^{13} \mathrm{cm}^{-3}$ is reported in literature [37, 44] for the upper limit of the acceptor concentration in a BIB, which is equivalent to a specific resistivity higher than 300 Ωcm. Crystalline Si of resistivities in this range can be fabricated by Czrochalski and float zone pulling without any serious technological difficulties, and bulk substrates of a purity sufficient for blocking layers in a BIB are commercially available. On the other hand, the growth of a Si layer of this purity by *epitaxial* means is technologically highly demanding. MBE chambers capable of this process are highly specialized and have to be kept absolutely free of any contamination by dopants. Thus, only few research groups worldwide have demonstrated competency to fabricate BIBs of high quality.

As far as alternative material systems like GaAs or Ge are concerned, the difficulties associated with the growth of sufficiently pure blocking layers have ultimately hindered the successful implementation of a BIB device. The realization of BIBs in these systems has attracted considerable research interest due to their potential for shifting the sensitivity onset further down the terahertz spectrum. In [38] the authors conclude, that a long-wavelength threshold of 190 μm can be achieved for Ge:Ga BIBs, while according to [40] onset wavelengths higher than 220 μm could be reached by GaAs BIBs. The sensitivity onset for Si:B BIBs is found around 42 μm. Thus, establishing an alternative fabrication process for BIBs involving non-epitaxial methods not only promises a higher accessibility of BIB production, but also holds the potential of opening novel strategies for realizing BIBs in alternative material systems. Put differently, an alternative, epitaxy-free fabrication strategy has the potential of granting a technologically simpler access to ultra-sensitive detectors for radiation deep in the terahertz gap of the electromagnetic spectrum.

BIBs are based on a highly doped semiconductor layer, within which an impurity band can form. Now, in semiconductor technology there are two conventional methods for incorporating dopants into a semiconductor. One is the epitaxial growth of material containing the desired impurity concentration. This has been the method of choice for BIB fabrication so far. The second method is given by ion-implantation of impurities into the pure substrate, a standard process in semiconductor industries. Ion-implantation as a means of introducing a high impurity concentration into a substrate intrinsically holds the advantage, that the highly doped layer interfaces with a region of ultra-pure substrate remaining beyond the reach of the implantation process and the subsequent dopant diffusion during annealing steps.

Recent efforts aimed at producing Ge BIBs by ion-implantation [42]. The devices presented by Beeman et al. were planar ones, for which the impurity layer region was defined lithographically. Even though this enabled the formation of a relatively sharp interface between the impurity layer and the blocking layer, the devices exhibited lateral inhomogeneities due to the depth profile of the implanted doping concentration. As a result, at an applied voltage an inhomogeneous generation of photocurrent has to be expected, where the detailed drift and gain properties of a photoexcited carrier depend on the depth, in which the carrier was generated by an incident photon. It can be assumed that the inhomogeneities of the device presented in [42] *perpendicular to the photocurrent drift direction* lead to responsivities of the device not even competitive with those featured by simple Ge:Ga photoconductors. Further, planar devices are unsuitable for fabrication of detector arrays, which was after all the main motivation behind the initial development of BIB devices.

As the insufficiencies of the device presented in [42] are directly correlated with the nature of a planar device, the work presented in this chapter concentrates on the fabrication of a vertical device. A vertical BIB based on ion-implantation brings about two technological challenges: Firstly, the implantation of a relatively deep, homogeneous and sharply confined doping profile, and secondly, establishing an electric contact to the blocking layer of the device, which is formed by the pure substrate *underneath* the impurity layer.

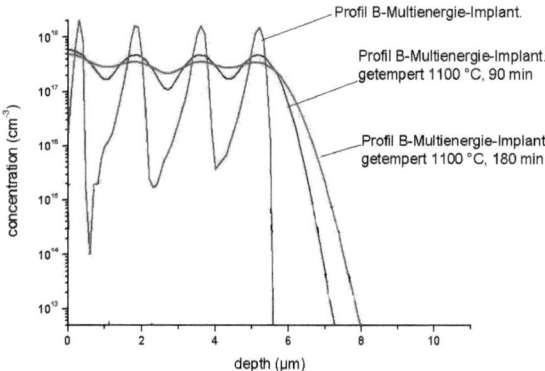

Figure 5.1: Simulated doping profile for **Linz 1 3 2008**. The red curves shows the doping profile directly after the four boron implantation steps given in table 5.1, but prior to the annealing process. The blue and magenta lines show the profile after a tempering step at 1100°C for 90 and 180 minutes, respectively.

Epitaxially grown photoconductive structures conventionally employ highly doped layers as bottom and top contacts of the vertical device. So do the BIB structures fabricated by MBE in e.g. [37]. This contacting approach allows the evaporation of metallic bonding pads on top of the highly doped bottom contact layer, which are laterally shifted from the etched device mesa. However, as the blocking layer of our structure consists of pulled intrinsic Si substrate material and is buried underneath the ion-implanted impurity region, this approach is not possible. In order to circumvent the necessity of a highly doped bottom contact layer, the blocking layer was directly accessed from the backside of the substrate wafer by a deep etching process through the handling wafer of an SOI substrate. This allowed the direct deposition of a metallic layer on top of the blocking layer totally covering the mesa area and thus creating an electric contact of a homogeneity unseen for planar devices. The etching is performed employing a Bosch-process, which is described in detail in the following.

In order to compare the devices fabricated in this work with conventionally fabricated BIBs, the choice of their target parameters was based on the BIB presented in [37]. The Si:B BIB discussed in this reference featured a blocking layer of 1 μm thickness and an acceptor concentration of less than $5 \cdot 10^{13} \text{cm}^{-3}$. The impurity layer had a thickness of 6 μm and was boron-doped with a concentration of $1 \cdot 10^{18} \text{cm}^{-3}$, which is just below the critical concentration for the semiconductor-metal transition. The counter-doping concentration, which determines the field profile within the BIB, was about $1 \cdot 10^{14}$cm^{-3}. Thus, for the sample **Linz 1 3 2008**, a targeted boron concentration of $1 \cdot 10^{18}$cm^{-3} was chosen for the impurity layer. For the counter-doping concentration, a value of $2 \cdot 10^{15}$cm^{-3} has been chosen. In order to be able to compare the device responsivities with those in reference [37], an impurity layer thickness of 6 µm was targeted.

The ion-implantation parameters required for meeting these specifications were determined by *Crystal-TRIM* simulations, which were performed by M. Zier and A. Kolitsch of the Institute of Ion Beam Physics and Materials Research, Forschungszentrum Dresden-Rossendorf. *Crystal-TRIM* simulates the implantation of ions into a single crystals and can derive depth distribution profiles. Figure 5.1 shows the simulation results of the boron distribution for a set of parameters finally used for the implantation process of sample **Linz 1 3 2008**. The dopants were induced in four implantation steps, whose parameters are given in table 5.1 for **Linz 1 3 2008**. This multi-implantation process results in four concentration peaks in approximately equidistant depths, as seen from the red curve in Fig. 5.1. However, the implanted ion dose is redistributed over the impurity layer of the sample during an annealing step, as was simulated for a temperature of 1100 °C and a duration of 90 (180) minutes. The resulting boron profiles are presented in Fig. 5.1 by the blue (magenta) curve. As concluded from the simulations, the boron doping profile achieved by the multi-step ion-implantation process with subsequent annealing should be sufficiently homogenous for BIB fabrication, and is expected to drop to values lower than the required blocking layer concentration at a sample depth of about 7 µm.

As the formation of a backside contact to the blocking layer requires etching through nearly the total thickness of the substrate material, and the etching is required to stop precisely in order to leave a well defined layer of pure substrate for the formation of the blocking layer, an efficient etch-stop is needed. This etch-stop is provided by the use of a commercial SOI (silicon-on-insulator) substrate, as thermal oxide is only weakly attacked by the Bosch-process. The substrate employed for the devices presented in this work consists of a handling wafer of 300 µm thickness and a 10 µm thick device layer, which are separated by 1 µm of thermal oxide.

5.2 Ion implantation

Based on the ideal parameters evaluated by the *Crystal-TRIM* simulations, the ion-implantation was carried out in the high-energy facilities of the Institute of Ion Beam Physics and Materials

Table 5.1: Ion-implantation sequence of sample **Linz 1 3 2008**:

Linz 1 3 2008	Boron		Phosphorus	
	Energy	Dose	Energy	Dose
Step 1	150 keV	$1 \cdot 10^{14}$ cm^{-2}	900 keV	$2 \cdot 10^{11}$ cm^{-2}
Step 2	1 MeV	$1 \cdot 10^{14}$ cm^{-2}	2.7 MeV	$2 \cdot 10^{11}$ cm^{-2}
Step 3	2 MeV	$1 \cdot 10^{14}$ cm^{-2}	5.5 MeV	$2 \cdot 10^{11}$ cm^{-2}
Step 4	3 MeV	$1 \cdot 10^{14}$ cm^{-2}	8 MeV	$2 \cdot 10^{11}$ cm^{-2}
Target concentration	$1 \cdot 10^{18}$ cm^{-3}		$2 \cdot 10^{15}$ cm^{-3}	

Research, Forschunsgzentrum Dresden-Rossendorf, where implantation energies of up to 10 MeV are available. The parameters for the two samples fabricated in the course of this work were chosen as given in tables 5.1 and 5.2. The target concentrations for sample **Linz 1 3 2008** were already commented on above. As suggested by the simulations, the implantation was carried out in four steps per dopant, featuring a constant dose but varying energies. The main doping was realized by implanting a boron dose of $1 \cdot 10^{14}$ cm^{-2} at energies of 150 keV, 1 MeV, 2 MeV and 3 MeV. The counter-doping was established by the implantation of $2 \cdot 10^{11}$ cm^{-2} of phosphorus at energies of 900 keV, 2.7 MeV, 5.5 MeV and 8 MeV.

For the implantation of sample **Linz 1 4 2008**, the implantation energies were chosen in analogy to **Linz 1 3 2008** in order to produce the same relative doping concentration profile. However, the doping doses were changed. For **Linz 1 4 2008**, the main doping concentration was reduced by a factor of two, while the counter-doping dose was decreased by a factor of four in comparison to **Linz 1 3 2008**.

The parameter expected to influence the BIB performance in the most critical way is the counter-doping concentration. As this concentration determines the width of the depletion zone, it decides whether establishing a strong avalanche gain succeeds and whether proper BIB operation is possible. The two samples studied in this chapter therefore differ in their counter-doping concentration in order to investigate, how drastically the operation of a BIB device is influenced by small deviations in the quantity of counter-dopants. This is of particular interest, as reaching a well defined counter-doping concentration in the extremely low range given for **Linz 1 3 2008** and **Linz 1 4 2008** is highly demanding in case of BIB fabrication by MBE growth.

Further, both changes from **Linz 1 3 2008** to **Linz 1 4 2008** should result in a reduction of the dark-current for a fixed bias. This is due to the fact that reducing the main doping concentration increases the impurity ionization energy by a decrease of the width of

Table 5.2: Ion-implantation sequence of sample **Linz 1 4 2008**:

Linz 1 4 2008	Boron		Phosphorus	
	Energy	Dose	Energy	Dose
Step 1	150 keV	$5 \cdot 10^{13}$ cm^{-2}	900 keV	$5 \cdot 10^{10}$ cm^{-2}
Step 2	1 MeV	$5 \cdot 10^{13}$ cm^{-2}	2.7 MeV	$5 \cdot 10^{10}$ cm^{-2}
Step 3	2 MeV	$5 \cdot 10^{13}$ cm^{-2}	5.5 MeV	$5 \cdot 10^{10}$ cm^{-2}
Step 4	3 MeV	$5 \cdot 10^{13}$ cm^{-2}	8 MeV	$5 \cdot 10^{10}$ cm^{-2}
Target concentration	$5 \cdot 10^{17}$ cm^{-3}		$5 \cdot 10^{14}$ cm^{-3}	

the impurity band. Thus the thermal generation of dark-current is hindered. Reducing the counter-doping concentration leads to an increase in the depletion zone width according to Eqn. 4.1 and a decrease in the maximum field value present in the structure. This makes the generation of dark-current less efficient. In addition, for a controlled flow of dark-current, a decreased main doping concentration and the consequential reduction of the hopping conductivity in the impurity band leads to a stronger field in the neutral region of the impurity layer. This is favorable in order to draw holes generated outside the depletion zone into the gain region. On the other hand, a reduced main doping concentration results in a reduction of the quantum efficiency of the layer. The two latter effects partly compensate each other in their influence on the responsivity of the BIB device.

Following the implantation process, the samples are tempered at 1100°C for 90 minutes in order to anneal the material damaged by the ion bombardment and incorporate the boron and phosphorus atoms into the Si monocrystal. The choice of the anneling parameters ensures, that the SOI compound remains undamaged, as the thermal oxide is stable up to 1300°C and is elastic at 1100°C. Thus, apart from a few surface sliplines no dislocations are expected [46]. As already stated above, according to the simulations the implantation and annealing processes are expected to result in a homogeneous doping concentration down to a substrate depth of about 6 μm, and at 7 μm the concentration of both impurities reaches values sufficiently low for a blocking layer. Therefore, the region of the device layer of the substrate, which serves as the blocking layer of the ion-implanted BIB, exhibits a thickness of about $3 - 4$ μm.

5.3 Processing and backside etching

Following the implantation and annealing steps given in the previous section, the BIB samples were processed into mesas and contacted by employing a deep backside etching process. A complete overview over the total fabrication process of the BIB samples is given in table

5.3 and illustrated in Fig. 5.2. Whenever the expressions *device layer* and *handling wafer* are mentioned in this table, they indicate the sample side, which is subject to the respective processing step. The backside etching sequence was chosen to close the processing order, as the handling of the free-standing, 10 μm thin mesas might have proven troublesome, if this sequence had been carried out earlier in the process.

The fabrication steps implemented at the Forschungszentrum Dresden-Rossendorf were followed by a plasma deposition step covering the handling wafer backside by 800 nm of silicon oxide, which served as an etching mask during the final Bosch-process step. The oxide was then structured by an reactive-ion-etching (RIE) step, which was masked by lithographically defined photoresist. The RIE process employed a plasma generated by a radio frequency (RF) field of 80 W in an atmosphere composed of 6.2 sccm CF_4 and 30 sccm H_2 at 40 mTorr. According to [45], the plasma generated in this gas combination preferentially etches silicon oxide with an selectivity ratio of approximately 2:1 in comparison to Si. By these means a silicon oxide mask featuring quadratic openings of 360×360 μm^2 (**Linz 1 4 2008**) and 320×320 μm^2 (**Linz 1 3 2008**) was defined.

Subsequently, a standard metal deposition and structuring step was performed on the device layer. The photolithographically structured resist mask was aligned to the silicon oxide mask on the sample backside using the backside alignment capability of the EVG 620 mask aligner. The results were quadratic frame contacts composed of 200 nm thick aluminum, which were directly opposing the holes in the oxide mask on the sample backside.

The subsequent processing sequence required an etching step through the device layer down to the thermal oxide of the SOI wafer, in order to define the detector mesas. This etching step No. 16, as well as that used to etch through the handling wafer in order to access the blocking layers of the BIB mesas for contacting, employed a Bosch-etching process implemented in a Oxford Instruments Plasmalab System 100 reactive ion etcher.

The Bosch-process [47, 48] allows the highly anisotropic etching of silicon with high etching rates of up to 10 μm per minute, where the aspect ratio can be up to 1:30. Additionally, photoresist and silicon oxide are highly inert to the process, with a selectivity ratio of 1:150 for oxide. The process is usually employed for MEMS fabrication. It consists of two steps, which are carried out alternately, one for etching Si and one for passivating the Si surface. The etching step involves an atmosphere of 100 sccm SF_6 and 3 sccm C_4F_8 at a pressure of 30 mTorr in the etching chamber. A plasma is generated by inductively coupling the gas to the magnetic component of a radio frequency field, while the electric component is shielded. The advantages of this coupling induced plasma (ICP) are its low-energetic ions, its low pressure and its high density. The ICP source employed for the fabrication of our BIBs was operated at a power of 700 W. An additional radio frequency field of 25 W was applied during the etching step. One etching step lasts 7 seconds and chemically removes 400 nm of unprotected Si. For the passivation step, an atmosphere of 3 sccm SF_6 and 100 sccm C_4F_8 at 30 mTorr is employed. Again, the ICP source is operated at 700 W, while the power of the additional RF is as low as 10 W. During this 5 s long passivation step a polymer is deposited, which protects the sample surface from further chemical etching during the etching step. However, during the etching step the polymer at the bottom of the etching trench is sputtered off by directional ions, exposing the trench bottom to the chemical etching process. The side walls of the trench are still protected against further chemical attacks. The highly anisotropic Bosch-process therefore enables the fabrications of extremely deep trenches without any underetch-blur of the structure defined by the oxide mask.

Employing 25 steps of this Bosch process, the samples were structured into 400×400 μm^2 mesas, where the previously deposited top contact metallization, in respect to which the photolithographic mask for the etching process was aligned, formed frame contacts on top of the quadratic mesas. The Bosch-etching stopped at the thermal oxide of the SOI substrate.

As already stated before, the backside contacting of our devices by deep etching through the handling wafer forms an essential part of the fabrication process. The oxide mask for this etching step had already been formed in previous processing steps. In order to support this oxide mask, an additional photoresist mask was spun on and structured, equal to that used for the etching of the oxide mask. To ensure that no native oxide hindered the chemical etching during the Bosch-process, the samples were subject to an RIE step lasting 10 minutes with the same parameters as used for the oxide mask structuring. This dry etching process was the equivalent to an HF dip, which could not be employed due to the presence of the front side metallization. The samples were then directly transferred to the load lock of the Plasma Lab System 100, in which 602 steps of the Bosch-process were carried out, giving a nominal etching depth of 350 μm in unprotected silicon. However, as the masks could not stand the whole duration of the etching process, those parts of the handling wafer originally protected by the mask were thinned down to about 170 μm, a thickness which still enables comfortable handling of the samples.

In order to evaporate a bottom metallization directly onto blocking layer laid plain during the etching process, the thermal oxide, which served as an etch-stop in the trenches, had to be removed. Employing a dry etching process proved unsuccessful due to its low rates for thermal oxide. Therefore a wet etching step by HF had to be employed, a process which was tricky to implement while preserving the front side metallization composed of aluminum.

Table 5.3: Processing steps for BIB samples.

Step No.	Process	Equipment	Parameters
1	ion-implantation	implantation facility FZD	see previous section
2	annealing	annealing oven FZD	1100°C, 90 min
3	cleaning	acetone, methanol	
4	deposition oxide handling wafer	plasma deposition chamber	800 nm, 100°C
5	cleaning	acetone, methanol	
6	resist deposition handling wafer	resist 1818, spinner	40 s, 4000/s
7	softbake	oven	90°C, 15 min
8	photolithography oxide mask	mask aligner EVG 620	4.5 s exposition
9	developing	developer	1 min
10	oxide etching handling wafer	reactive-ion-etcher	29.4% CF_4, 30% H_2, 40 mT, 15% RF, 50 min, 800 nm
11	cleaning	acetone, methanol	
12	resist deposition device layer	resist 1818, spinner	40 s, 4000/s
13	softbake	oven	90°C, 15 min
14	photolithography top contact	mask aligner EVG 620	4.5 s exposition
16	developing	developer	1 min
17	native oxide removal	hydrofluoric acid	
18	vapor deposition	evaporation chamber, Al target	200 nm Al
19	lift-off	acetone	
20	cleaning	acetone, methanol	
21	resist deposition device layer	resist 1818, spinner	40 s, 4000/s
22	softbake	oven	90°C, 15 min
23	photolithography mesa	mask aligner EVG 620	4.5 s exposition
24	developing	developer	1 min
25	mesa etching	RIE, Bosch process	10 μm, 25 steps

Step No.	Process	Equipment	Parameters
26	cleaning	acetone, methanol	
27	resist deposition handling wafer	resist 1818, spinner	40 s, 4000/s
28	softbake	oven	90°C, 15 min
29	photolithography oxide mask	mask aligner EVG 620	4.5 s exposition
30	developing	developer	1 min
31	native oxide etching handling wafer	reactive-ion-etcher	29.4% CF_4, 30% H_2, 40 mT, 15% RF, 10 min
32	backside etching	RIE, Bosch process	350 μm, 602 steps
33	cleaning	acetone, methanol	
34	gluing to carrier wafer, sealing of device layer	resist 1818	
35	thermal oxide removal	hydrofluoric acid	several minutes
36	vapor deposition handling wafer	evaporation chamber, Al target	400 nm Al
37	contact alloying	rapid thermal annealing oven	380°C, 20 sec $N_2 + H_2$
38	backside contacting	conductive silver	
39	bonding	bonder, Al wire	force: 3.5, 5 time: 7, 8 power: 9.1, 8.1

The sample was placed on a piece of silicon exceeding the sample in size, where the sample's device layer side faced this carrier wafer. Then the edges of the sample were carefully sealed with photoresist using a syringe. Consequentially, the front side of the BIB device was protected by the carrier wafer during HF etching, while the photoresist used to glue the sample to the carrier prevented an intrusion of HF from the sample edges. The whole sample-carrier-compound was dipped into HF for several minutes, until the thermal oxide at the bottom of the etched trenches had been completely removed. The sample was then detached from the carrier wafer by acetone.

Directly after removing the thermal oxide, 400 nm of aluminum were evaporated to the handling wafer side of the samples, establishing a contact to the blocking layer of the BIB

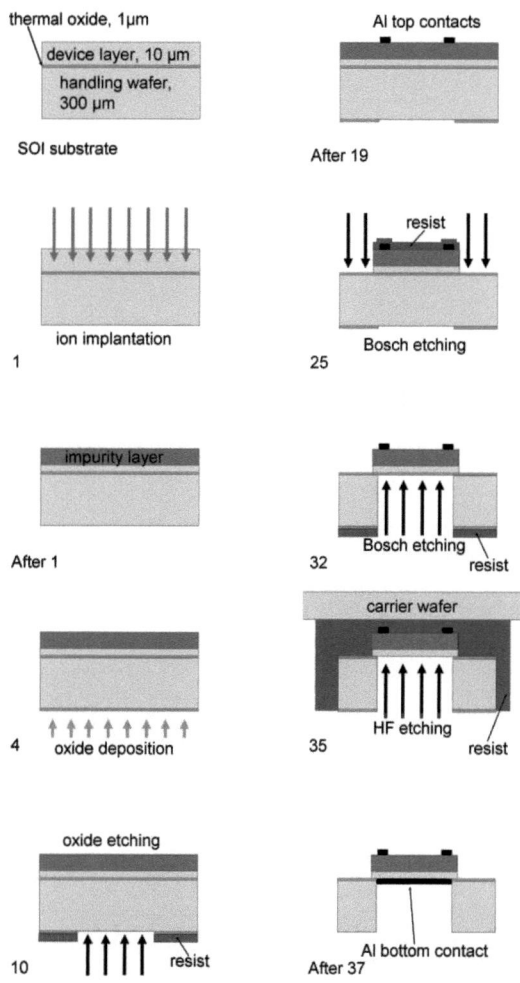

Figure 5.2: An illustration of a selection of the BIB processing steps given in detail in table 5.3. The numbers in the lower left corner of the individual sketches correspond to the processing step number in this table.

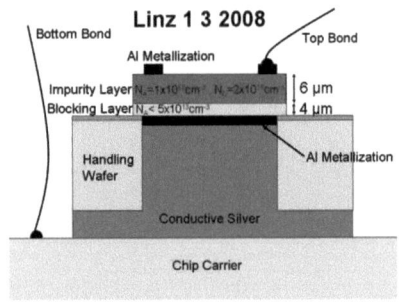

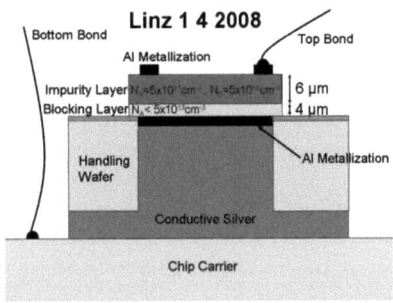

Figure 5.3: Schematic cross sections of the ion-implanted BIB devices **Linz 1 3 2008** and **Linz 1 4 2008**. Note that the blocking layer is located at the bottom of the device, in contrast to the device presented in [37] and shown in Fig. 5.4.

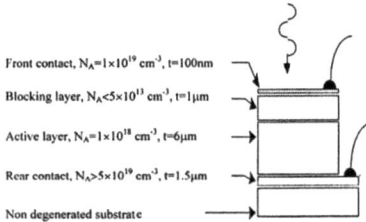

Figure 5.4: Schematic cross section of a BIB-device fabricated by molecular beam epitaxy [37].

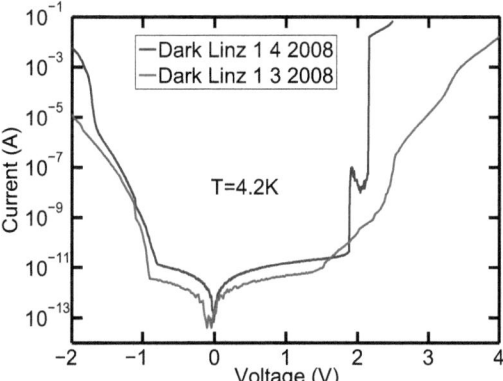

Figure 5.5: Dark-I-V characteristics. The red and blue lines show the dark-current curves of **Linz 1 3 2008** and **Linz 1 4 2008**, respectively, exhibiting a strong asymmetry and a breakthrough-behavior for high negative voltages. Note the threshold-like behavior of **Linz 1 4 2008** for positive voltages, which is associated with the onset of dark-current gain.

mesas in the trenches. The contacts were afterwards alloyed into the silicon by a rapid thermal annealing step (380°C, 20 s) in reducing atmosphere. After the contact alloying, the handling wafer side of the sample was connected to a gold-covered chip carrier by conductive silver. The conductive silver filled the etching trenches of the handling wafer and established a direct electric contact between the chip carrier and the blocking layer metallization of the mesas. The top contacts of the respective BIB mesas were connected to the chip carrier pins by Al wire bonding.

Figure 5.3 presents a schematic cross section of the final devices. Note that the devices fabricated in the course of this work are built up fundamentally differently than the MBE fabricated device presented by Leotin in [37] and shown in Fig. 5.4.

5.4 Photocurrent experiments

5.4.1 Dark-current characteristics

General observations

The dark-current characteristics of the samples **Linz 1 3 2008** and **Linz 1 4 2008** were measured under low background conditions at a temperature of 4 K by placing the devices in a liquid helium vessel. Employing a Keithley 236 SMU, a voltage sweep was performed, where a long stabilization time of 60 seconds was granted to every recorded point. The data were read out by a PC. The results for the dark-current-voltage characteristics are presented in Fig. 5.5 for both samples. The black and blue lines, showing the dark-current curves, exhibit a strong asymmetry due to the structure of the devices. The dark-current characteristics of sample **Linz 1 4 2008** exhibit features commonly associated with a BIB device.

For negative applied bias, a situation which is equivalent to biasing in reverse to the operational direction of the BIB, the dark-current characteristics of **Linz 1 3 2008** and **Linz 1 4 2008** feature a multi-exponential behavior including several kinks, similar to the observations made for a Si:As BIB in [49]. In this publication, Esaev states that the initial exponential section of the I-V characteristics can be associated with dark-current-injection over a barrier generated by the sample contacts for a case, in which the total voltage drops over the blocking layer of the BIB. The latter is true for low negative voltages due to the formation of an ultra-thin space charge region in the impurity layer next to its interface with the blocking layer, where the width of the space charge region is given by the high concentration of majority dopants in the impurity layer. The assumption that all of the applied bias drops over the blocking layer is based on the high hopping conductivity in the impurity band. Under negative biasing, the dark-current is produced by injection of electrons directly into the conduction band (for the Si:As BIB in [49]). Now, according to Esaev, a significant fraction of these carriers is captured by the impurity atoms and is then transported via hopping within the impurity band. If this hopping current reaches the maximum which is possible due to the amount of doping, the hopping conductivity is reduced, the ultra-thin space charge region cannot be sustained any more and the voltage ceases to drop entirely over the blocking layer. The voltage drop consequentially extends into the impurity layer. This, so Esaev, introduces a kink into the dark-current characteristics.

Dark-current of Linz 1 4 2008

For positive applied bias, the dark-current of **Linz 1 4 2008** remains below 10^{-10} A, which is equivalent to a current density of $1.6 \cdot 10^{-13}$ A/cm^2, up to high applied voltages around 1.9 V,

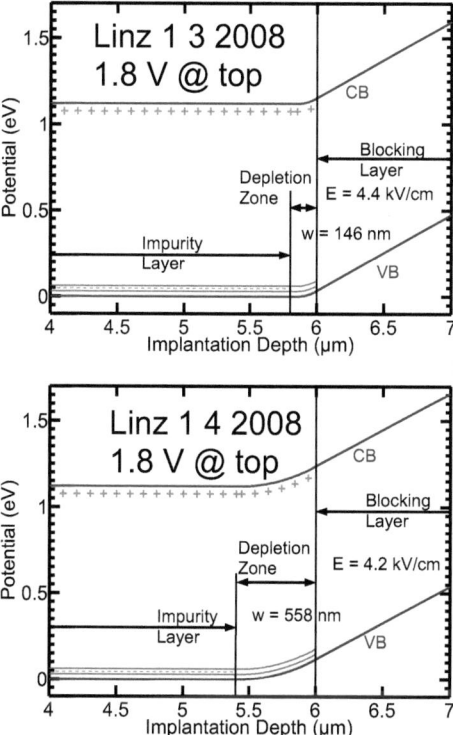

Figure 5.6: Bandstructure of both BIB samples for an applied voltage of 1.8 V. The plotted region was chosen in a way to highlight the depletion zone. The potential profile was calculated in absence of current flow, thus no field is present in the neutral region of the impurity layer. The blue lines indicate the band edges of the valence (VB) and conduction bands (CB). The red lines indicate the impurity band, where the + and - symbols represent charged counter- and main doping sites, respectively. Note that while the field in blocking layer, which is equivalent to the maximum field in the structure, is only slightly higher for **Linz 1 3 2008** (4.4 kV/cm) than for **Linz 1 4 2008** (4.2 kV/cm), the widths of their gain regions differ crucially. These differences originate from the unequal counter-doping concentrations of the two structures.

before rising in a sharp step over three orders of magnitude. The behavior of the BIB sample for positive applied voltages is fundamentally different from that at negative bias, which is a direct consequence of the asymmetric structure of the device. As already mentioned, at negative applied bias, charge carriers are injected from highly conductive (in our case metallic) contacts into the valence band (for Si:B BIBs) due to the absence of an impurity band in the blocking layer. On the other hand, for positive applied bias, which in our case resembles the bias direction of the operating BIB, the holes injected from the contact are trapped and freeze out in the impurity band at low temperatures of 4.2 K. And even though holes can be efficiently transported within the impurity band by hopping, at low bias the blocking layer prevents the carriers from reaching the sample's blocking layer contact. In order to generate a dark-current, electrons have to tunnel into the impurity band. This process is very inefficient at low applied voltages, leading to the observed low dark-current for a bias below 1.89 V in Fig. 5.5.

Next to the blocking layer/impurity layer interface a depletion region is established. The positively charged phosphorus ions lead to a significant voltage drop over the depletion region, where the depletion width decreases with increasing counter-doping concentration as seen from Eqn. 4.1. In an optimized BIB structure the counter-doping concentration is therefore chosen as low as possible while maintaining homogeneity, which is required to prevent local high-field regions from forming dark-current leaks.

At an applied voltage around 1.9 V, the dark-current of **Linz 1 4 2008** rises by orders of magnitude. A jump-like behavior of the dark-current characteristics analogous to that of **Linz 1 4 2008** was reported for the Si:As BIB in [49]. In [50, 51], this jump in the photocurrent is attributed to the impact ionization of additional impurity sites by the dark-current holes. The jump in the dark-current characteristics of **Linz 1 4 2008** for positive applied voltages can thus be associated with the onset of strong dark-current gain by avalanche multiplication. Esaev observed a dependence of the threshold voltage for this dark-current amplification by impact-ionization on the temperature of the device [49]. This dependence was again attributed to a change in the space charge profile and the consequential difference in the field distribution within the BIB. During the dark-current measurements performed on the BIB samples subject of this work, the precise threshold voltage for the jump in the dark-current was extremely sensitive to the speed, at which the voltage was scanned by the SMU. Only by choosing long stabilization delay times in order to overcome capacitive effects, reproducible dark-current characteristics could be obtained. These findings are thus in line with the explanation for the temperature dependence of the voltage threshold given in [49].

Dark-current of Linz 1 3 2008

For negative applied bias, **Linz 1 3 2008** behaves in analogy to **Linz 1 4 2008**. However, the dependence of its dark-current on positive applied bias differs from that of **Linz 1 4 2008**. The reason for this can be found in the band structure of both devices, which is shown in Fig. 5.6 for an applied bias of 1.8 V.

As can be calculated from Eqn. 4.1, at this voltage the depletion zone of **Linz 1 4 2008** exhibits a width of 558 nm, assuming the absence of dark-current, which is not strictly correct but serves the purpose of highlighting the difference between the two samples. Under these conditions, the field present within the blocking layer amounts 4.2 kV/cm. Due to the higher counter-doping concentration of **Linz 1 3 2008**, its depletion zone is at 1.8 V only 146 nm wide, while the field in the blocking layer exhibits a value of 4.4 kV/cm.

The reason for the nearly identical field in the blocking layer of both samples is the fact that its thickness exceeds that of the depletion zone by one order of magnitude. Therefore, the field present in the blocking layer, E_B is mainly determined by its width t_B and the applied voltage V_b, and is thus given approximately by $E_B = \frac{V_b}{t_B}$. The relatively thick blocking layer of our samples pins the maximum field in the device at a value nearly independent of the field distribution within the impurity layer, which should increase the stability of the dark-current behavior of the BIBs.

As expected from both the higher boron (stronger impurity band broadening) and phosphorus (higher electric fields) concentrations, the dark-current of **Linz 1 3 2008** starts to increase at lower voltages of 1.5 V than for **Linz 1 4 2008** at 1.89 V. Apart from the differing onset voltages for significant dark-current, the shape of the dark-current rise is fundamentally different. For **Linz 1 3 2008** the dark-current originates from valence band electrons tunnelling into the impurity band at the high-field interface between impurity layer and blocking layer, and rises slowly with the applied bias. Along with the tunnelling of electrons, holes are left back in the valence band. For **Linz 1 4 2008** the dark-current remains below 10^{-10} A up to 1.89 V and then rises abruptly, indicating that the hole current generated by tunnelling experiences gain by impact ionization of impurity band sites. As already mentioned, the latter was also observed in [49–51] for BIB samples grown by MBE. The reason for the absence of a jump-like increase in dark-current of **Linz 1 3 2008**, which is interpreted as the lack of dark-current amplification by impact ionization, can be found in the width of the gain region of this sample. For **Linz 1 3 2008** the width of the depletion zone, within which gain by impact ionization is possible, is by nearly a factor 4 lower than for **Linz 1 4 2008**. The conclusion can be drawn, that for the sample **Linz 1 3 2008** the initial dark-current generated by tunnelling does not experience avalanche gain. This might imply, that this sample does not exhibit avalanche gain properties at all due to the strong counter-doping and the consequentially short depletion zone, which would be unfavorable for BIB operation in the high-gain regime. Another possible explanation for the lack of dark-current gain is the generation of the majority of dark-current holes close to the interface with the blocking layer, as in this case these holes would not pass the gain region of the device. The latter would be highly favorable for low-noise BIB operation. The final answer to this question is delivered by experiments under illumination, which are discussed in the following section.

Dips in the dark-current characteristics

An additional remarkable feature of the dark-current characteristics of **Linz 1 4 2008** is given by the dip in the current around 2.1 V. This dip is reproducible, when keeping the stabilization delay during the measurement long and might be due to a change in field distribution similar to that responsible for the temperature dependence of the threshold voltage for the dark-current gain observed by Esaev [49] and/or a self-stabilization of the gain process [52,53]. Shadrin states in [53], that the spatially varying avalanche gain $M(x)$, which originally depends strongly on the externally applied bias, has to saturate. The maximum of the avalanche gain for a given applied voltage is spatially located at the impurity layer/blocking layer interface, as there the field in the structure is strongest. The saturation of the gain at this interface, $M_s(w)$, is reached, when the current generated in the valence band by avalanche multiplication is too high to enable redelivering of the excited carriers by hopping transport in the impurity band. From this point on, the additionally applied voltage drops solely at the charge-neutral region of the impurity layer, and does therefore not influence the depletion zone and the gain. The condition for the saturation of the avalanche gain is given by Eqn. 5.1.

$$\sigma_p \cdot M_s(x) = \sigma_n$$
$$\mu_p \cdot p \cdot M_s(x) = \mu_n N_D \qquad (5.1)$$

In this equation, σ_p resembles the hole conductivity, which depends on the amount of charge carriers in the valence band p and the hole mobility μ_p. The hopping transport of holes within the impurity band is based on the transport of electrons in the opposite direction. Thus, the hopping conductivity σ_n is given by the electron mobility, multiplied by the number of electrons in the impurity band of the neutral region, which is equal to the amount of counter-doping sites N_D. Trivially, p rises monotonically along the structure due to avalanche multiplication and the lack of re-trapping of valence band holes into the impurity band. Thus, as long as N_D is constant, the saturation gain is reached at the impurity layer/blocking layer interface and has a well-defined value, namely $M_s(w) = \frac{\mu_n}{\mu_p} \frac{N_D}{p(w)}$. However, in our structure the counter-doping concentration might vary along the growth direction over the impurity layer, and Eqn. 5.1 can therefore be fulfilled for more than one value of $M(x)$, and at different

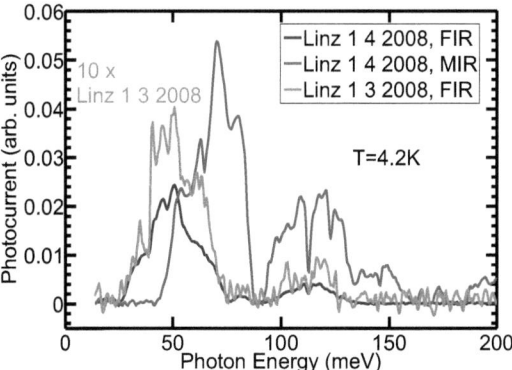

Figure 5.7: Photocurrent spectra of **Linz 1 3 2008** and **Linz 1 4 2008**. The blue and red curves present the spectral response of **Linz 1 4 2008** to the radiation source in a far-infrared and a mid-infrared spectrometer, respectively, at liquid helium temperature and an applied bias of 1.7 V. The green line shows the spectrally resolved photocurrent of **Linz 1 3 2008** in the far-infrared region at a voltage of 2 V. Note that for both samples the onset of photoresponse lies around 30 meV.

lateral positions x. This probably causes a meta-stability in the saturation gain and poses a possible explanation for the observed dip in the dark-current characteristics as seen in Fig. 5.5.

Dark-current conclusions

From the dark-current characteristics of both samples the conclusion can be drawn, that the blocking layer serves its purpose of suppressing the dark-current sufficiently well up to high voltages. Their dark-current characteristics clearly exhibit features, which are typically associated with BIBs. This becomes particularly clear by the comparison to the I-V curves of samples merely consisting of contacted impurity layers without any blocking layer, as done by Esaev in [49] for conventionally fabricated BIBs. While **Linz 1 4 2008** exhibits a strong dark-current gain for voltages above 1.89 V, **Linz 1 3 2008** did not show any sign of avalanche multiplication in its dark-current characteristics.

5.4.2 Photocurrent spectra

Setup

As Si:B BIBs are sensitive to radiation in the terahertz frequency regime, the photocurrent experiments on the ion-implanted BIB devices had to be carried out in a helium cryostat featuring diamond inner and polyethylene outer windows. As the backside metallization of the BIB samples is directly connected to their chip carrier, the carrier had to be electronically isolated from the cryostat sample holder via a sapphire plate providing good thermal contact. A Bruker IFS113 Fourier-transform-interferometer was employed to determine the spectral shape of the BIB response to far-infrared radiation. The radiation of a mercury lamp passed the interferometer with a 6 μm mylar beamsplitter, and was cleared from spectral components with energies above the mid-infrared by a black polyethylene low pass filter. It then directly illuminated the BIB mesa, which was cooled to liquid helium temperature, under normal incidence. A voltage was applied to the top contact of the mesa under investigation in respect to its bottom contact by using the biasing capability of a Femto DLPCA-200 low-noise current amplifier. The photocurrent through the biased sample generated by the far-infrared radiation was measured by this current amplifier, where the voltage signal at its output was fed into the spectrometer. Further data processing involved the acquisition processor of the interferometer and a PC. While the sample adjustment was carried out with the help of a chopper and a lock-in-amplifier, the interferograms were recorded in fast-scan mode without chopping the radiation.

The spectral range of the setup described in the previous paragraph was limited to photon energies below approximately 70 meV. Thus, the response of our BIB to mid-infrared radiation had to be measured separately in a Bruker Vertex 80 FTIR featuring a KBr beam splitter. The interferometer radiation from a globar source was coupled out of the spectrometer using one of its exits and focused to the position of the BIB mesa by a parabolic 90° off-axis gold mirror with a diameter of 58 mm and a parent focal length of 76.2 mm. The spectrometer was controlled by a PC. The Femto DLPCA-200 amplifier was employed in a way analogous to the far-infrared measurements. Again, the interferograms were recorded in fast-scan mode, while the sample temperature was controlled by a Lakeshore 331 Temperature Controller.

Spectral results

The results of the spectrally resoved experiments are presented in Fig. 5.7. In this plot, the photocurrent response to the radiation from the FTIR is given without any division by the references or window transmission characteristics. The curves for **Linz 1 4 2008** were measured at a voltage of 1.7 V, that for **Linz 1 3 2008** at a bias of 2 V. For both samples, the response to terahertz radiation sets in around 30 meV. As the ionization energy for isolated boron is about 45 meV, this red-shift of the responsivity onset is a direct evidence for the broadening of the impurity levels and thus for the formation of an impurity band, which is essential for the operation of a BIB. This result is consistent with that reported for a MBE grown Si:B BIB in [37]. Note that the scales of the red and the blue curves are uncorrelated due to the fact that they were recorded by two different spectrometers. However, the photocurrent spectra of the two samples exhibit the same features. It shall be mentioned, that for sample **Linz 1 4 2008** viable photocurrent spectra could only be recorded for voltages up to approximately 1.8 V. For higher applied biases, the spectral photocurrent measurements were dominated by noise. Possible explanations for this observation will be discussed in section 5.4.4.

In order to determine the spectral shape of the actual responsivity of the samples, the spectra presented in Fig. 5.7 had to be related to the spectral characteristics of the radiation source and the transmission properties of both the optical components in the spectrometer and the cryostat windows. Further, the quantitative determination of the responsivity of the BIB samples requires black-body experiments, which are presented in the next section.

5.4.3 I-V characteristics under illumination

Setup

After observing clear indications for BIB operation of **Linz 1 4 2008** in both its dark-current curves (sharp jumps) and the spectral shape of its response (red-shift of the onset), I-V characteristics under illumination are employed to finally determine whether our samples qualify for BIB operation in the terahertz regime. For this purpose, radiation emitted by a black-body radiator at 300°C was collimated by a spherical mirror, and illuminated the respective BIB mesa under investigation. The mirror had a focal length of 150 mm and a diameter of 60 mm. The black-body radiation was modulated by a chopper, operated at a frequency of 364 Hz. The top contact of the mesa was connected to the output-high-pin of a Keithley 236 Source Measure Unit, while the sample bottom contact was connected to the input of a Femto DLPCA-200 low-noise current amplifier. The DLPCA-200 thus measured the current through the sample for an applied bias given by the Keithley 236. The output of the amplifier was fed into the voltage-input of a Stanford Research System DSP SR830 lock-in amplifier, which was triggered by the chopper signal. Both the Keithley 236 and the SR830 were connected to a PC via IEEE interfaces. The Keithley SMU was used to perform a voltage sweep of the mesa's top contact, where for each bias value the output of the LIA was recorded. Further, the measurement was split into eleven parts, where the appropriate amplification setting of the DLPCA-200 was chosen separately between 10^3 V/A

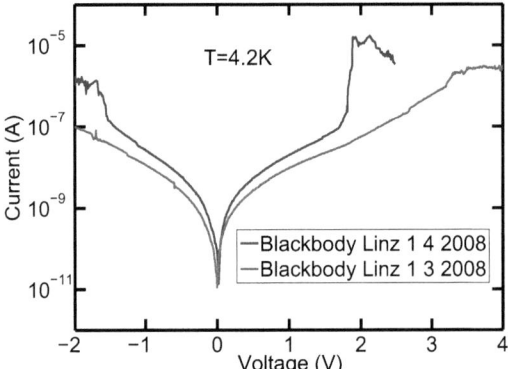

Figure 5.8: The red (**Linz 1 3 2008**) and blue (**Linz 1 4 2008**) lines present both samples' responses to 300°C black-body radiation. Similar to the dark-current in Fig. 5.5, the photocurrent response of **Linz 1 4 2008** shows threshold characteristics and features a sharp rise over two orders of magnitude, associated with the onset of high photocurrent gain by impact ionization.

(L) and 10^8 V/A (L) for each section of the sweep. Before each recorded value, the system was allowed to stabilize for 7 s (**Linz 1 3 2008**) or 20 s (**Linz 1 4 2008**). The resulting curves were scaled according to the amplification factor, and multiplied by a factor $2 \cdot \sqrt{2}$. The latter accounts for the LIA-specific scaling factor between the LIA output voltage and the peak-to-peak value of a sinus-shaped input signal at the chopper frequency. The value of interest for the data evaluation is the peak-to-peak voltage of the LIA input equivalent to the signal difference between the sample response under black-body illumination and that under background conditions. The scaled curves were stitched together, and the results are presented in Fig. 5.8.

I-V characteristics of Linz 1 3 2008

For **Linz 1 3 2008**, the photocurrent response to the black-body radiation increases rapidly with small voltages below 0.2 V. In this regime, the transport is limited by hopping. The initial rise was identified in [44] as a consequence of the field-dependence of the Poole-Frenkel effect, on which the hopping transport in the impurity band is based. The activation conductivity σ_i, as Aronzon [44] calls it, depends on the electric field E as stated in Eqn. 5.2.

$$\sigma_i = \sigma_{i0} \cdot \mathrm{e}^{-\frac{\epsilon_3}{kT}} \mathrm{e}^{\sqrt{\frac{0.69}{kT} \frac{4e^3 E}{\chi}}} \qquad [44] \qquad (5.2)$$

In this equation, ϵ_3 is the activation energy for the hopping transport. Following the steep, in the logarithmic plot square-root shaped initial increase induced by the field-dependence of the impurity band conductivity via the Poole-Frendel-effect, the photoinduced current rises slowly and continuously up to a high bias of 3.3 V. In [44], this flattening of the initial increase was attributed to a situtation, in which the dependence of the photocurrent on the applied bias is not any more limited by the hopping transport, but by the process of photoexcitation. A thorough discussion of this case is given in the next paragraph. The consideration of the dark-current characteristics presented in Fig. 5.5 leads to the conclusion that **Linz 1 3 2008** exhibits BIB operation up to a voltage of 1.5 V, as below this bias the dark-current via hopping transport in the impurity band is efficiently suppressed by the blocking layer, while the sample is highly sensitive to far-infrared radiation. However, within this voltage range the sample does not exhibit any signs of photocurrent gain via avalanche multiplication. The reason for this can be found in the extremely narrow depletion zone of **Linz 1 3 2008**. Put differently, the counter-doping concentration of this sample is too high to result in a gain region of sufficient width for avalanche multiplication. This finding it consistent with the conclusions drawn in section 5.4.1 from the absence of any jumps in the dark-current characteristics. Only at a high bias of 3.3 V a current jump of a factor of 2 can be identified as the onset of weak avalanche gain. However, at this voltage the dark-current is far too high for proper BIB operation.

I-V characteristics of Linz 1 4 2008

The photocurrent response of **Linz 1 4 2008** exhibits a similar behavior to that of **Linz 1 3 2008** up to a voltage of 1.8 V: After an initial steep increase up to a bias of 0.5 V it rises slowly with the applied bias. This initial, flat increase of the photocurrent response can be attributed to a widening of the zone in the impurity layer, from which photogenerated charge carriers reach the blocking layer by drifting along the applied field. There are two contributions to this increase in photogenerated carriers reaching the sample contact. First, the increase in the width of the depletion zone with the applied bias, and second, the increase in the field along the neutral region of the impurity layer due to a rise in the dark-current.

From an applied voltage of 1.8 V on, **Linz 1 4 2008** behaves fundamentally differently than **Linz 1 3 2008**. Between 1.8 and 1.9 V the photocurrent through **Linz 1 4 2008** features a sharp increase by two orders of magnitude, a rise which cannot be attributed to the increase in the amount of photogenerated charge carriers reaching the blocking layer. This abrupt rise can only be explained by the onset of avalanche gain, as this gain is expected to exhibit a threshold-like behavior: Above 1.8 V a hole generated by the absorption of black-body radiation gains sufficient energy to cause a charge avalanche by impact ionization while drifting along the field of the depletion region. Concluding from the height of the jump, the photoconductive gain by impact ionization reaches values of about 100. Petroff states in [36], that a detector device with a capacitance of 1 pF has to feature a built-in gain of about 200 to 300 in order to qualify for single-photon detection. The capacity of our devices approximately amounts 4 pF, as concluded from a given blocking layer thickness of 4 μm, a mesa area of $1.6 \cdot 10^{-7}$ m^2 and a dielectric constant of 12 for pure silicon. Thus, the avalanche gain factor of sample **Linz 1 4 2008** is of an order, which qualifies for single photon detection in the terahertz regime.

To recapitulate, the I-V characteristics of **Linz 1 4 2008** under illumination exhibit three regimes, namely that of an initial steep rise due to the activation characteristics of the hopping transport in the impurity band for low voltages, followed by that of a slow increase in the photogenerated current with the applied bias, and finally that of a jump-like rise with a subsequent saturation behavior for high voltages caused by avalanche multiplication. Esaev also observed three regimes for the total current under illumination through an epitaxially fabricated Si:As BIB [54]. In this work, Esaev attributes the final sharp rise around 2 V to a punch-through of the depletion region and the subsequent increase in dark-current. In contrast to the rise in the *total* current reported in [54], the jump observed in the characteristics of **Linz 1 4 2008** occurs in the *current originating from the modulated black-body radiation* measured in lock-in technique. This jump thus occurs in the voltage dependence of the responsivity itself and therefore clearly marks the onset of strong photocurrent gain. Further, remarkably the profile of the I-V characteristics under illumination above 1.9 V with its dips is reproducible. It is assumed that the dips in the characteristics originate from a bi-stability in the saturation gain, as was already discussed for dips in the dark-current in section 5.4.1.

The I-V-curves recorded under illumination by a black-body radiator together with the photocurrent spectra presented in section 5.4.2 could be used for determining the quantitative responsivity characteristics of both BIB samples fabricated in the course of this work.

5.4.4 Responsivity

Responsivity spectra

The spectral shape of the responsivity of the BIB samples was gained from the photocurrent spectra presented in Fig. 5.7. In this process, the photocurrent spectra recorded in the mid- and far-infrared at a certain bias V_b^{fix}, $S_{PC}^{MIR}(\nu, V_b^{fix})$ and $S_{PC}^{FIR}(\nu, V_b^{fix})$, were divided by the spectral characteristics of the respective spectrometer radiation sources, $S_{ref}^{MIR}(\nu)$ and $S_{ref}^{FIR}(\nu)$ measured as a reference using the internal DTGS detector of the FTIR. During

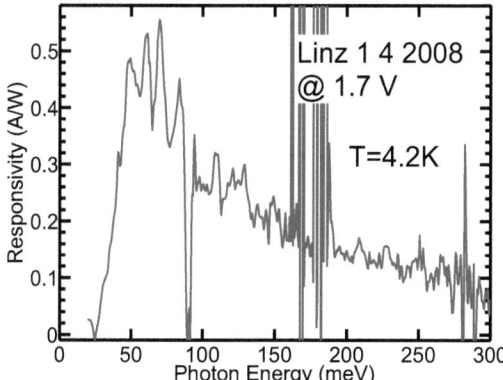

Figure 5.9: Responsivity spectrum of **Linz 1 4 2008** measured at 1.7 V. Note the onset of responsivity at low energies of 30 meV, which is shifted in respect to that for isolated boron at 45 meV due to impurity band broadening. The fringes are caused by resonances resulting from the device thickness of 10 μm.

the measurement of the spectral source characteristics in fast-scan mode a spectrally flat detector response had to be ensured by employing sufficiently slow mirror speeds. The resulting relative spectra for both infrared regions were scaled aiming at a matching overlap and stitched together. Additionally, the transmission characteristics of the cryostat windows, $T_{cryo}(\nu)$, had to be taken into account. This was necessary due to the fact that the polyethylene windows show a strong cut-off in the mid-infrared region from a photon energy of 150 meV on, as evident from the photocurrent spectra shown in Fig. 5.7. The transmission characteristics of one diamond/polyethylene window pair were experimentally determined by measuring the transmission through two pairs of cryostat windows in both the far- and the mid-infrared region, again employing the internal detectors of the spectrometers, and then taking the square-root of the resulting spectra. In analogy to the photocurrent spectra, the transmission spectra of the cryostat window pair recorded in two different spectral regions were scaled and stitched together. In this way, the uncalibrated responsivity spectrum $\tilde{R}(\nu)$ is gained, as given in Eqn. 5.3.

$$\widetilde{R}(\nu, V_b^{fix}) = \frac{S_{PC}^{MIR+FIR}(\nu, V_b^{fix})}{S_{ref}^{MIR+FIR}(\nu)\, T_{cryo}(\nu)}$$

$$C_{cal}(V_b^{fix}) = \frac{I_{blackbody}(V_b^{fix})}{\zeta(V_b^{fix}) \cdot A_{mesa}\left(1 - \cos\left(2\arctan(\tfrac{d}{4f})\right)\right)}$$

$$\zeta(V_b^{fix}) = \int_0^\infty \widetilde{R}(\nu, V_b^{fix}) \frac{\pi h \nu^3}{c^2}\left(\frac{1}{e^{h\nu/kT}-1} - \frac{1}{e^{h\nu/kT_{bg}}-1}\right) d\nu$$

$$R(\nu, V_b^{fix}) = \widetilde{R}(\nu, V_b^{fix}) \cdot C_{cal}(V_b^{fix}) \tag{5.3}$$

The calibration factor C_{cal}, which is also given in this equation, quantifies this experimentally obtained responsivity characteristics. In Eqn. 5.3, A_{mesa} is the mesa area of $16 \cdot 10^{-8}\,\text{m}^2$, d the diameter of the spherical mirror of 60 mm, f its focal length of 150 mm, T the temperature of the black-body radiator, T_{bg} the chopper wheel temperature of 300 K, and $I_{blackbody}(V_b^{fix})$ the value of the black-body induced photocurrent at the respective applied bias, as given in Fig. 5.8. The final result for the spectrally resolved responsivity $R(\nu)$ of sample **Linz 1 4 2008** is shown in Fig. 5.9 for an applied bias of 1.7 V.

As already discussed for Fig. 5.7, the sensitivity of the sample to infrared radiation sets on at about 30 meV, a value which is associated with the formation of impurity bands in Si:B BIBs according to [37]. The responsivity of **Linz 1 4 2008** peaks at around 70 meV, exhibiting fringes with a period of about 18 meV due to interference oscillations in the 10 μm thin device. The strong noise above 100 meV is due to the polyethylene window's weak transmission at these energies. The absence of any spectral information within a narrow band around 170 meV is attributed to a minimum in the beam splitter characteristics of the FTIR.

External quantum efficiency

Remarkably, when plotting $\frac{R(\nu) \cdot h\nu}{e}$, which gives the number of electrons reaching the sample contacts per incident photon, this quantity is found to be independent of the photon energy within the spectral range of our experiments. This experimental observation, which is presented in Fig. 5.10, lacks a simple or elegant explanation, as the detailed shape of the absorption coefficient of the impurity band depends on the dispersion and energetic width of the impurity band as well as on its symmetry and that of the heavy- and light-hole bands. Additionally, as seen in Eqn. 5.4, the external quantum efficiency scales with the featured photocurrent gain, where a dependence of the gain on the photon energy cannot be excluded. Nevertheless, this experimental finding implies that in the operational mode of solid state photomultiplication, the probability that an incoming photon produces an electron burst in

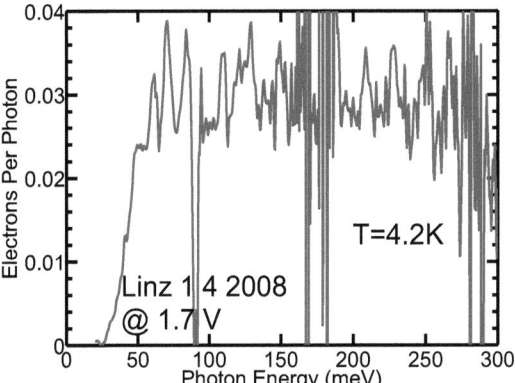

Figure 5.10: Spectrally resolved external quantum efficiency of **Linz 1 4 2008** measured at 1.7 V. Note that after a sharp onset around 30 meV, the number of charge carriers injected through the sample contacts per incident photon remains at a constant value of about 0.03 up to high photon energies. Again, the fringes are due to resonances resulting from the device thickness of 10 μm.

the sample is independent of the energy of the photon. In other words, our device detects photons of all wavelengths within a spectral region from 50 meV to 300 meV equally well.

Internal quantum efficiency

In order to compare the experimentally determined external quantum efficiency value of 0.03 with the absorption coefficient expected from the impurity concentration present in the sample, the definition of the external quantum efficiency η_{ext} as given in Eqn. 5.4 has to be considered.

$$\eta_{ext}(\nu) = \eta(\nu) \cdot g(\nu)$$

$$\eta(\nu) = \frac{\alpha(1-R_1)}{1-R_1 R_2 \cdot e^{-2\alpha(\nu)d_{IL}}} <\Theta>(\nu)$$

$$<\Theta>(\nu) = \frac{e^{-\alpha(\nu)d_{IL}}}{\alpha(\nu)}[(e^{\alpha(\nu)w}-1) + R_1(1-e^{-\alpha(\nu)w})]$$

$$\alpha(\nu) = N_A \cdot \sigma_{photoion}(\nu) \tag{5.4}$$

This equation follows [50, 51], where $\eta(\nu)$ is the internal quantum efficiency, $g(\nu)$ represents the photocurrent gain, N_A the boron doping concentration, $\sigma_{photoion}$ the photo-ionization cross section of the boron impurities, d_{IL} the thickness of the impurity layer, w the width of the depletion zone, and R_1 and R_2 the bottom and top reflectivities of the silicon sample (both 0.3). For sample **Linz 1 4 2008** a value of 700 cm^{-1} is gained for $\alpha(\nu_0)$ at the absorption onset, when using $\sigma_{photoion}(\nu_0) = 1.4 \cdot 10^{-15}$ cm^2, as found in [55]. Solving Eqn. 5.4 gives an internal quantum efficiency of $\eta(\nu_0) = 0.023$, a value close to the external quantum efficiency of 0.03. However, the model in Eqn. 5.4 is based on the assumption, that only impurity band carriers excited within the depletion region of the device contribute to a photocurrent. This assumption is only valid in absence of a significant voltage drop over the neutral impurity layer zone. In presence of a significant current flow within the impurity band, a field builds up within the neutral region of the impurity layer, driving charge carriers generated in this region into the depletion zone. If impurity band carriers contribute homogeneously to a photocurrent independent of the exact location of the absorption process within the impurity layer, w in Eqn. 5.4 has to be replaced by d_{IL}. In this case, the predicted internal quantum efficiency $\eta(\nu_0)$ is 0.3, implying a gain of 0.1 in case of sample **Linz 1 4 2008** at a bias of 1.7 V. As the actual generation process of photocurrent is to be found between the two extreme cases just discussed, the internal quantum efficiency of our sample lies between 0.02 and 0.3, resulting in gain values between 0.1 and 1.5 as derived from the measured external quantum efficiency and the calculated internal quantum efficiency.

Experimental challenges for BIBs under high radiation background

Note that Fig. 5.9 presents the spectrally resolved responsivity of sample **Linz 1 4 2008** at an applied bias of 1.7 V, a value which is below the onset of high photocurrent gain. For stronger biasing, the acquired photocurrent spectra suffered from heavy noise, which seemed to exceed that expected from the dark-current characteristics presented in Fig. 5.5. A possible explanation for this behavior is the fact that during the photocurrent experiments carried out in this work, the BIB samples were operated under high photon flux conditions, induced by both the radiation background at 300 K and the intensity of the FTIR light source. This operating condition results in two different mechanisms, which degrade the signal-to-noise ratio of the obtained measurement results. Both are discussed in the following.

Excess noise in a BIB

First of all, the noise generated in a BIB operated in the avalanche gain mode exceeds that of a photodetector based on conventional QWIP gain mechanisms. In a conventional QWIP, noise is generated by the statistical processes of photoexciting charge carriers from a quantum well into the continuum, and of re-trapping a fraction of these carriers, resulting in a gain g_{photo}. Equation 5.5 gives the spectral noise density $\frac{<i^2_{photo}>}{\Delta f}$ for a QWIP device, where $<i_{photo}>$ is the photocurrent generated by incident radiation.

$$\frac{<i^2_{photo}>}{\Delta f} = 4eg_{photo} <i_{photo}> \qquad \textbf{QWIP} \qquad (5.5)$$

In case of background-limited operation, that means under conditions, where the noise generated by the radiation background exceeds the dark-current noise, the detectivity $D^*(\nu)$ is given by Eqn. 5.6.

$$D^*(\nu) = \frac{e\,\eta(\nu)\,g_{photo}(\nu)\,\sqrt{\Delta f}\sqrt{A_{mesa}}}{h\nu\sqrt{<i^2_{bg}>(f)}} \qquad \textbf{QWIP} \qquad (5.6)$$

As the noise generated by the background radiation flux density $\Phi_{bg}(\nu)$ can be derived from Eqn. 5.5 using Eqn. 5.7, the detectivity of a QWIP under background limited conditions is independent of the photoconductive gain, as stated in Eqn. 5.8.

$$<i_{bg}> = e\,\Phi_{bg}(\nu)\,\eta(\nu)\,g_{photo}\,A_{mesa} \qquad \textbf{QWIP} \qquad (5.7)$$

$$D^*(\nu) = \frac{1}{h\nu}\frac{1}{2}\sqrt{\frac{\eta(\nu)}{\Phi_{bg}(\nu)}} \qquad \textbf{QWIP} \qquad (5.8)$$

Thus, for a QWIP a decrease of the signal-to-noise ratio, or detectivity, with the photocurrent gain could not be explained by Eqn. 5.8. However, in **Linz 1 4 2008** the detectivity decreases, when increasing the applied bias above 1.7 V. This was indicated by the extremely weak signal-to-noise ratio during photocurrent experiments at voltages above 1.7 V, which prevented the experimental determination of photocurrent spectra in this bias region. As concluded from Fig. 5.5, dark-current noise should not dominate the current signal through the device for voltages below 1.9 V. Thus, the drop in detectivity has to be related to the increase in gain between 1.7 V and 1.8 V, as highlighted in Fig. 5.12.

In BIB devices, an additional source of noise is introduced by the process of avalanche multiplication, which is built up by a series of statistic events depending on each other. This causality of avalanche processes is reflected in the integral form of the photocurrent present in a BIB structure, as given in Eqn. 5.9 following [50] and discussed in detail in chapter 5.5.1.

$$<i_{bg}> = e \int_0^w u_{bg}(x)M(x)dx \qquad \textbf{BIB} \qquad (5.9)$$

$$\frac{<i^2_{bg}>}{\Delta f} = 2e^2 \int_0^w u_{bg}(x)M(x)[2M(x)-1]dx \qquad \textbf{BIB} \qquad (5.10)$$

In this equation, $u_{bg}(x)$ represents the excitation rate of charge carriers by the radiation background, depending on the sample depth. $M(x)$ is the impact multiplication factor. The fact that avalanche multiplication obeys fundamentally different statistics than gain in a QWIP is reflected in the dependence of the background noise on the background current itself, which for a BIB cannot be expressed in a simple linear relation like that in Eqn. 5.5. As shown in Eqn. 5.10, the background current noise exhibits an integral dependence on the excitation rate and multiplication factor functions. As a consequence, the detectivity of a BIB under background limited conditions depends on the detailed gain mechanisms. In Eqn. 5.11, this is accounted for by the excess noise factor $\gamma\{M(x), G(x)\}$, which is actually a functional depending on the impact multiplication factor $M(x)$ and the gain factor $G(x)$, which is defined by Eqn. 5.20 and is discussed in detail in chapter 5.5.1.

$$D^*(\nu) = \frac{1}{h\nu}\sqrt{\frac{\eta(\nu)}{2\Phi_{bg}(\nu)\gamma\{M(x),G(x)\}}} \qquad \textbf{BIB} \qquad (5.11)$$

As shown in [50], this behavior ultimately leads to a decrease in a BIB's detectivity with increasing avalanche gain, a phenomenon which might explain the weak signal-to-noise ratio observed for **Linz 1 4 2008** at applied voltages above 1.7 V. Below the onset of strong avalanche multiplication, the radiation of the interferometer probably is above the detection limit imposed by the background noise. When the avalanche multiplication process sets in, the noise originating from the strong radiation background eventually exceeds the current signal through the BIB originating from the FTIR source, and the experimental determination of the spectral response becomes cumbersome.

Additional noise sources

Under conditions of strong illumination, our BIB samples might be subject to additional processes hindering the measurement of photocurrent spectra for high applied bias. The performance of the BIB samples might be suffering from non-linear photocurrent response caused by a saturation behavior of the gain as described in section 5.4.1 and 5.4.3 together with consequential bi-stabilities in the gain as discussed in section 5.4.1. Thus, the observed ultra-high noise in the photocurrent spectra recorded for voltages above 1.7 V is probably not only generated by conventional noise mechanisms, but could be a result of the flipping of the avalanche gain between different meta-stable saturation values due to the presence of strong photon flux in the interferometer. However, for biases below the high-gain regime, the recording of spectra with low noise is possible due to the absence of gain saturation.

The gain bi- or multi-stabilities would be especially detrimental for experiments carried out in the fast-scan mode of a FTIR. This is due to the fact, that during the scanning process through an interferogram, the intensity of radiation hitting the BIB is changed at a high frequency. Thus, considering the meta-stable gain states of the system, the detector is not given sufficient time to acquire an equilibrium. Therefore, for the proper recording of photocurrent spectra in the high-gain regime, a spectrometer featuring a step-scan mode would have a crucial advantage over the employed IFS113. Determining the spectral dependence of the responsivity in step-scan mode would enable the use of long sample stabilization times of the order of those employed during the measurement of the I-V characteristics under illumination as presented in Fig. 5.8. For these measurements, the stabilization time granted to the system between two recorded data points was as high as 7 or 20 s. However, the ideal conditions for the characterization of a BIB are given by low background and low signal photon flux, which would require dipping the sample into liquid helium and illuminating it by radiation from a light source located in the helium vessel.

Bias-dependence of responsivity and external quantum efficiency

The spectral shape of the absorption coefficient originating from valence band-impurity band transitions is expected to be independent of the voltage applied to the sample. Thus, even though photocurrent spectra could not be measured at voltages above 1.7 V, the voltage dependence of the sample responsivity at a certain photon energy could be determined by employing the spectrum recorded at $V_b^0 = 1.7$ V for the calculation of the (now) bias dependent calibration factor $C_{cal}(V_b)$, as given in detail in Eqn. 5.12.

$$\widetilde{R}(\nu, V_b^0) = \frac{S_{PC}^{MIR+FIR}(\nu, V_b^0)}{S_{ref}^{MIR+FIR}(\nu)\, T_{cryo}(\nu)}$$

$$C_{cal}(V_b) = \frac{I_{blackbody}(V_b)}{\int_0^\infty \widetilde{R}(\nu, V_b^0) A_{mesa}\left(1 - \cos\left(2\arctan(\frac{d}{4f})\right)\right) \frac{\pi h \nu^3}{c^2} \frac{1}{e^{h\nu/kT}-1}\, d\nu}$$

$$R(\nu^{fix}, V_b) = \widetilde{R}(\nu^{fix}, V_b^0) \cdot C_{cal}(V_b) \tag{5.12}$$

Responsivity of Linz 1 3 2008

Figure 5.11 presents the result for the voltage-dependence of the responsivity of **Linz 1 3 2008** to radiation with a photon energy of 50 meV as a solid red line. The broken red line compares this responsivity to the profile of the dark-current through the sample, shown on an arbitrary ordinate axis. The responsivity of **Linz 1 3 2008** lies about one order of magnitude below that of epitaxially grown Si:B BIBs as reported in [54], where a value of 8 A/W is reached for a sample temperature of 7 K. In comparison, at liquid helium temperature **Linz 1 3 2008** reaches values of 0.1 A/W at an applied voltage of 1.5 V, a voltage at which dark-current is as low as $9 \cdot 10^{-12}$ A. The sample presented by Esaev in [54] and **Linz 1 3 2008** exhibit a similar dependence of the responsivity on the applied bias. For both devices, the responsivity rises in good approximation exponentially, without any significant jump. As

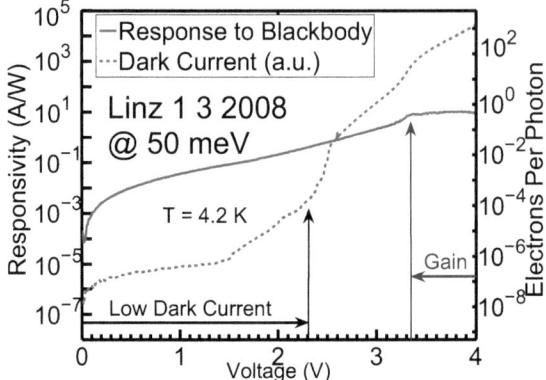

Figure 5.11: Responsivity and number of charge carriers per incident photon (external quantum efficiency) of **Linz 1 3 2008** at a photon energy of 50 meV. **Linz 1 3 2008** exhibits peak responsivity values of 10 A/W. In contrast to sample **Linz 1 4 2008**, the dark-current of **Linz 1 3 2008** (shown on an arbitrary ordinate axis) rises before the structure exhibits any avalanche gain. This is due to the fact that the counter-doping concentration is too high to enable the formation of a sufficiently wide depletion zone. Therefore **Linz 1 3 2008** does not feature strong avalanche gain, even though proper BIB operation is given.

already stated before, the counter-doping concentration of **Linz 1 3 2008** is too high to enable the formation of a depletion region sufficiently wide for a high multiplication gain before the onset of unacceptably high dark-current, resulting in the low responsivity values for this sample within this low-dark-current regime. At higher voltages of 3.5 V, the sample exhibits competitive responsivities of 10 A/W. However, in this bias region the dark-current generated by tunnelling from the valence band into the impurity band of the blocking layer rises up to $1.3 \cdot 10^{-3}$ A, disqualifying sample **Linz 1 3 2008** from proper BIB operation. Put differently, the bias regions of low dark-current and competitive photocurrent gain do not overlap.

Responsivity of Linz 1 4 2008

As seen in Fig. 5.12, for sample **Linz 1 4 2008** responsivity values rise tremendously after a slow initial exponential increase below 1.7 V. Between a bias of 1.82 V, which is identified as the onset of heavy avalanche gain, and 1.89 V, at which the dark-current rises rapidly, the responsivity increases jump-like from 1.1 A/W to 13.5 A/W, which corresponds to a rise in

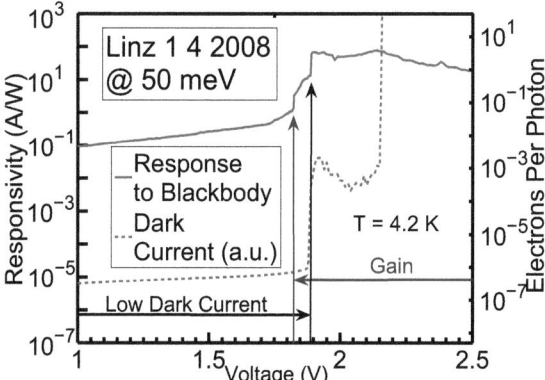

Figure 5.12: Responsivity and number of charge carriers per incident photon (external quantum efficiency) of **Linz 1 4 2008** at a photon energy of 50 meV. **Linz 1 4 2008** exhibits high peak responsivity values of 65 A/W and external quantum efficiencies of up to 3, although in this regime dark-currents (shown on an arbitrary ordinate axis) are high due to the presence of dark-current gain. However, the onset of high dark-current is shifted from the onset of photocurrent gain by 70 mV. Within this work-point window, external quantum efficiencies exceed 0.5, while dark-currents are still insignificant, qualifying **Linz 1 4 2008** for competitive BIB operation in the strong-gain-regime.

external quantum efficiencies from 0.06 to 0.7 eletrons per photon. These highly impressive values for the responsivity and external quantum efficiency of **Linz 1 4 2008** are reached within a bias window of 70 mV, within which the dark-current remains at remarkably low values of less than $5 \cdot 10^{-11}$ A. This voltage window represents the working point region of our sample, within which highly competitive BIB operation featuring strong avalanche gain is observed. For comparison, the responsivity value of 8 A/W shall be mentioned, as reported for conventionally fabricated Si:As BIB structures in [54].

The rise in the responsivity occurs in several jumps, which are reproducible and appear to originate from field-redistribution process and multi-stable gain saturation effects, as already discussed for the dark-current in 5.4.1.

At higher applied bias of 1.9 V, the responsivity of **Linz 1 4 2008** even reaches an impressive maximum value of 65 A/W, equivalent to an external quantum efficiency of 3.3 electrons per photon. Concluding from the abrupt rise of the responsivity over three orders of magnitude between 1.7 and 1.9 V, the avalanche multiplication gain reaches values of 100 or more at high voltages, where the exact figure remains unknown due to the lack of experimental data on the internal quantum efficiency. According to [35] and as already mentioned in detail in section 5.4.3, gain values of this order of magnitude qualify for single-photon detection. However, at these high voltages the dark-current has already risen dramatically to 10^{-7} A, a value which disqualifies from proper BIB operation in this bias region.

5.5 Simulation of figures of merit

In this section, an analytical model presented by Szmulowicz in [50] and [51] is applied to the BIB structures of **Linz 1 3 2008** and **Linz 1 4 2008** in order to compare the figures of merit determined experimentally in the previous section with the values predicted by this theory. However, due to the lack in precise figures for several parameters required for the definition of the model, the relevance of the detailed interpretation of the simulation results is restricted. The present section shall give a more detailed illustration of the BIB operation and the relevance of structural parameters for the figures of merit *on the basis of model calculations*.

5.5.1 Analytical model

The models by Szmulowicz as reported in [50] and [51] are based on the assumption, that the absorption of radiation relevant for the generation of a photocurrent occurs exclusively in the depletion zone of the impurity layer. The number of carriers excited by incident photons is subsequently enhanced by impact ionization in the depletion zone. Put differently, the processes included in this model take place exclusively within the charged region of the impurity layer. This simplification makes the model numerically highly accessible due to the fact that neither a current dependent field across the neutral region of the impurity layer, nor the presence of dark-current has to be considered. However, this assumption limits the use of the model for simulating BIB devices under conditions of significant current flow. Reference [50] reports on a model for front-side illuminated, conventionally fabricated BIBs, for which incident radiation enters the device from the blocking layer side. In [51], this model is expanded for backside-illuminated BIBs. As our devices are fabricated "upside down", with a blocking layer located deep within the sample, radiation enters the structure from the impurity layer side for front-side illumination and the latter model has to be used.

Analytical expression for the BIB responsivity

In the following, a BIB device with an impurity-band-induced absorption coefficient $\alpha(\nu)$ and an impurity layer thickness of d_{IL} is considered. The fraction of photons hitting the BIB surface, which reaches a sample depth x, where $x = 0$ corresponds to the blocking layer/impurity layer interface, is given in Eqn. 5.13.

$$\Pi(\nu, x) = \theta(\nu) \cdot \Theta(\nu, V, x)$$

$$\theta(\nu) = \frac{1 - R_2}{1 - R_1 R_2 \, e^{2\alpha(\nu)d_{IL}}}$$

$$\Theta(\nu, x) = e^{-\alpha(\nu)d_{IL}} \left(e^{\alpha(\nu)x} + R_1 e^{-\alpha(\nu)x} \right) \tag{5.13}$$

In this equation, R_1 and R_2 account for the reflection coefficient at the sample back and front surface, respectively. The number of photons absorbed per incident photon within a region dx at a sample depth x is thus given by $\alpha(\nu) \cdot \Pi(\nu, x)$.

At sufficient electric fields in the device, the charge carriers excited at a sample depth x experience gain by impact ionization of additional carriers. According to [50], the impact ionization coefficient follows Eqn. 5.14, where N_A is the main doping concentration in the impurity layer, σ_I the impact ionization cross section of the boron impurities, E_c the critical field for impact ionization and $E(x)$ the electric field at position x.

$$\xi(x) = N_A \sigma_I e^{-\frac{E_c}{E(x)}} \tag{5.14}$$

$$E(x) = \frac{eN_D}{\epsilon \epsilon_0} x \tag{5.15}$$

The impact ionization coefficient gives the average number of impact ionization incidents induced by a single valence band hole (for p-type BIBs) per unity length drifted along the applied field. Let us assume that a charge carrier is present at the location x, where it experiences a multiplication factor of $M(x)$. As the charge carrier as well as the carriers subsequently ionized by it drift by dx along the field, each generates additional $\xi(x)\,dx$ carriers. The multiplication factor thus has to increase with x by the number of additionally generated carriers and therefore obeys Eqn. 5.16.

$$dM(V, x) = \xi(x) M(V, x)\, dx \tag{5.16}$$

Partial integration leads to the solution shown in Eqn. 5.17.

$$M(V, x) = e^{N_A \sigma_I \Xi(V,x)}$$

$$\Xi(V, x) = w(V) e^{-\frac{W}{w(V)}} - (w(V) - x) e^{-\frac{W}{w(V)-x}} - W \int_{\frac{W}{w(V)}}^{\infty} \frac{e^{-z}}{z} dz +$$

$$+ W \int_{\frac{W}{w(V)-x}}^{\infty} \frac{e^{-z}}{z} dz$$

$$W = \frac{\epsilon \epsilon_0 E_c}{e N_D} \tag{5.17}$$

In this equation, $w(V)$ is the depletion width as given by Eqn. 4.1. In order to obtain an expression for the external quantum efficiency, the sum of all charge carriers generated by the absorption of a photon and the subsequent impact ionization, $\alpha(\nu) \cdot \Pi(\nu, x) \cdot M(V, x)$, between $x = 0$ and $x = w(V)$ has to be determined, as shown in Eqn. 5.18.

$$\eta_{ext}(\nu, V) = \alpha(\nu) \cdot \theta(\nu) \cdot \int_0^{w(V)} \Theta(\nu, x) \cdot M(V, x) \, dx$$
$$= \alpha(\nu) \cdot \theta(\nu) \cdot < \Theta M > (\nu, V) \tag{5.18}$$

The internal quantum efficiency $\eta(\nu, V)$ of the structure is given by Eqn. 5.19, summing up over the fraction of photons absorbed throughout the whole depletion zone.

$$\eta(\nu, V) = \alpha(\nu) \cdot \theta(\nu) \cdot \int_0^{w(V)} \Theta(\nu, x) \, dx$$
$$= \alpha(\nu) \cdot \theta(\nu) \cdot < \Theta > (\nu, V) \tag{5.19}$$

The expressions for the external and internal quantum efficiencies allow the definition of a gain factor $G(\nu, V)$ according to Eqn. 5.17, which has to be interpreted as the *average* number of charge carriers reaching the sample contacts per excited carrier.

$$G(\nu, V) = \frac{\eta_{ext}}{\eta}$$
$$= \frac{<\Theta M>}{<\Theta>} \tag{5.20}$$

Given the definitions introduced in Eqns. 5.13 and 5.17, the responsivity of a BIB device can be written as in Eqn. 5.21.

$$R(\nu, V) = \frac{e}{h\nu} \eta_{ext}(\nu, V)$$
$$= \frac{e}{h\nu} G \cdot \eta$$
$$= \frac{e}{h\nu} \alpha(\nu) \cdot \theta(\nu) < \Theta M > (\nu, V) \tag{5.21}$$

Analytical expression for noise in a BIB

This section deals with the fundamental difference between noise in a QWIP and in a BIB and explains the analytical expression for BIB noise as given in detail in [50]. This topic has already been brought up in section 5.4.4, but shall in the following be illuminated from a more mathematical point of view. In a quantum well infrared photodetector, gain is based on the fact that in order to re-trap as many continuum carriers as generated by an optical absorption process, significantly more carriers have to be injected via the device contacts than have been photoexcited. This gain effect is necessary to reach a stationary number of carriers trapped in the structure's quantum wells during illumination. The noise in a QWIP is the result of two statistical, independent events, namely the excitation of a charge carrier by a photon, and the re-trapping of a carrier by a quantum well. The emission current noise of a QWIP is thus given by two times the expression for shot noise, as given in Eqn. 5.22.

$$\frac{<i^2>}{\Delta f} = 2e <i_{em}> \qquad \textbf{shot noise} \qquad (5.22)$$

However, each fluctuation in the number of emitted carriers has to be balanced by a gain process, in order to maintain stationary conditions in the structure. Thus the noise current also is enhanced by the gain factor g_{photo}, and therefore obeys Eqn. 5.23. Note that $<i_{photo}> = g_{photo} <i_{em}>$, when comparing Eqns. 5.5 and 5.23.

$$\frac{<i^2_{photo}>}{\Delta f} = 4eg^2_{photo} <i_{em}> \qquad \textbf{QWIP} \qquad (5.23)$$

In a BIB, the gain is based on a completely different mechanism. BIB devices exhibit gain due to the impact ionization of additional impurity sites, where each ionization process is a statistical event and thus a source for noise. Furthermore, each charge carrier generated by impact ionization may cause an additional excitation of impurity band carriers, thus increasing the number of statistical events. The noise generated by an emission current $i_{em}(x)$ from the valence band into the impurity band at location x is subject to a multiplication gain $M(x)$, thus the increase in the noise current density within dx is given by $2eM^2(V,x) \cdot \frac{di_{em}(x)}{dx} dx$. As any charge carrier generated by photo- or impact-ionization in the valence band reaches the sample contact, the total noise in a BIB device is given by Eqn. 5.24. Note the difference between Eqns. 5.24 and 5.23.

$$\frac{<i^2>}{\Delta f} = 2e \int_0^w M^2(x) \cdot \frac{di_{em}(x)}{dx} dx \qquad (5.24)$$

$$\frac{di_{em}(x)}{dx} = \xi(x) \cdot i_p(x) + e \cdot u(x) \qquad (5.25)$$

The emission current in a BIB is increased within dx by carriers generated by the absorption of photons, $e \cdot u(x)$, and by carriers generated by impact ionization, $\xi(x) \cdot i_p$, as seen in

Eqn. 5.25. $i_p(x)$ is the hole current flow within the valence band, composed of carriers generated by one of the two mechanisms. Note that in a p-type BIB only valence band holes experience gain by avalanche multiplication. Equation 5.24 is solved by inserting 5.25 and substituting $M(x) \cdot \xi(x)$ by $\frac{dM(x)}{dx}$. Further, $i_p(x)$ is substituted by $i - i_{em}(x)$, where i is the total current through the device, which is independent of x due to the continuity condition. The latter substitution is valid, as the total current is built up by the transport within the valence band, $i_p(x)$, and the hopping transport within the impurity band, which has to be equivalent to the emission current $i_{em}(x)$ under equilibrium conditions. The result of these substitutions is given in Eqn. 5.26.

$$\frac{<i^2>}{\Delta f} = 2e \int_0^w \left[-\frac{dM(x)}{dx} M(x) \cdot i_{em} + \xi(x) M^2(x) \cdot i + e \cdot u(x) \cdot M^2(x) \right] dx \tag{5.26}$$

The first term is then integrated in parts, resulting in Eqn. 5.27.

$$\frac{<i^2>}{\Delta f} = -e[M(x)^2 \cdot i_{em}(x)]_0^w + e \int_0^w M^2(x) \cdot \frac{di_{em}(x)}{dx} dx + e \cdot i \int_0^w \frac{dM^2(x)}{dx} dx +$$
$$+ 2e^2 \int_0^w u(x) \cdot M^2(x) \, dx \tag{5.27}$$

The second term in Eqn. 5.27 is equivalent to $\frac{1}{2}\frac{<i^2>}{\Delta f}$, and $M(w)^2 \cdot i_{em}(w)$ in the first term can be substituted by $M(w)^2 \cdot i - M(w)^2 \cdot i_p(w)$.

$$\frac{<i^2>}{\Delta f} = 2eM(0)^2 \cdot i_{em}(0) - 2eM(w)^2 \cdot i + 2eM(w)^2 \cdot i_p(w) + 2eM(w)^2 \cdot i - 2eM(0)^2 \cdot i +$$
$$+ 4e^2 \int_0^w u(x) \cdot M^2(x) \, dx \tag{5.28}$$

As $i_p(w)$ is the valence band current at the interface between the depletion zone and the neutral region of the impurity layer, this quantity is expected to be negligible. The quantity $i_{em}(0)$, the emission current at the blocking layer/impurity layer interface, can as well be neglected in case of efficient blocking of a direct current between the impurity band and the blocking layer. Additionally, the multiplication factor at the impurity layer/blocking layer interface $M(0)$ is trivially one. These simplifications result in Eqn. 5.29.

$$\frac{<i^2>}{\Delta f} = -2e \cdot i + 4e^2 \int_0^w u(x) \cdot M^2(x) \, dx \tag{5.29}$$

Now, the total current through a BIB device i is built up by the contributions of carriers generated by incoming photons within dx around x, which experience a multiplication gain $M(x)$, and can thus be written as in Eqn. 5.30

$$i = e \int_0^w u(x) \cdot M(x) \, dx \tag{5.30}$$

Inserting Eqn. 5.30 into Eqn. 5.29 leads to the final expression for the noise current in a BIB, as given by Eqn. 5.31.

$$\frac{<i^2>}{\Delta f} = 2e^2 \int_0^w u(x) \cdot M(x) \cdot [2M(x) - 1] \, dx \qquad (5.31)$$

In case of background-limited operation, the charge carrier generation by the absorption of photons, $u(x)$, is dominated by background radiation, and with $u(x) = u_{bg}(x)$ Eqn. 5.31 transforms into Eqn. 5.10. The amount of charge carriers generated by the background flux density $\Phi_{bg}(\nu)$ follows Eqn. 5.32, where A_{mesa} is the device area.

$$u_{bg}(x) = \alpha(\nu)\theta(\nu)\Theta(\nu,x)\Phi_{bg}(\nu) \cdot A_{mesa} \qquad (5.32)$$

Inserting this expression into Eqn. 5.31 gives the background induced current noise, as presented in Eqn. 5.33

$$\begin{aligned}\frac{<i_{bg}^2>}{\Delta f} &= 2e^2 \alpha(\nu)\theta(\nu)\Phi_{bg}(\nu) \cdot A_{mesa} \int_0^w \Theta(x) \cdot M(x) \cdot [2M(x) - 1] \, dx \\ &= 2e^2 \alpha(\nu)\theta(\nu)\Phi_{bg}(\nu) \cdot A_{mesa} \left[2<\Theta M^2> - <\Theta M>\right] \end{aligned} \qquad (5.33)$$

As seen from Eqn. 5.33, the expression for the background limited current noise in a BIB has an analogous structure to that for the background noise current through a QWIP device, and these two expressions can thus be related. In [50], Szmulowicz defines an excess noise parameter $\gamma(\nu, V)$, which is calculated according to Eqn. 5.34 and gives the ratio between the spectral noise current density of a BIB and that of a QWIP, assuming that both devices exhibit the same gain factor.

$$\gamma(\nu,V) = \frac{(2<\Theta M^2> - <\Theta M>)<\Theta>}{<\Theta M>^2} \qquad (5.34)$$

The expression for the spectral background noise density thus reads as in Eqn. 5.35.

$$\frac{<i_{bg}^2>}{\Delta f} = 2e^2 G^2(\nu,V) \cdot \eta(\nu,V) \cdot \gamma(\nu,V) \cdot \Phi_{bg}(\nu) \cdot A_{mesa} \qquad (5.35)$$

Inserting $\frac{<i_{bg}^2>}{\Delta f}$ into the definition of the detectivity $D^*(\nu, V)$, which is presented in Eqn. 3.1, leads to the final expression for the detectivity within the model of Szmulowicz as presented in Eqn. 5.36.

$$\begin{aligned} D^*(\nu,V) &= \frac{R(\nu,V)\sqrt{\Delta f \, A_{mesa}}}{\sqrt{<i_{bg}^2>}} = \\ &= \frac{1}{h\nu}\sqrt{\frac{\alpha(\nu)\theta(\nu)}{2\Phi_{bg}(\nu)}} \frac{<\Theta M>}{\sqrt{2<\Theta M^2> - <\Theta M>}}(\nu,V) = \\ &= \frac{1}{h\nu}\sqrt{\frac{\eta}{2\Phi_{bg}\gamma}} \end{aligned} \qquad (5.36)$$

97

Table 5.4: BIB simulation parameters:

Parameter	Linz 1 3 2008	Linz 1 4 2008
d_{IL}	6 µm	6 µm
d_{BL}	4 µm	4 µm
N_A	$1 \cdot 10^{18} \text{cm}^{-3}$	$5 \cdot 10^{17} \text{cm}^{-3}$
N_D	$2 \cdot 10^{15} \text{cm}^{-3}$	$5 \cdot 10^{14} \text{cm}^{-3}$
$\alpha(50\text{meV})$	1400 cm^{-1}	700 cm^{-1}
R_1	0.3	0.3
R_2	0.3	0.3
σ_I	$1 \cdot 10^{-12} \text{cm}^2$	$1 \cdot 10^{-12} \text{cm}^2$
E_c	7000 V/cm	7000 V/cm
F_{loss}	0.3	0.3

As seen from the comparison between Eqn. 5.8 and Eqn. 5.36, the detectivities of a QWIP and a BIB device of equal gain factors and quantum efficiencies differ by a factor $\sqrt{\gamma}$, where the BIB detectivity is reduced by the excess noise originating from the statistical properties of the gain mechanism.

5.5.2 Simulation results

The parameters employed in order to simulate the respective figures of merit of **Linz 1 3 2008** and **Linz 1 4 2008** are presented in table 5.4. The value for E_c has been directly taken from reference [50], where it was used in the simulation of a Si:As BIB. However, in this reference the simulated quantities are not compared to experimental results. Furthermore, it has not been shown, that the parameters in reference [50] are valid for the Si:B system. The value of σ_I has been changed from $1.6 \cdot 10^{-13} \text{cm}^3$, as found in [50], to $1 \cdot 10^{-12} \text{cm}^3$ in order to fit our experimental results with higher accuracy. As no accurate value for this parameter can be found in literature, this figure is within the physically possible range. The uncertainty concerning the values of E_c and σ_I poses one of the shortcomings of using the model of Szmulowicz for our Si:B structures. A series of designated experiments aiming at the determination of these material parameters would be necessary in order to closely correlate simulation with experiment.

The simulation parameter values for $\alpha(\nu)$ were calculated from 5.4 by inserting the value of $\sigma_{photoion} = 1.4 \cdot 10^{-15} \text{cm}^2$, as found in [55]. The loss factor F_{loss} was introduced to consider intensity losses by e.g. scattering in the optical setup.

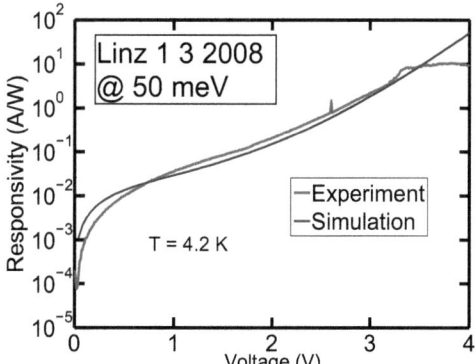

Figure 5.13: Comparison between the simulated and experimentally determined responsivity characteristics of **Linz 1 3 2008** at a photon energy of 50 meV. Note that the simulation results were scaled by a loss factor of 0.3 in order to fit the experiment in absolute values. However, the simulated curve shape reproduces the measurements remarkably well.

Simulated responsivity

Figure 5.11 presents the simulation results for the responsivity of sample **Linz 1 3 2008**. The blue curve in this graph shows $F_{loss} \cdot R(\nu_0, V)$ for $h\nu_0 = 50$ meV, as calculated from Eqn. 5.21 and scaled by the loss factor. F_{loss} was treated as a fit parameter, the absolute values of the simulated curve are thus by a factor of three higher than that of the measured characteristics. However, the shape of the experimentally determined bias-dependence of the responsivity is reproduced by the simulation with remarkable accuracy. There are two regions with significant deviations of the simulation from the experiment, one for voltages below 0.5 V, one above 3.5 V. The reason for the discrepancy between model and experiment for low voltages can be found in the activation energy for hopping transport (see Eqn. 5.2), as discussed in section 5.4.3 following [44]. The activation behavior of the hopping transport within the impurity band dominates the curve shape of the I-V characteristics of a BIB at low voltages. However, this process is not included in the model of Szmulowicz. The deviation of the simulated curve from the experimental one for high voltages is caused by the flattening of the experimental curve due to saturation effects, as discussed in detail in the paragraph covering the dips in the I-V characteristics in chapter 5.4.1. These effects are not accounted for in the model of Szmulowicz.

Figure 5.12 compares the simulation to the experimental results for **Linz 1 4 2008**. In

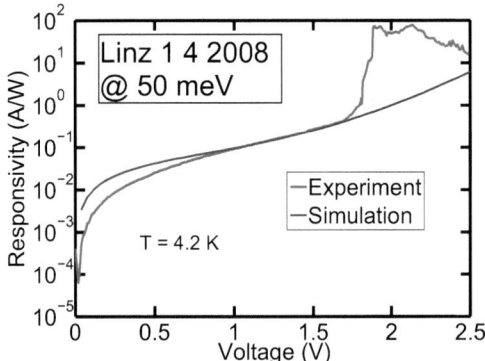

Figure 5.14: Comparison between the simulated and experimentally determined responsivity characteristics of **Linz 1 4 2008** at a photon energy of 50 meV. Note that the simulation results were scaled by a loss factor of 0.3 in order to fit the experiment in absolute values. Note the strong deviation of the simulation from the measured curve in the high-gain-region above 1.8 V.

contrast to those of **Linz 1 3 2008**, the measured and simulated characteristics for **Linz 1 4 2008** differ substantially. For voltages below 1 V, this discrepancy can be attributed to the unaccounted activation characteristics of hopping transport. Between 1 and 1.7 V, simulation and experiment agree well, given the same loss factor as used in Fig. 5.11 for sample **Linz 1 3 2008**. Around 1.7 V, both simulation and experiment feature a change in slope. For voltages higher than 1.7 V, the measurement curve rises tremendously faster than the simulated characteristics, leading to differences of two orders of magnitude between the two curves. The reason for this disagreement is not understood in detail. Notably, the discrepancy between model and experiment sets in at the voltage, which is the threshold for strong gain by avalanche multiplication. Thus, it can be assumed, that the model fails to deliver correct values for the responsivity of a BIB operated in high-gain mode.

One aspect of a BIB operated under conditions of strong photon flux not included in the model by Smulowicz are effects caused by a strong current flow through the device. The model assumes that there is no significant voltage drop over the neutral region of the impurity layer, which therefore does not contribute to the generation of a photocurrent. In the model, the single influence of this neutral zone on the responsivity of the device lies in its reduction of the intensity reaching the electro-optically active depletion zone, as seen in Eqn. 5.13. However, in a BIB featuring a significant current flow, a field develops within the neutral region of the impurity layer, driving photogenerated holes into the depletion region.

Thus, the neutral zone *does* contribute to the photocurrent through the device. The stronger the current flow gets, the more influence this effect has on the responsivity. At the onset of strong avalanche gain, the current flow through the sample rises drastically, resulting in an additional field within in the neutral region of the impurity layer, which in turn boosts the generation of photocurrent, and so on. For high applied voltages the model thus definitely underestimates the slope of the responsivity characteristics. However, it is doubtful that the underestimation due to a widening of the region contributing to the photocurrent accounts for the total difference between the curves shown in Fig. 5.12.

In order to come to a definite conclusion regarding the reason for the difference in curve shape between the experimental and simulated data, further experiments and a more thorough analysis of the model are required.

Even though the model of Szmulowicz does not live up to a full reproduction of the experimental results, it does predict the relation between the responsivities of **Linz 1 3 2008** and **Linz 1 4 2008** with high accuracy within the voltage region of low multiplication gain. This becomes evident when considering the fact that the simulated curves in Figs. 5.11 and 5.12 were gained employing the same simulation parameters σ_I and E_c as well as the same loss factor F_{loss}. The following subsections illustrate, how the difference in structural parameters of our BIB samples influences their figures of merit according to the model.

Simulated gain, internal quantum efficiency

The main doping (acceptor) concentrations of samples **Linz 1 3 2008** and **Linz 1 4 2008** differ by a factor of two. As a consequence, **Linz 1 3 2008** features a by a factor two higher absorption coefficient $\alpha(\nu)$ in the far- and mid-infrared than **Linz 1 4 2008**. This means that in the photocurrent-generating depletion zone of **Linz 1 3 2008** radiation is converted into a current signal with higher efficiency. However, in the model of Szmulowicz the neutral region of the impurity layer does not contribute to a photocurrent, and thus just acts as a medium damping the radiation actually reaching the depletion zone. The neutral region of **Linz 1 3 2008** absorbs incoming radiation to a higher content than that of **Linz 1 4 2008**. Therefore, as calculations imply, in case of an equally wide depletion zone, the internal quantum efficiency of **Linz 1 3 2008** is only by a factor of 1.3 higher than that of **Linz 1 4 2008**.

For a given voltage, even though the amount of absorbing impurities is higher for **Linz 1 3 2008**, its internal quantum efficiency $\eta(V,\nu_0)$ is lower than that of **Linz 1 4 2008**, as shown in Fig. 5.15 and calculated according to Eqn. 5.19. This is due to the assumption of the model that the generation of photocurrent is restricted to the depletion zone, as a

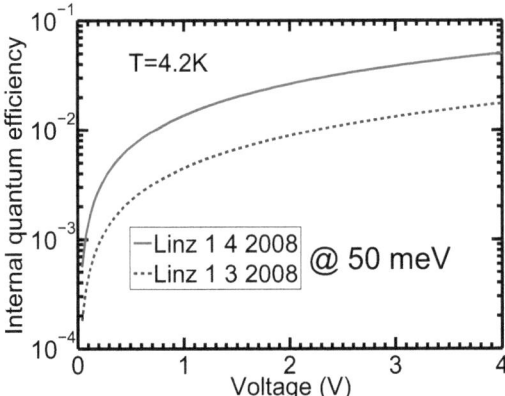

Figure 5.15: Simulated internal quantum efficiency $\eta(V, \nu_0)$ calculated for **Linz 1 3 2008** and **Linz 1 4 2008** at $h\nu_0 = 50$ meV. Note that even though the main doping concentration of **Linz 1 3 2008** is higher, its internal quantum efficiency is lower than that of **Linz 1 4 2008** due to its inferior depletion width, which is seen in Fig. 5.16.

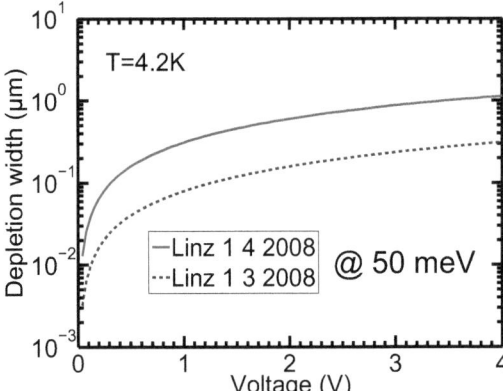

Figure 5.16: Simulated depletion width $w(V)$ calculated for **Linz 1 3 2008** and **Linz 1 4 2008**. Note that the curve shape of the internal quantum efficiency in Fig. 5.15 is directly related to that of the depletion width.

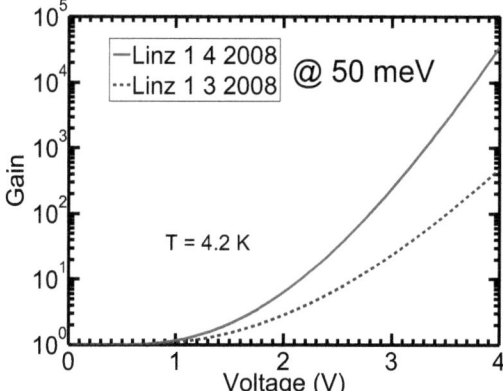

Figure 5.17: Simulated gain $G(\nu_0, V)$ calculated for **Linz 1 3 2008** and **Linz 1 4 2008** at $h\nu_0 = 50$ meV. Note that the gain starts to rise significantly from 1.5 V on, where its rise is due to both the increase in the depletion width and the field within the depletion zone with the applied bias.

consequence of which the internal quantum efficiency rises with the depletion width $w(V)$.

Figure 5.16 presents the dependence of the depletion width, which was calculated using Eqn. 4.1, on the applied voltage for both samples. Trivially, the curve $w(V)$ shows a square-root dependence. As can be easily seen, the curve shape of the internal quantum efficiency in Fig. 5.15 directly reflects the depletion width characteristics.

When comparing the bias dependence of the responsivity in Figs. 5.11 and 5.12 with that of the depletion width, it becomes evident that up to the kink in the responsivity characteristics around 1.5 V, the simulated curves for $R(\nu_0, V)$ and $\eta(V)$ (and thus $w(V)$) are closely correlated. Put differently, up to a voltage of about 1.5 V, the increase in the simulated responsivity is based on a bias-induced widening of the depletion region and therefore the electro-optically active zone. This finding is consistent with the conclusion drawn from the experimentally obtained data in 5.4.3.

For a given voltage, the depletion width of **Linz 1 3 2008** is, due to the higher counter-doping concentration N_D, about a factor four below that of **Linz 1 4 2008**. This difference in the depletion width results in a ratio of 2.5 between their measured responsivities below 1 V, which rises for higher applied voltages due to the onset of strong avalanche gain in **Linz**

1 4 2008. The difference between **Linz 1 3 2008** and **Linz 1 4 2008** in their responsivity for low applied voltages is slightly reduced from the value expected from the difference in the counter-doping concentration by the higher main doping concentration of **Linz 1 3 2008**.

The observation, that the increase in the responsivity of our samples is below 1.5 V exclusively caused by a rise in the internal quantum efficiency, is consistent with the simulated data presented in Fig. 5.17. This plot shows the simulated dependence of the gain $G(\nu_0, V)$ on the applied bias, which was calculated using Eqn. 5.20. Below 1.5 V, the gain for both samples remains below two, and thus only insignificantly influences the responsivity characteristics in Figs. 5.11 and 5.12. However, above 1.5 V the gain rises strongly, where above 2 V an exponential increase is reached. As a consequence, for higher applied voltages the shape of the responsivity characteristics is dominated by the bias dependence of the gain factor. The rise of the gain factor with the applied bias is not only induced by the increase of the depletion width, but is additionally influenced by the activation behavior of the impact ionization coefficient given in Eqn. 5.14. The latter dominates the shape of the gain characteristics for voltages above 2 V. While in both the experimentally obtained and the simulated responsivity characteristics a dominant influence of the gain curve shape is observed for high voltages, the rise of the responsivity of **Linz 1 4 2008** in 5.12 is strongly underestimated by the model of Szmulowicz. The reasons for this have been discussed above.

In Fig. 5.17, the gain factor of **Linz 1 4 2008** rises remarkably faster than that of **Linz 1 3 2008**, a simulated behaviour which can be exclusively attributed to the by a factor four lower counter-doping concentration of **Linz 1 4 2008** and to the resulting increased depletion width. The main doping concentration is expected to also strongly effect the gain factor and therefore the responsivity for high applied bias. A higher main dopant concentration is predicted to lead to a higher gain factor. The model predictions regarding the dependence of the gain coefficient on the main and counter-doping concentrations of a BIB device will be discussed in more detail in the following paragraphs.

Simulated detectivity

Figure 5.18 presents the simulation results for the normalized detectivity $\sqrt{\Phi_{bg}(\nu_0)} \cdot D^*(\nu_0, V)$ calculated for both **Linz 1 3 2008** (red line) and **Linz 1 4 2008** (blue line) by using Eqn. 5.36. The normalization procedure aims at giving a measure independent of the actual background flux density $\Phi_{bg}(\nu)$ for the background limited detectivity. As seen in Fig. 5.18, the model of Szmulowicz predicts the background limited detectivity of both BIB samples to rise up to a maximum value (around 1.2 V for **Linz 1 4 2008** and 1.6 V for **Linz 1 3 2008**), and to subsequently drop again for higher voltages. This behavior of the detectivity of a BIB is due

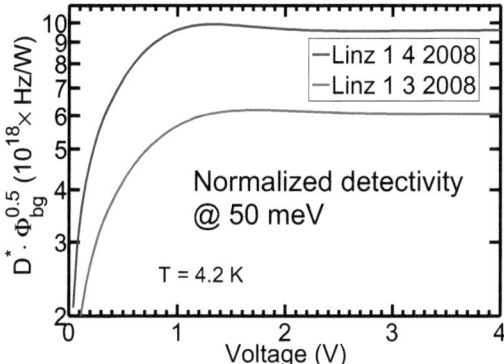

Figure 5.18: Simulated normalized detectivity $\sqrt{\Phi_{bg}(\nu_0)} \cdot D^*(\nu_0, V)$ calculated for **Linz 1 3 2008** and **Linz 1 4 2008** at $h\nu_0 = 50$ meV. Note that while the simulated responisivity in Figs. 5.11 and 5.12 rises monotonically, the simulated detectivity reaches a maximum and drops again.

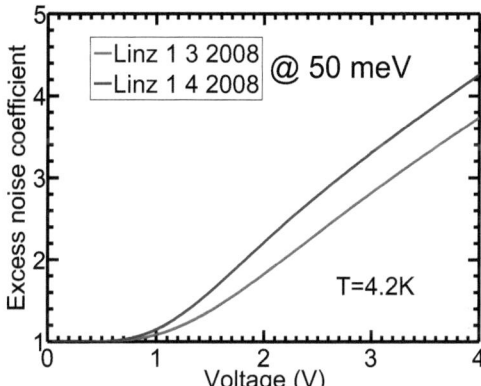

Figure 5.19: Simulated excess noise coefficient $\gamma(\nu_0, V)$ calculated for **Linz 1 3 2008** and **Linz 1 4 2008** at $h\nu_0 = 50$ meV. Note that the excess noise coefficient reaches values significantly exceeding one with the onset of gain, as seen by comparing to Fig. 5.17.

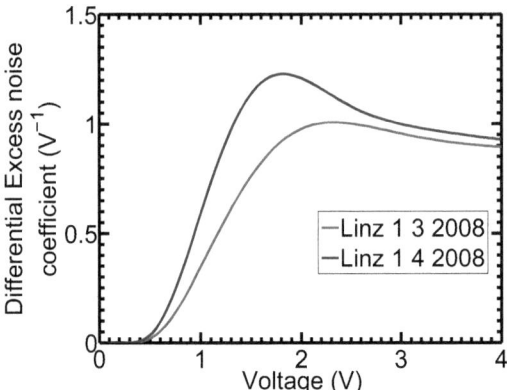

Figure 5.20: Simulated differential excess noise coefficient $\frac{d\gamma(\nu_0, V)}{dV}$ calculated for **Linz 1 3 2008** and **Linz 1 4 2008** at $h\nu_0 = 50$ meV. For **Linz 1 4 2008** the curve reaches its maximum exactly at the experimentally observed onset of noise too strong for the measurement of reliable photocurrent spectra.

to the generation of excess noise current by the avalanche multiplication process, as already discussed in sections 5.4.4 and 5.5.1. Figure 5.19 presents the excess noise coefficient for both BIB samples, as calculated following Eqn. 5.34. When comparing Figs. 5.17 and 5.19, it can be concluded that the excess noise factor starts to rise strongly with the onset of avalanche gain.

Remarkably, when plotting the derivative $\frac{d\gamma(\nu, V)}{dV}$, as is done in Fig. 5.20, its maximum for **Linz 1 4 2008** can be found around 1.8 V. This corresponds exactly to the bias value, from which on no proper photocurrent spectra could be measured for this sample due to the strong influence of noise, as already addressed in section 5.4.4. The bias, from which on **Linz 1 4 2008** was subject to heavy noise during the measurements of photocurrent spectra is thus well predicted by the simulations.

Even if the biasing point, from which on strong noise hinders the use of the device under high background conditions, is well understood within the model of Smulowicz, the use of the detectivity characteristics presented in Fig. 5.18 is very limited. First of all, as the model fails to quantitatively predict the responsivity characteristics for high applied voltages, it is not expected to treat the excess noise correctly within this bias region. Furthermore, as BIB devices unfold their full potential under low background conditions, which is also the

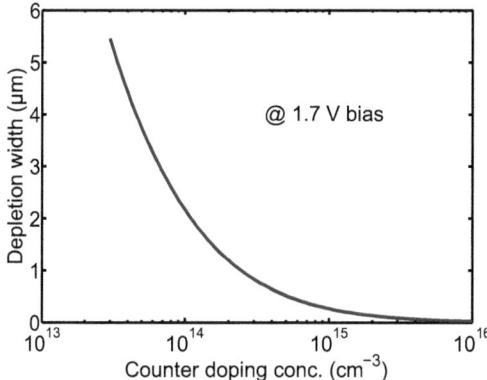

Figure 5.21: Calculated dependence of the depletion width on the counter-doping concentration at an applied voltage of 1.7 V and a blocking layer width of 4 μm.

operational mode for single photon counting, the relevant detectivity figure is that evaluated under dark-current limitation rather than background limitation. However, dark-current is not included in the model of Szmulowicz, leaving this important figure of merit unaccessible by our calculations.

Dependence of the depletion width on the doping concentrations

This section addresses the dependence of the figures of merit of a BIB on its two main device parameters, namely the main and counter-doping concentrations, as predicted by the model of Szmulowicz. Even though Szmulowicz reported on the dependence of the BIB performance on the counter-doping concentration in detail in reference [50], the main doping concentration was kept constant within his considerations. In the following, the model predictions for the BIB characteristics are mapped by varying both the main and counter-doping, employing the simulation parameters in table 5.4. Despite the observed underestimation of the gain properties of our devices by the model, the simulated data in this section shall be employed for once more highlighting the differences between **Linz 1 3 2008** and **Linz 1 4 2008** as well as suggesting changes in the device parameters for optimizing the performance.

As already mentioned on numerous occasions, the width of the depletion zone within the impurity layer of a BIB is one of the essential parameters for the optimization of both the internal quantum efficiency and the avalanche gain. Trivially, as seen from Eqn. 4.1, the

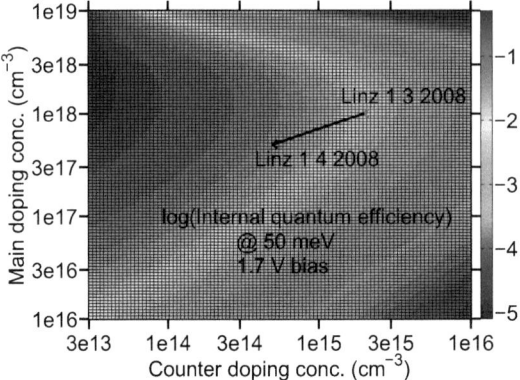

Figure 5.22: Logarithmic surface plot of the dependence of the **internal quantum efficiency** on both main and counter-doping concentrations. The plot shows $\log(\eta(N_a, N_d))$ for an applied voltage of 1.7 V, where red surface parts correspond to high quantum efficiency values, and blue areas to low efficiencies.

depletion width is independent of the main doping concentration. Figure 5.21 presents the theoretical dependence of the depletion width on the counter-doping concentration for an applied voltage of 1.7 V and a blocking layer thickness of 4 μm. The depletion widths for **Linz 1 3 2008** and **Linz 1 4 2008** differ by a factor of four.

Trivially, in order to maximize the depletion width of a BIB, the counter-doping concentration has to be chosen as low as possible. However, the possibility of keeping the counter-doping concentration as low as possible during fabrication is, in case of an ion-implanted BIB, fundamentally limited by the purity of the substrate. Furthermore, for a given workpoint voltage, the maximum in the depletion zone is anyhow limited to the total thickness of the impurity layer, thus in case of an applied voltage of 1.7 V and a 6 μm thick impurity layer, a counter-doping concentration below $3 \cdot 10^{13}$ cm^{-3} is futile regarding the depletion width.

Dependence of the internal quantum efficiency on the doping concentrations

The increase in depletion width with decreasing counter-doping concentration leads to a monotonic dependence of the internal quantum efficiency on the amount of counter-doping for any concentration of the main dopant, as seen in Fig. 5.22. This figure presents a surface plot of the logarithm of $\eta(\nu_0, V_0)$ for varying doping concentrations, calculated by using Eqn. 5.19 for $h\nu_0 = 50$ meV and $V_0 = 1.7$ V. With the width of the depletion zone, the thickness of the impurity layer region generating photocurrent and therefore the internal quantum efficiency increases. However, the dependence of the internal quantum efficiency on the main doping concentration N_d is less trivial. As seen in Fig. 5.22, the internal quantum efficiency reaches a maximum for a main doping value N_a around $1 \cdot 10^{18}$ cm^{-3}. The reason for this behavior can be found in the fact that increasing N_a not only increases the amount of radiation absorbed within the depletion zone, but also the loss of incoming intensity by absorption within the neutral region of the impurity layer. For main doping concentrations above $1 \cdot 10^{18}$ cm^{-3}, the absorption losses in the neutral region of the impurity layer, which does not contribute to the photocurrent in the model of Szmulowicz, are too strong to be compensated by the enhanced local quantum efficiency in the depletion zone.

The positions of both **Linz 1 3 2008** and **Linz 1 4 2008** in the parameter space are indicated in Fig. 5.22 by crosses, and both coordinates are connected by a black line in order to give a guide for the eye regarding the change of the internal quantum efficiency when changing the parameters along this line. This line visualizes the observance that even though the main doping concentration of **Linz 1 3 2008** is optimal regarding the internal quantum efficiency, the influence of the change in counter-doping from **Linz 1 3 2008** to **Linz 1 4 2008** overrules the influence of the changed main doping concentration, and **Linz 1 4 2008** exhibits the superior quantum efficiency. The latter is at least true within the model of Szmulowicz, where the neutral region does not contribute to a photocurrent. As this is not totally true in reality, the conclusions drawn from Fig. 5.22 concerning the optimal (N_a, N_d) coordinate for a maximal internal quantum efficiency have to be treated with caution.

Dependence of the gain on the doping concentrations

As there is a clear strategy for the choice of the counter-doping concentration as far as the optimization of the depletion width is concerned, namely to choose it as low as technologically possible, the same holds true for the main doping concentration, when it comes to optimizing the gain of a BIB. As gain is based on the impact ionization of impurity sites by holes generated in the valence band, its optimization is achieved by chosing the number of impurity atoms as high as possible. This conclusion can be drawn from Fig. 5.23, which visualizes the dependence of the gain $G(\nu_0, V_0)$ on the respective doping concentrations, as derived from Eqn. 5.20 for $h\nu_0 = 50$ meV and $V_0 = 1.7$ V. As far as the counter-doping concentration is concerned, the simulated gain factor does not show a monotonic behavior, but reaches a maximum around $5 \cdot 10^{13}$ cm^{-3}. The reason for this behavior, which has in a similar form been reported in [50], can be found in the fact that the avalanche gain depends on both the depletion width and the field present in the depletion region. Now, for high values of the counter-doping concentration, the depletion width is small in respect to the thickness of

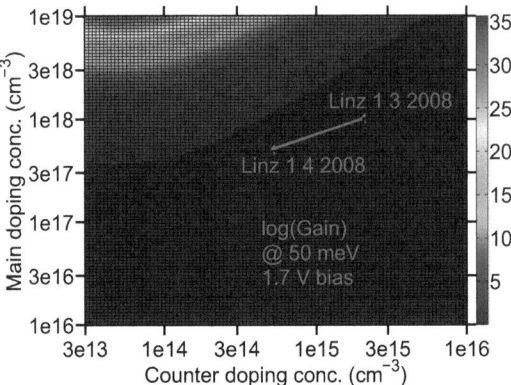

Figure 5.23: Logarithmic surface plot of the dependence of the **gain** on both main and counter-doping concentrations. The plot shows $\log(G(N_a, N_d))$ for an applied voltage of 1.7 V, where red surface parts correspond to high gain values, and blue areas to low gain factors.

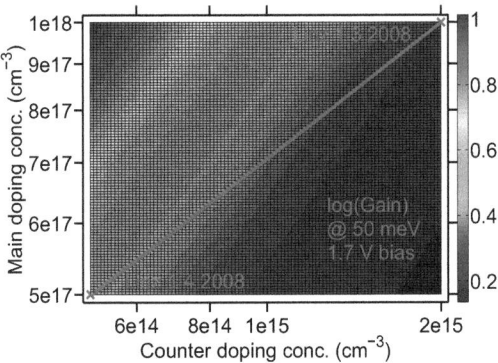

Figure 5.24: Zoom into the parameter region of Fig. 5.23 relevant for our BIB samples. The plot shows $\log(G(N_a, N_d))$ for an applied voltage of 1.7 V.

the blocking layer, and the maximum value of the electric field in the structure is pinned to $\frac{V}{d_{BL}}$. In this region of the parameter space the gain thus rises with decreasing counter-doping due to the increase in the depletion width. As N_d is now reduced further and the depletion width grows to dimensions similar to that of the blocking layer thickness, the maximum

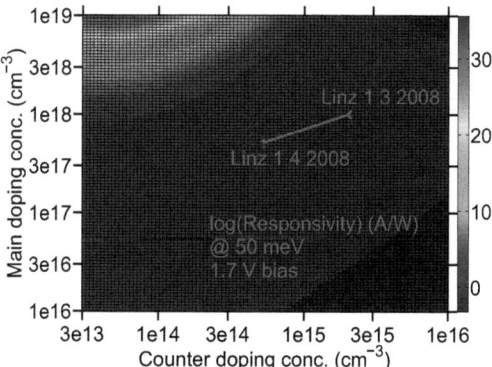

Figure 5.25: Logarithmic surface plot of the dependence of the **responsivity** on both main and counter-doping concentrations. The plot shows $\log(R(N_a, N_d))$ for an applied voltage of 1.7 V, where red surface parts correspond to high responsivity values, and blue areas to low responsivities.

field in the structure is reduced significantly and induces a drop in the gain factor. Below $N_d = 5 \cdot 10^{13}$ cm^{-3}, the influence of the reduction in the present fields overrules that of the increasing depletion width, and the gain starts to drop towards smaller counter-doping concentrations. For a desired operation bias of 1.7 V, a maximum gain is thus predicted by the model for a counter-doping concentration of $5 \cdot 10^{13}$ cm^{-3} and a main doping concentration as high as possible.

Dependence of the reponsivity on the doping concentrations

For an improvement of the performance of a BIB, not the individual figures for gain and internal quantum efficiency but rather their product has to be optimized. Figure 5.25 shows the theoretical dependence of the responsivity $R(\nu_0, V_0)$ on the respective doping concentrations, as derived from Eqn. 5.21 for $h\nu_0 = 50$ meV and $V_0 = 1.7$ V. From a comparison of Figs. 5.23 and 5.25 the conclusion can be drawn that the dependence of the responsivity on the doping concentrations is dominated by the gain characteristics. Thus, for optimizing the responsivity at an operation voltage of 1.7 V, according to the model the main doping concentration should be maximized, while the counter-doping should be kept at $5 \cdot 10^{13}$ cm^{-3}.

As far as our samples are concerned, the optimum condition for N_d is more closely met by **Linz 1 4 2008**, that for N_a by **Linz 1 3 2008**. However, as seen in Fig. 5.26, the change

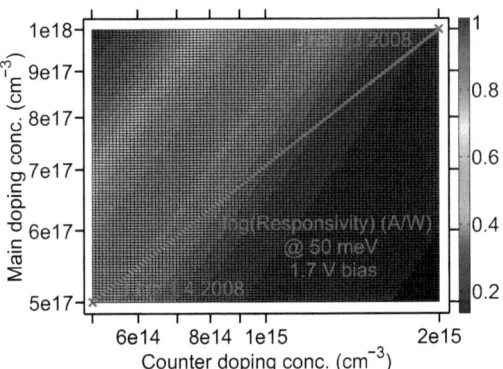

Figure 5.26: Zoom into the parameter region of Fig. 5.25 relevant for our BIB samples. The plot shows $\log(R(N_a, N_d))$ for an applied voltage of 1.7 V, where red surface parts correspond to high responsivity values, and blue areas to low responsivities.

in the counter-doping influences the responsivity figure more strongly than that in the main doping, resulting in **Linz 1 4 2008** exhibiting the higher responsivity. This theoretical prediction was confirmed by our experiments, as mentioned previously on several occasions.

Dependence of the detectivity on the doping concentrations

Besides the responsivity, the detectivity is the major figure of merit for any detector. The model of Szmulowicz is only able to make predictions concerning the background-limited detectivity, a surface plot of which is presented in Fig. 5.27. This plot shows the theoretical dependence of the detectivity $D(\nu_0, V_0)$ on the respective doping concentrations, as derived from Eqn. 5.36 for $h\nu_0 = 50$ meV and $V_0 = 1.7$ V. While the shape of the responsivity characteristics is dominated by the doping dependence of the gain factor, the detectivity characteristics follow that of the internal quantum efficiency, a fact which becomes evident when comparing Figs. 5.22 and 5.27.

This behavior can be understood by considering the definition of the background-limited detectivity, which is given by dividing the responsivity by the square-root of the noise current density. As the square-root spectral noise density is proportional to $G\sqrt{\gamma \cdot \eta}$ (see Eqn. 5.35), the influence of the gain factor on the responsivity vanishes during this division, as evident from Eqn. 5.36. Thus, with the dominant influence of the gain factor divided out, the de-

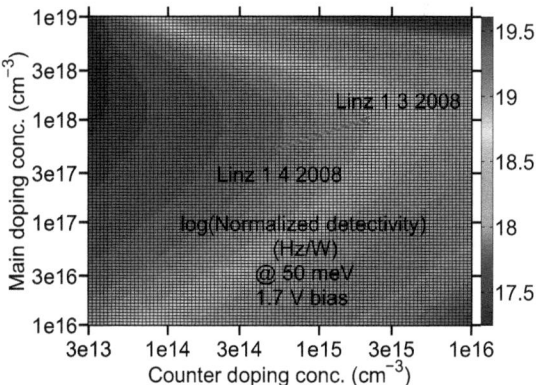

Figure 5.27: Logarithmic surface plot of the dependence of the background-limited **detectivity** on both main and counter-doping concentrations. The plot shows $\log(D^*(N_a, N_d))$ for an applied voltage of 1.7 V, where red surface parts correspond to high responsivity values, and blue areas to low responsivities.

tectivity characteristics follow the doping dependence of the internal quantum efficiency. In order to reach the maximum background-limited detectivity, the model by Szmulowicz implies that the counter-doping has to be chosen as low as possible. The ideal main doping concentration for a workpoint bias of 1.7 V is predicted to be $1 \cdot 10^{18}$ cm^{-3}.

However, the independence of the detectivity from the gain factor is only true in case of background-limited operation. Under dark-current-limited, low background conditions, which are the favorable mode of operation for ultra-sensitive applications, the detectivity *does* strongly depend on the gain factor, particularly as noise gain and photocurrent gain are based on different mechanisms. As dark-current is not included in the model of Szmulowicz, it does not deliver predictions on the dark-current-limited detectivity, which would be essential for the optimization of BIB devices for single-photon detection.

Summary of the optimal parameter range and fabrication limits

In the previous paragraph, the optimization of the doping concentration parameters was discussed for a *fixed desired bias* of operation. This enabled a direct comparison of the characteristics of **Linz 1 3 2008** and **Linz 1 4 2008**. However, usually the bias workpoint of a device is flexible. Under this condition, the optimization criteria for the BIB parameters are as follows.

The main issue for the optimization of parameters for the fabrication of future ion-implanted BIB samples should be an increase in detectivity. As our currently used substrate has a specified resistivity of higher than 400 Ωcm, it is guaranteed to feature a concentration of n-type dopants below $2 \cdot 10^{13}$ cm^{-3}. Therefore, reducing the implanted counter-doping concentration below this value is futile. Now, in case of a flexible voltage of operation, the counter-doping concentration has to be selected as low as possible in order to increase the BIB detectivity. For future BIB fabrication experiments, the substrate could be left at its intrinsic counter-doping concentration, without employing any additional phosphorus implantation steps. Reducing the number of implantation steps would substantially reduce the fabrication efforts and would pose a possibility of reducing the counter-doping concentration to a substrate-induced limit, thus increasing the background-limited detectivity.

However, for a well-defined amount of counter-doping, which is not given in the intrinsic substrate due to fabrication-induced variations, a counter-doping of $5 \cdot 10^{13}$ cm^{-3} should be chosen for further experimental samples. This concentration holds the advantages of being significantly above the maximum impurity concentration guaranteed for the plain substrate.

The lower limit for the main doping is defined by the concentration, at which an impurity band forms and efficient hopping transport within this band sets in. Below this concentration, the impurity sites are isolated, and carriers frozen out in these sites are immobile, a situation unsuitable for BIB operation. The upper limit for the main doping concentration results from the width of the impurity band, where the ultimate upper boundary for the choice of this fabrication parameter is defined by the concentration value, for which the impurity band starts to overlap with the valence band. However, for a given choice of the operating temperature of a BIB device the main doping concentration has to be chosen low enough to ensure, that the electrons freeze out in the valence band.

If the main doping concentration is chosen too high, a strong dark-current is expected to appear already at low voltages due to thermally assisted tunnelling from the impurity band into the valence band, an effect not considered in the dark-current-free model of Szmulowicz. This increase in dark-current results in the reduction of the *dark-current limited* detectivity, while the model is merely able to treat the *background limited* detectivity. However, for BIB operation under low background conditions, which are essential for the employment as a single photon detector, the dark-current limits the BIB's figures of merit rather than the background, and a careful optimization of the performance of the BIB in the dark has to be performed experimentally or by Monte-Carlo simulation, as this task is beyond the scope of the analytical model by Szmulowicz.

However, in case of an counter-doping concentration of $5 \cdot 10^{13}$ cm^{-3}, which is suggested for further sample development, the ideal main doping concentration is predicted by the simulations to be around $1 \cdot 10^{18}$ cm^{-3} in case of an impurity layer width of 6 μm and in a bias range similar to that employed during the experiments on **Linz 1 3 2008** and **Linz 1 4 2008**. This value is well within the parameter window feasible for fabrication. Furthermore, as has been demonstrated experimentally in this work, this value is low enough to ensure low dark-current through the device for the range of electric fields required for avalanche multiplication. The latter is crucial for optimizing the dark-current performance of the device.

To summarize and conclude this section, for future experiments aiming at optimizing both the performance and fabrication efficiency of ion-implanted BIBs, the main doping concentration should be chosen around $1 \cdot 10^{18}$ cm^{-3}, which is within the range of **Linz 1 3 2008** and **Linz 1 4 2008**. The counter-doping concentration has to be significantly reduced to $5 \cdot 10^{13}$ cm^{-3}, or even left at the low value intrinsic to the substrate itself, which would additionally simplify the fabrication process. The latter is, however, only possible if the intrinsic impurities of the substrate include the type required for the counter-doping of the BIB device (in our case p-type). These suggested changes from the sample parameters experimentally characterized within this work are expected to improve both the responsivity and detectivity figures of the BIB device.

5.6 Summary and conclusion

In conclusion, this chapter reported on the concept, fabrication and experimental characterization of ion-implanted blocked impurity band detectors. The experimentally obtained data were additionally compared to the results of an analytical model by Szmulowicz.

In search for an alternative to the conventional BIB fabrication process by epitaxial growth, which is technologically highly cumbersome due to the purity requirements for the growth of the blocking layer and the precise incorporation of a very low counter-doping in the impurity layer, a main and counter-doping profile was implanted into the ultra-pure device layer of an SOI substrate. The optimal implantation parameters were evaluated by *Crystal-TRIM* simulations, and the implantation was carried out in four steps in the high-energy facilities of the Institute of Ion Beam Physics and Materials Research, Forschunsgzentrum Dresden-Rossendorf. The implanted concentrations of both the boron and phosphorus dopants are nominally homogenous down to a sample depth of 6 μm, leaving 4 μm of pure substrate as a blocking layer. A contact to the blocking layer was established by etching through the carrier wafer from the sample backside, where etching stopped at the thermal oxide of the

SOI wafer. The etching through the 300 μm thick handling wafer employed the so-called Bosch-process, where the openings were defined by photolithography and were aligned to the BIB mesas on the front side of the wafer. The fabrication of the BIB devices was thus free of any epitaxial processes, which poses a tremendous increase in the technological accessability of BIB fabrication.

The BIB mesas of two samples differing in main and counter-doping were characterized by obtaining current-voltage characteristics under low background as well as under illumination by a black-body radiator. Further, their spectral characteristics were obtained by Fourier-transform-interferometry in both the far- and mid-infrared regions. The measured data allowed the determination of the figures of merit of the ion-implanted BIB devices like responsivity, quantum efficiency and gain, as dependent on both the applied electric field and the wavelength of the incoming radiation.

The experiments showed that for both samples the pure substrate layer exhibits excellent blocking characteristics, as seen from the featured low dark-currents up to high applied voltages. Additionally, clear evidence for the formation of an impurity band, within which hopping transport occurs, was obtained. The latter was given by a significant reduction of the photo-ionization energy of the impurity atoms when compared to separated boron sites incorporated into silicon. The samples are at liquid helium temperature highly sensitive to radiation from the far- to the mid-infrared, where the responsivity sets on at a photon energy of 30 meV. Summed up, both samples exhibit the well-understood characteristics of a BIB in the operational mode of low gain, within which the responsivity reaches competitive values between 0.1 and 0.5 A/W.

Futhermore, for sample **Linz 1 4 2008** strong avalanche gain up to a factor of 100 was observed for applied voltages above 1.8 V. The capacity of **Linz 1 4 2008** to operate as a solid-state photomultiplier was correlated to its low counter-doping concentration, which substantially increases the width of the depletion zone, within which gain by impact ionization can occur due to the presence of high fields. This strong gain by avalanche multiplication leads to highly impressive responsivity values of up to 65 A/W and external quantum efficiencies of 3 electrons per incoming photon. However, this exceptionally high figures of merit are only reached at electric fields, for which the observed dark-current is unacceptably high and the detectivity is thus too low for proper BIB operation.

Nevertheless, within a bias window of 70 meV, **Linz 1 4 2008** reaches highly competitive responsivity values of 13.5 A/W and external quantum efficiencies up to 0.7. These impressive figures of merit are reached within a biasing region, for which dark-currents remain below

$5 \cdot 10^{-11}$ A. The performance of the ion-implanted sample **Linz 1 4 2008** compares well with BIB devices fabricated by conventional, epitaxial methods.

The dependence of the figures of merit of the BIB samples on their fabrication parameters of main and counter-doping concentrations was simulated using an analytical model by Szmulowicz. Within this model, the experimentally observed behavior of the BIB samples under low-gain conditions can be understood and reproduced in detail. In contrast, experiment and model differ significantly under conditions, where high avalanche gain is observed. Nevertheless, the analytical model can be employed to further improve the figures of merit of ion-implanted BIBs by optimizing the fabrication parameters.

By further optimizing the implantation parameters for future BIB fabrication, both the accessability of the fabrication process and the BIB performance can be improved. By reducing the width of the impurity layer and thus the implantation energies required for introducing the impurity band into the pure substrate, conventional keV implantation facilities could qualify for BIB fabrication. This would resemble a huge step towards cheap and broadly accessible fabrication of ultra-sensitive detectors featuring responsivity onsets in the terahertz region of the infrared spectrum. The fabrication process could be additionally simplified by omitting any implantation step for the counter-doping and thus employing the intrinsic impurity concentration in the substrate material as counter-doping. Additionally, by optimizing the doping concentrations in order to enable high avalanche gain under low-dark-current conditions, gain values of 100, which have already been observed in this work in presence of strong dark-currents, could be employed for BIB operation. Gain figures of this order of magnitude would even qualify the structures for single photon detection.

The demonstration of ion-implanted vertical Si:B BIBs with competitive figures of merit including the capability of operating in the solid-state photomultiplication mode not only opens a highly promising road to technologically simple and cheap BIB array fabrication with implantation substituting the delicate MBE growth. The concepts demonstrated in this work for devices realized in the silicon system could also be adapted to material systems, for which the realization of an operating BIB failed due to impregnable technological difficulties concerning the epitaxial growth of sufficiently pure blocking layers. The concept of replacing the epitaxial growth of the layer sequences of a BIB by an ion-implantation of the required impurity concentration into intrinsically pure substrate could be adapted for the Ge and GaAs systems, in order to expand the wavelength detection range up to 200 μm.

Chapter 6

Intersubband relaxation times: An introduction

6.1 Intersubband silicon optoelectronics

6.1.1 Challenges and motivation for the Si system

Silicon is an indirect semiconductor. This fundamental property crucially restricts the use of silicon for the implementation of detectors and emitters based on optical transitions between the valence and conduction bands. The reasons for the extensive interest in pushing the silicon system into optoelectronics despite its inherent flaws are found in the enormous possibilities that would open up with the successful integration of all-silicon optoelectronics onto a chip. On the side of more conventional applications the monolithic integrability of infrared detectors and emitters with the signal processing microelectronics would enable the fabrication of environmental sensors on a single chip. However, the crucial driving force for research interest in this field is induced by the challenges met by the progress in integrated circuit scaling itself. With multicore processors of constantly increasing core numbers as well as system-on-chip products on the rise, metallic chip-to-chip interconnects meet their bandwidth limits. Intel representatives list chip-to-chip interconnects among the five future scaling challenges, and trade the transition from electrical to optical interconnects as the possible solution [56]. On a long term prospect, the electronics-to-optics transition could be extended from chip-to-chip interconnects down to the level of on-chip-interconnects. However, in order to perform this transition, the development of integrable optoelectronics is crucial.

As already demonstrated in the previous chapters, despite the indirect band gap of silicon detector devices can be realized in this material system. However, in order to circumvent the disadvantages of the material regarding optical conduction- to valence-band transitions, alternative optical processes have to be employed. For the detector structures presented in chapters 2 and 3, intersubband transitions within the valence band were exploited, where the devices studied in chapter 5 are based on transitions between a band formed by boron impurities and the valence band. Thus, by using intersubband transitions as a basis for optical activity, detector devices can be realized in the group-IV system. Furthermore, recent breakthroughs like the demonstration of a high-speed optical modulator in Si [57] bring the transition from electrical to optical interconnects closer to realization.

Despite the numerous successful developments providing step by step the components required for substituting electrons by photons for inter-chip communications, the base of silicon optoelectronics is still missing: An electrically pumped group-IV laser source. Up to now, the only stimulated emission processes reported in the silicon system are Raman lasing [58] and stimulated emission from bismuth impurity atoms [59]. Both processes are unsuitable for the development of an integrated laser source, as a pumping laser is required to drive lasing in the silicon.

Given the successful implementation of quantum well infrared detectors based on intersubband transitions in the SiGe system, enabled by the maturing of molecular beam epitaxy in this material system within the last 15 years, the most promising concept for realizing a group-IV laser source is that of a device also based on intersubband transitions: The quantum cascade laser.

6.1.2 The quantum cascade laser: An intersubband device

Since their first demonstration in the InGaAs/AlInAs/InP system by Faist et al. in 1994 [60], quantum cascade lasers (QCLs) have kept a huge scientific community busy, with numerous groups devoting their work to the successive improvement and extension of this laser concept. As a consequence of this extensive research efforts, QCLs have been implemented in various III-V material systems (AlGaAs/GaAs [61], InAs/AlSb [62]), CW operation at room temperature has been achieved [63], and threshold currents, maximum operation temperatures and output powers have been successively optimized. These improvements have been made possible by thorough band structure design, progress in growth technology, as well as innovative cavity design. In 2002 the frequency of QCL radiation was pushed into the terahertz region of the spectrum [64], in 2006 short wavelengths down to 3.3 μm could be reached [62]. In other words, from its invention on the QCL has developed into a highly optimized commercial product within very short time, while still attracting extensive interest from numerous groups of basic researchers. The latter is due to the design flexibility and the resulting tremendous control over the energetic positions and wave functions of the quantum

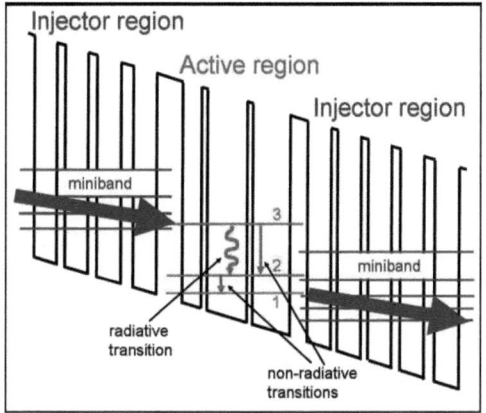

Figure 6.1: Schematic of the QCL operation. The black curve shows the conduction band edge of a QCL structure, while the blue, red and green horizontal lines indicate the quantized energy levels. Electrons are induced via the miniband into state 3, and create a photon during the transition to level 2. The latter is depopulated into level 1 by phonon scattering, from where the electron is transferred to the excited state of the subsequent period's active region via the next injector miniband.

states involved. Reviews of the field of QCLs can be found in references [65, 66]. Despite the highly successful realization and exploitation of the QCL concept in III-V materials, the silicon system remains unconquered.

On the road to a silicon-based QCL, the first step was to realize a quantum cascade emitter, which was successfully completed by Dehlinger in 2000 [12]. Since then, electroluminescence in SiGe quantum cascade structures has been reported for numerous wavelengths [13, 14]. Nevertheless, lasing in such a device still has to be reported. Details on the progress in the development of a SiGe QCL will be given in section 6.1.4.

Quantum cascade laser basics

QCLs are composed of hundreds of layers of compound semiconductor material with thicknesses in the nanometer regime, and are fabricated by epitaxial growth. The fabrication of these devices therefore makes high demands on growth technology regarding precise layer thickness control over several hundred growth cycles, which are nevertheless met by state-of-the-art epitaxial systems. QCLs in the III-V material systems operate in the conduction band and are therefore built up by n type doped heterostructures, which form up to hundred periods of a sequence of conduction band quantum wells.

The basic idea behind the QCL is to generate stimulated emission by the transition between excited states and lower states within the quantum wells formed by the structure, thus exploiting transitions *within one and the same band*. In contrast to interband semiconductor lasers, which generate radiation during the recombination of electrons and holes, the emitting charge carrier in a QCL is not annihilated by the radiative process. It can be transferred to the excited state of the neighboring active region, and thus be used for the generation a series of photons. The energy, which is emitted by the lasing process is provided to the electron by externally applying an electric field to the QCL structure. The charge carriers drift along the field, gain energy and emit photon by photon in the active regions. As the electrons therefore run down a potential profile like water down a cascade, the name quantum cascade laser was given to the device.

Figure 6.1 illustrates the operation of a QCL. The solid black line represents the conduction band edge of a period of an example structure, which could e.g. be realized in InGaAs/AlInAs [67]. The energetic depths of the quantum wells depend on the band offset between the two materials employed for the well- and the barrier layers, respectively. The additional tilt of the band edge is induced by applying an external voltage to the structure. The blue, green and red horizontal lines indicate the quantized energy levels formed in the quantum wells. Now, in an operating QCL electrons are injected into the structural period from the left via states of the injector region forming a miniband, within which an efficient carrier transport (indicated by the thick blue arrow) along the field is possible. The transport properties of the injector region are crucial for the operation of a QCL, as the region serves the purpose of filling the upper state of the lasing transition (state 3 in Fig. 6.1). The structure illustrated in Fig. 6.1 features a so-called three-well vertical-transition active region [65], which shall serve as an example in order to illustrate QCL operation. Levels 1, 2 and 3 are the weakly confined ground states of the three wells of the active region. Now, by stimulating a radiative 3→2 transition, coherent radiation is generated in the structure. 3→2 forms the lasing transition of the QCL (indicated by the red arrow), where population inversion between the two states is a fundamental requirement for lasing. For the achievement of gain, it is thus crucial to rapidly depopulate state 2 by carrier relaxation into state 1 (indicated by a green arrow). This relaxation process usually involves the resonant emission of longitudinal optical (LO) phonons. From state 1, the electron is transported to the active region of the subsequent period by drifting within the miniband of the injector located in the right half of Fig. 6.1, ready to undergo further emission processes.

Gain in a quantum cascade laser

In this subsection, the amount of gain expected within a QCL is considered. Equation 2.3 gives the absorption coefficient induced by an arbitrary quantum well structure. For photons in resonance with the 3→2 transition of our sample QCL structure in Fig. 6.1, only those two states are to be included in the calculation of the absorption coefficient, simplifying Eqn. 2.3 into Eqn. 6.1.

$$\alpha(\nu_{res}) = \frac{e^2}{\pi \, m_{effav}^2 \, \nu_{res} \, n_r \, c \, \epsilon_0 \, V} |<\Psi_2 \, |\boldsymbol{\pi}| \, \Psi_3> \cdot \boldsymbol{\epsilon}|^2 \cdot \frac{1}{\Gamma} \cdot (f_2 - f_3) \tag{6.1}$$

In this equation, m_{effav} represents the average effective mass in the structure, n_r the average refractive index, $\boldsymbol{\pi}$ the momentum operator as defined in Eqn. 2.1, $\boldsymbol{\epsilon}$ the polarization vector of the radiation, Γ the energetic broadening of the transition, and $f_3 - f_2$ the difference in occupation between states 2 and 3. In order to have domination of the stimulated emission over the absorption of light, which is the basic requirement for lasing, the absorption coefficient has to turn negative, which demands population inversion $f_3 > f_2$. Let now be τ_3 the total decay time of state 3, τ_{32} the scattering time between states 3 and 2, and τ_{21} the scattering time between states 2 and 1. In our simple model, τ_3 depends on both the relaxation rate from state 3 into states 1 and 2. Assuming that charge carriers are injected into the active region at a rate I_{inj} leads to the rate equations as given in Eqn. 6.2.

$$\begin{aligned} \frac{df_3}{dt} &= -\frac{f_3}{\tau_3} + I_{inj} \\ \frac{df_2}{dt} &= -\frac{f_2}{\tau_{21}} + \frac{f_3}{\tau_{32}} \end{aligned} \tag{6.2}$$

In case of stationarity, the difference in occupation between states 2 and 3 follows Eqn. 6.3.

$$f_3 - f_2 = I_{inj} \cdot \tau_3 (1 - \frac{\tau_{21}}{\tau_{32}}) \tag{6.3}$$

Substituting Eqn. 6.3 into Eqn. 6.1 with $j_{inj} = \frac{I_{inj}}{A_0}$ and $V = A_0 \cdot L_p$, where A_0 is the device area and L_p the period thickness, results in Eqn. 6.4.

$$\begin{aligned} \alpha(\nu_{res}) &= -\frac{e^2}{\pi \, m_{effav}^2 \, \nu_{res} \, n_r \, c \, \epsilon_0 \, L_p} |<\Psi_3 \, |\boldsymbol{\pi}| \, \Psi_2> \cdot \boldsymbol{\epsilon}|^2 \cdot \frac{1}{\Gamma} \cdot \tau_3 (1 - \frac{\tau_{21}}{\tau_{32}}) \cdot j_{inj} \\ &= -g_L \cdot j_{inj} \end{aligned} \tag{6.4}$$

In this equation the gain-coefficient g_L is defined. Equation 6.4 demonstrates that the sign of the absorption coefficient in the active zone of a QCL exclusively depends on the relation between the relaxation times τ_{21} and τ_{32}. As long as level 2 can be depopulated faster than it is refilled by charge carriers from level 3, i.e. $\tau_{21} < \tau_{32}$, population inversion builds up independently of the amount of carriers injected into the excited state 3, and independently of the overall decay time of state 3. However, this is merely true for a QCL in equilibrium, whereas the transient effects concerning the individual state populations *do* depend on τ_3, $|<\Psi_f\,|\boldsymbol{\pi}|\,\Psi_i>\cdot\boldsymbol{\epsilon}|$ and j_{inj}. Note that in Fig. 6.1 most relaxation processes are indicated by arrows, were τ_{21} is associated with the green arrow, and τ_{32} with the red arrow. However, τ_3 does not have a simple representation, as it is dependent on the relaxation of state 3 into both state 1 and 2.

As any lasing device, QCLs are not merely composed of the optically active material, but require an optical resonator in order to sufficiently enrich the radiation field for the increase of stimulated emission over spontaneous emission and the non-radiative decay of the excited state 3. In the simplest case, the resonator is composed of a ridge etched into the optically active material, which is terminated by two cleaved semiconductor facets acting as the resonator mirrors. More sophisticated resonators employ high-reflection coating in order to reduce mirror losses α_m.

QCLs employing a waveguide ridge etched into the active material as a resonator are called Fabry-Perot type QCLs. Due to the high refractive index of the semiconductor material QCLs are composed of, the laser modes are intrinsically well confined in a ridge etched out of the grown structure. However, within any waveguide ridge resonator fractions of the confined laser mode exist, which do not overlap with the optically active material, in our case the quantum cascade structure, and therefore do not experience gain. The fraction of the mode overlapping with the active material is given by the confinement factor Γ_c. As QCLs are electrically pumped and therefore require highly doped semiconductor contact layers as well as metallic contacts, laser modes in the mid-infrared are subject to waveguide losses α_w due to free-carrier absorption. There are several concepts to deal with this losses, one strategy being the separation of the metallic contacts from the optical modes, which tend to couple to surface plasmons, using dielectric claddings [68]. Another way of dealing with confinement losses is to exploit the coupling of the radiation to the surface plasmon for a stronger confinement of the resonator mode to the active material in so-called plasmon-enhanced waveguides [68]. Such waveguides enable the confinement of the mode without cladding the QCL core with thick dielectric layers by directly evaporating a metallic layer (usually gold) on top of the active structure [69,70]. In order to achieve single mode operation, which is not possible for simple Fabry-Perot resonators, periodic structures can be etched into the waveguide ridge. QCLs employing this kind of resonator are called distributed feedback QCLs [71]. Further sophisticated means of narrowing the number of modes in a cavity are provided by microresonators, e.g. in disk shaped whispering gallery mode resonators [72].

Whatever concept the resonator of a QCL is based on, a fundamental condition for lasing in this structure is given in Eqn. 6.5.

$$\Gamma_c \cdot g_L \cdot j_{inj} > \alpha_m + \alpha_w \qquad (6.5)$$

This equation expresses the necessity of the radiative gain in the active material to compensate the losses due to resonator mirrors and free carriers in the cavity, leading to the definition of the threshold current of a QCL as given in Eqn. 6.6.

$$j_{thresh} = \frac{\alpha_m + \alpha_w}{\Gamma \cdot g_L} \qquad (6.6)$$

For an operating device, a threshold current as low as possible is desired. However, it is absolutely crucial that j_{thresh} is low enough to still be able to cool the device below the critical temperature for lasing operation.

Summing up, there are two fundamental requirements for laser operation of a quantum cascade structure. The first, most basic requirement is that of population inversion in the active region, equivalent to $\tau_{21} < \tau_{32}$. The second necessary condition is that of a sufficiently low threshold current, which is in turn linked with the heat properties of the semiconductor material. Now, the development of a QCL in the SiGe system requires the thorough optimization of cavity losses as well as the gain factor. The latter is fundamentally dependent on two properties, the strength of the optical transition, and the efficiency of the non-radiative carrier relaxation between the states involved in the lasing process.

Charge carrier relaxation is therefore the most fundamental issue, when it comes to laser development, and gaining experimental access to the intersubband relaxation times in SiGe heterostructures is a major step towards mastering the up to now unravelled difficulties on the road to a SiGe QCL.

Optical phonon-induced transitions in k-space

The most important depopulation mechanism in III-V QCLs is the emission of optical phonons. Optical phonons possess a finite energy at a zero wave vector, $h\nu_0$, as e.g. shown for phonons in bulk silicon in Fig. 6.4. Therefore, in contrast to scattering by acoustic phonons with a wavevector around zero, the emission of optical phonons can result in a significant loss of energy without a change in the momentum of the scattered charge carrier. On the other hand, as the dispersion of optical phonons is rather flat, the difference in the transferred energy between the emission of an high-q phonon and one with zero wave vector is low, which is once more in strong contrast to the behavior of acoustic phonon scattering.

In the following, it shall be assumed that charge carriers populate an excited state, which

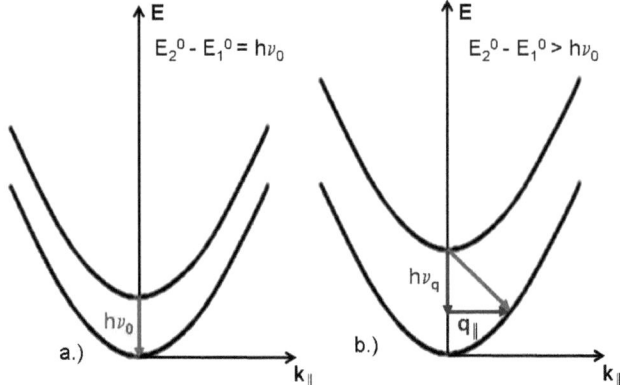

Figure 6.2: Position of optical phonon scattering processes in k-space. Graphs (a) and (b) show the in-plane dispersion of quantum states confined in z-direction in case of a transition energy $E_2^0 - E_1^0$ (defined as the energetic spacing between the states at zero in-plane wave vector) equal to and exceeding the optical phonon energy, respectively. Note that for optical phonon scattering in (a) direct transitions in k-space are possible, whereas for higher transition energies in (b) phonons of a significant $q_\parallel$ are emitted.

is confined in z-direction, in a quantum well of a heterostructure. Further, their in-plane momentum shall be relaxed ($k_\parallel = 0$). In Fig. 6.2, the in-plane dispersion of two quantum well levels is shown for two different transition energies at zero wave vector $E_2^0 - E_1^0$, which is throughout this work referred to as *transition energy* of the respective states. This is, strictly speaking, not fully correct, as the actual transition energy depends on the trajectory of the process in k-space. Nevertheless, as the energetic position of quantum well levels is in this work consistently given for $k_\parallel = 0$, the transition energy between two states is hereby defined at zero in-plane wave vector.

Now, if the transition energy between states 1 and 2 is below the optical phonon energy, the emission of such a phonon is prohibited by energy conservation. If the transition energy $E_2^0 - E_1^0$ matches the optical phonon energy, a scattering event vertical in k-space is possible, as indicated in Fig. 6.2 (a). As the transition energy between the states increases above the optical phonon energy, phonons of a finite in-plane wave vector are required to allow transitions from the excited level at $k_\parallel = 0$ to high $k_\parallel$-states of the lower level. This process is shown in Fig. 6.2 (b).

The dependence of the scattering efficiency between two quantum well states on their transition energy reflects the dependence of this transition rate on the wave vector of the phonon involved. The further the transition energy increases beyond the optical phonon energy, the higher the required in-plane wave vector of the scattering phonon is. If the scattering rate decreases with the length of $q_\parallel$ and thus with increasing transition energy between the quantum well states, the scattering event is called *resonant*, as its efficiency is highest in case of a transition energy matching the phonon energy. The question, whether a specific scattering mechanism behaves resonantly, will be addressed in the following sections. The next subsection covers the case of polar optical phonon scattering.

Interaction potential of polar optical phonons in III-V materials

III-V materials exhibit a zinc-blende crystal structure with two atoms of different electronegativity in the unit cell, therefore forming a polar crystal. In such a crystal, longitudinal optical (LO) phonons can induce a polarization on a macroscopic level. The field caused by a LO phonon of wave vector q, $\boldsymbol{E}_{LO}^q$, is proportional to the displacement between the atoms of the unit cell, $\boldsymbol{u}_{LO}^q(\boldsymbol{r},t)$, and reads as in Eqn. 6.7 ([73], pp. 125).

$$\begin{aligned}\boldsymbol{E}_{LO}^q &= \sqrt{4\pi N\, M\, \omega_{LO}^{q\,2}(\epsilon_\infty^{-1} - \epsilon_0^{-1})} \cdot \boldsymbol{u}_{LO}^q(\boldsymbol{r},t)\\ \frac{1}{M} &= \frac{1}{M_1} + \frac{1}{M_2}\end{aligned} \qquad (6.7)$$

In this equation, N represents the number of unit cells per unit volume in the crystal, M_1 and M_2 the masses of the unit cell atoms, ω_{LO}^q the LO phonon frequency, and ϵ_∞ and ϵ_0 the two limits of the dielectric constant. The scalar potential associated with this electric field is presented in Eqn. 6.8.

$$\Phi_{LO}^q = -\sqrt{4\pi N\, M\, \omega_{LO}^{q\,2}(\epsilon_\infty^{-1} - \epsilon_0^{-1})}\, \frac{u_{LO}^q(\boldsymbol{r},t)}{iq} \qquad (6.8)$$

$$(6.9)$$

In order to obtain the Fröhlich Hamiltonian, $u_{LO}^q(\boldsymbol{r},t)$ is expressed by the phonon creation and annihilation operators, c_q^+ and c_q, and the contributions for all different phonon wave vectors $\boldsymbol{q}$ are summed up.

$$\begin{aligned}H_{Fr} &= \sum_q \sqrt{4\pi N\, M\, \omega_{LO}^{q\,2}(\epsilon_\infty^{-1} - \epsilon_0^{-1})}\, \frac{i\, e\, u_{LO}^q(\boldsymbol{r},t)}{q}\\ &= \sum_q \sqrt{\frac{2\pi\, \hbar\, \omega_{LO}^q}{N\, V}(\epsilon_\infty^{-1} - \epsilon_0^{-1})}\frac{i\, e}{q} \cdot (c_q^+\, e^{i(\boldsymbol{q}\, \boldsymbol{r}-\omega_{LO}^q t)} + \\ &\quad + c_q\, e^{-i(\boldsymbol{q}\, \boldsymbol{r}-\omega_{LO}^q t)})\end{aligned} \qquad (6.10)$$

The Fröhlich interaction exclusively depends on macroscopic parameters, and can therefore be easily calculated. Note that the strength of the interaction decays with the absolute phonon wave vector q like q^{-1}. As will be shown in more detail in chapter 6.2.2, scattering by Fröhlich interaction is thus a resonant mechanism. Its efficiency decreases with increasing q-values and thus with increasing transition energies (see previous subsection). In a polar material system, this dependence of the Fröhlich interaction on the transition energy can be exploited for designing a QCL structure with optimized $\tau_{32} : \tau_{21}$ ratio, as will be discussed shortly on an example given in [65].

Time scales in a QCL

Due to the highly efficient Fröhlich interaction between electrons and LO phonons, intersubband relaxation times in III-V heterostructures are extremely short for transition energies above the LO phonon energy. Prior to the demonstration of the first QCL device by Faist et al. [60], these ultra-short relaxation times on a picosecond scale were expected to hinder the realization of an intersubband laser. However, as already mentioned, it turned out that this ultrafast relaxation could be employed for the highly efficient depopulation of the ground state of the lasing transition (state 2 in Fig. 6.1) by designing the $2 \rightarrow 1$ transition to be resonant with the LO phonon energy. In such a configuration, which involves short phonon wave vectors, the Fröhlich interaction has its maximum efficiency due to the q^{-1} dependence seen in Eqn. 6.10. On the other hand, the scattering process between excited state 3 and ground state 2 typically involves longer phonon wave vectors due to the higher transition energy, and is therefore less efficient. As long as the $2 \rightarrow 1$ transition energy is closer to the LO phonon energy than the $3 \rightarrow 2$ energy, the ground state can be emptied more quickly by the emission of an LO phonon than it is refilled from state 3. This enables population inversion despite the short lifetime of carriers in state 3. Due to a sufficiently high transition matrix element in III-V heterostructures, the radiative lifetime can be easily decreased below the non-radiative decay time by increasing the radiation field through a ridge cavity. In GaAs, the LO phonon energy is 36 meV, for InAs it values 30 meV.

In [65], Gmachl et al. present an example of an InGaAs/AlInAs QCL with an energetic spacing between ground state 2 and excited state 3 of 207 meV, and a transition energy between the ground state 2 and the extraction state 1 of 37 meV. In this work, relaxation times were simulated using the Frölich interaction model, giving $\tau_{32} = 2.2$ ps, $\tau_{31} = 2.1$ ps and $\tau_{21} = 0.3$ ps. Therefore, even though the decay time of excited state 3 is ultra-short, the depopulation time from the ground state is substantially shorter due to thorough design of the heterostructure, which gives a depopulation transition energy close to the LO phonon energy resulting in extremely efficient phonon emission.

Thus, careful design can turn the potentially cumbersome ultra-fast intersubband electron relaxation times in III-V heterostructures by polar optical phonon scattering into a useful tool to depopulate the ground state of the lasing transition. The question, whether this is in an analogous way possible for the intersubband hole relaxation times in SiGe heterostructures, directly leads to the need for detailed experimental studies on these decay times, which are subject of this work. The theoretical aspects of the difference in relaxation mechanisms between III-V electrons and SiGe holes will be addressed later on in this chapter.

6.1.3 Potential advantages of SiGe over III-V materials

Lattice absorption in III-V materials

Due to the polar nature of the unit cell of III-V compound semiconductors, electromagnetic waves of a field vector $\boldsymbol{E}$ can interact with transverse optical (TO) phonons, whose displacement vector $\boldsymbol{u}_{TO}(\boldsymbol{r},t)$ follows Eqn. 6.11 ([73], pp. 288).

$$M\frac{d^2\boldsymbol{u}_{TO}(\boldsymbol{r},t)}{dt^2} - M\gamma_d \frac{d\boldsymbol{u}_{TO}(\boldsymbol{r},t)}{dt} = -M\omega_{TO}^2 \boldsymbol{u}_{TO}(\boldsymbol{r},t) + Q \cdot \boldsymbol{E} \qquad (6.11)$$

In this equation, which includes local field corrections of the phononic eigenfrequency ω_{TO} due to the interaction of neighboring dipoles ([74], p. 353), M represents the reduced mass of the oscillation, γ_d its damping factor, and ω_{TO} the TO phonon frequency. Q represents the effective ionic charge of the oscillation, corrected by a local-field-induced factor as given by Eqn. 6.12.

$$Q = \frac{q^*}{1 - \frac{N\varrho_{el}}{3}} \qquad (6.12)$$

In this equation, ϱ_{el} is the electronic polarization factor, N is again the number of unit cells per crystal volume, and q^* represents the uncorrected effective charge attributed to the oscillation. The polarization, which is caused by the displacement between the two unit cell atoms and their electronic shells is presented in Eqn. 6.13.

$$\boldsymbol{P} = N \cdot Q \cdot \boldsymbol{u}_{TO}(\boldsymbol{r},t) + \epsilon_0 \frac{N}{1 - \frac{N\varrho_{el}}{3}} \boldsymbol{E} \qquad (6.13)$$

By combining Eqns. 6.11 and 6.13, Eqn. 6.14 is obtained for the dielectric function in a III-V crystal.

$$\epsilon(\nu) = \epsilon_\infty + \frac{\epsilon_0 - \epsilon_\infty}{1 - \frac{(2\pi\nu)^2}{\omega_{TO}^2} - i\frac{2\pi\nu\,\gamma_d}{\omega_{TO}^2}} \qquad (6.14)$$

ϵ_∞ and ϵ_0 represent the limits of the material's dielectric function for low and high frequencies of the electromagnetic wave. By considering the Gauss equation $\nabla \boldsymbol{D} = 0$ for the case of a

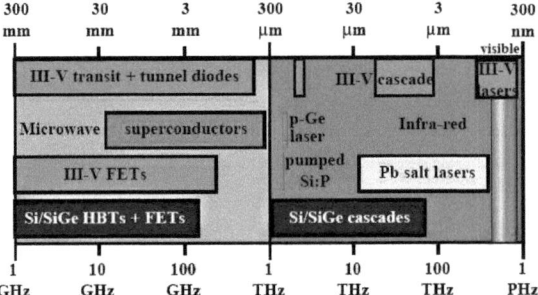

Figure 6.3: Coverage of spectral regions by the indicated radiation sources. Note the spectral region, which cannot be accessed by III-V QCLs due to the reststrahlen band in these materials. Further, there is a spectral region uncovered by any emitting devices between 1 and 10 THz, the so-called terahertz gap. The SiGe material system holds the potential to access both regions devoid of III-V and any other sources, respectively, as indicated by the blue bar.

longitudinal electric field, the so-called Lyddane-Sachs-Teller relation, as given in Eqn. 6.15, can be derived.

$$\frac{\epsilon_0}{\epsilon_\infty} = \frac{\omega_{LO}^2}{\omega_{TO}^2} \qquad (6.15)$$

When calculating the reflection coefficient for III-V materials from Eqn. 6.14 ([73], pp. 289), it becomes evident that radiation with frequencies between those of the TO and LO phonons cannot propagate in these material systems. The spectral region around the TO phonon frequency is called the reststrahlen band.

Due to the discussed interaction between polar optical phonons and far-infrared radiation, QCLs fabricated in III-V material systems cannot be realized for wavelengths within the reststrahlen band. The spectral region, which cannot be covered by GaAs devices, ranges from 20 to 35 μm.

Lattice absorption in the SiGe system

The unit cell of a SiGe crystal is composed of two group-IV atoms. As a consequence, an optical phonon is unable to induce a polarization in this material, resulting in the absence of polar optical phonon scattering in SiGe heterostructures [75, 76]. Even though first-order absorption of radiation by optical phonons is rendered impossible by the crystal structure of SiGe, weak lattice absorption is observed for Si and Ge crystals [77]. According to an explanation by Lax and Burstein [78], these absorption bands are originating from a second order interactions between phonons and electromagnetic waves. After the first phonon breaks the symmetry of the crystal lattice, a second phonon is able to induce a polarization in the crystal, which in turn couples to electromagnetic waves. However, the high order of this perturbation results in the effect being extremely weak. It can therefore be neglected during the considerations of the SiGe system for the realization of a QCL.

Therefore, the absence of a reststrahlen band in the SiGe system poses a clear advantage of this material over III-V materials. The high transparency of undoped SiGe for infrared radiation covers the whole range from 300 μm far into the mid-infrared beyond 2 μm. Thus, the SiGe systems offers potential as a basis for devices operating in the wavelength region inaccessible by III-V QCLs, as illustrated in Fig. 6.3.

Thermal properties of the SiGe system

One of the most important figures of merit of a QCL device is its threshold temperature. In order to achieve a device temperature sufficiently low for functionality while reaching a sufficiently high peak output power, many III-V QCLs have to be operated in pulsed mode. The demonstration of a QCL achieving CW operation at room temperature was a huge breakthrough within the QCL community [63]. Nevertheless, efficient cooling and critical temperatures continue to be major issues in the race for record figures of merit.

The first potential advantage of the silicon system regarding the matter of operation temperature lies in its intrinsically high thermal conductivity of 1.3 W cm^{-1}°C^{-1}, which exceeds that of III-V materials by more than a factor of two. These excellent cooling properties of the silicon system are exploited in integrated circuit technology. Thus, in case of equal ohmic losses, the cooling of SiGe devices below the critical lattice temperature is expected to be significantly less cumbersome than for III-V devices.

The second fundamental difference in material properties between SiGe and III-V systems with the potential to form a crucial advantage concerning the temperature dependence of the device operation, regards the dominant mechanisms for charge carrier relaxation. As pointed out in several publications [13,79], the absence of polar optical phonons should keep the intersubband relaxation times in SiGe heterostructures independent of the device temperature up to more than 100 K. This can be concluded from the observation, that carrier relaxation between states with a transition energy below the optical phonon energy is dominated by interface roughness scattering, which does not exhibit any temperature dependence up to 100 K [80]. On the other hand, for III-V structures the lifetime-shortening influence of polar optical phonon scattering becomes a critical issue from 40 K on even for transition energies below the LO phonon energy [81]. However, the scattering processes in SiGe heterostructures as well as the associated intersubband relaxation times will be covered in more detail in the following sections.

Selection rules and surface emission

The considerations presented in this section rather discuss the potential advantages of p-type devices over n-type structures than the advantages of SiGe over III-V material systems. However, as III-V devices are commonly realized in the valence band, while the SiGe devices discussed in this work are based on transitions within the valence band, this section presents the potential advantages of p-type SiGe structures over n-type devices in the III-V system.

Due to quantum mechanical selection rules, the emission of light from n-type III-V QCLs is restricted in polarization as well as in direction. Intersubband transition within the conduction band of III-V heterostructures can only produce light with an electric field vector parallel to the growth direction of the structure, that means in direction of the confinement potential of the charge carriers (TM polarization). This fact can be easily realized by the following considerations as found in basic textbooks ([82], pp. 299).

In an envelope function approach for conduction band states, the initial and final wavefunctions ϕ_i and ϕ_f of a transition can be written as in Eqn. 6.16.

$$\begin{aligned}
\phi_i &= S^{-\frac{1}{2}} u_i(\boldsymbol{r}) e^{i\boldsymbol{k}_\parallel \cdot \boldsymbol{r}} \Phi_i(z) \\
&= S^{-\frac{1}{2}} u_i(\boldsymbol{r}) \chi_i(\boldsymbol{r}) \\
\phi_f &= S^{-\frac{1}{2}} u_f(\boldsymbol{r}) e^{i\boldsymbol{k}'_\parallel \cdot \boldsymbol{r}} \Phi_f(z) \\
&= S^{-\frac{1}{2}} u_f(\boldsymbol{r}) \chi_f(\boldsymbol{r})
\end{aligned} \quad (6.16)$$

In this equation, $u_{i,f}(\boldsymbol{r})$ are lattice-periodic Bloch functions, $\boldsymbol{k}_\parallel$ and $\boldsymbol{k}'_\parallel$ are in-plane electronic wave vectors with $k_z = 0$, and $\Phi_{i,f}(z)$ are the envelope functions accounting for the carrier confinement in growth direction of the heterostructure. The envelope functions are gained by solving the one-dimensional Schrödinger equation for the confinement potential using the effective masses m_i^* and m_f^* of the respective bands, between which the transition takes place. As seen from Eqn. 2.3, the matrix element relevant for interactions of charge carriers with the radiation field is given by.

$$<\phi_f |\boldsymbol{p}| \phi_i> \cdot \boldsymbol{\epsilon} = S^{-1} <u_f \chi_f |\boldsymbol{p}| u_i \chi_i> \cdot \boldsymbol{\epsilon} \quad (6.17)$$

$\chi_{i,f}(r)$ are nearly constant on the scale of an unit cell and vary only slowly in comparison to $u_{i,f}(r)$. By taking this into account when writing Eqn. 6.17 in integral form, Eqn. 6.18 can be obtained.

$$\begin{aligned}<\phi_f |p| \phi_i> \cdot \epsilon &= S^{-1} <\chi_f | \chi_i><u_f |p| u_i>_{cell} \cdot \epsilon + \\ &+ S^{-1} <u_f | u_i>_{cell}<\chi_f |p| \chi_i> \cdot \epsilon \end{aligned} \quad (6.18)$$

In case of a III-V QCL, which employs intraband transitions within the conduction band for the generation of radiation, $\Phi_i(z)$ and $\Phi_f(z)$ are solutions to the same Schrödinger equation for one and the same effective mass and are thus orthogonal. As a consequence, the first term in Eqn. 6.18 vanishes during the evaluation of the z-integral for $<\chi_f | \chi_i>$. In the remaining second term, the polarization dependence of the optical transition is given by the term in Eqn. 6.19.

$$\begin{aligned}<\chi_f |p| \chi_i> &= \int d^3r \, \Phi_f^*(z) \, \Phi_i(z) \, e^{i(k_\parallel - k'_\parallel) \cdot r} \, k_\parallel + \\ &+ \int d^3r \, \Phi_f^*(z) \, e^{i(k_\parallel - k'_\parallel) \cdot r} \, p \, \Phi_i(z) \\ &= \delta_{k'_\parallel k_\parallel} \, k_\parallel <\Phi_f | \Phi_i> + \delta_{k'_\parallel k_\parallel} <\Phi_f |p_z| \Phi_i> e_z \end{aligned} \quad (6.19)$$

The last step in the evaluation of the equation is valid, as $\Phi_{i,f}(z)$ are exclusively dependent on z. For intraband transitions, the first term in Eqn. 6.19 vanishes due to reasons already mentioned. As a result, $<\chi_f |p| \chi_i>$ is composed of a non-vanishing z-component only, which implies that devices based on optical intrasubband transitions can exclusively interact with TM polarized photons only.

As a consequence of the selection rules discussed above, the transitions within the quantum wells of n-type QCLs lead to the emission of radiation exclusively directed parallel to the growth surfaces. The laser radiation of such devices therefore conventionally has to be coupled out of the side facetts. As this poses a disadvantage for numerous applications, complex gratings for surface emission are employed to overcome this shortage [83].

Devices implemented in p-type material do not exhibit this shortage. By using appropriate design, devices realized in the valence band are capable of interacting with radiation propagating parallel to the growth direction of the structure. This has been demonstrated for SiGe quantum well infrared photodetectors (see chapters 2 and 3 of this work and reference [16]) as well as for SiGe quantum well [84] and cascade emitters [13].

Analogous to intra-conduction band transitions in III-V devices, transitions within the heavy- or light-hole bands enable the absorption and emission of TM polarized radiation in p-type

SiGe devices. In addition to intraband transitions within the respective valence bands induced by TM polarized light, p-type heterostructures allow optical transitions between the different hole bands. When considering hole states *near the Γ-point*, the second term in Eqn. 6.18 vanishes for Bloch functions $u_{i,f}(\boldsymbol{r})$ belonging to different hole bands due to their orthogonality. In contrast to the case of intraband transitions, the first term is finite for intersubband transitions due to the non-orthogonality of $\Phi_i(z)$ and $\Phi_f(z)$. The latter holds true for states of the one-dimensional potential wells with the same quantum numbers belonging to different hole bands, which is trivial, but as well for different states due to the different effective masses of the respective hole bands. As seen from Eqn. 6.18, in case of optical intersubband transitions near the zone center, the allowed polarization is given by $< u_f \, |\boldsymbol{p}| \, u_i >_{cell}$. The selection rules for this kind of interaction with the radiation field are therefore analogous to that in bulk material. For relaxed SiGe (100) bulk material, transitions between HH and LH states are forbidden for TM polarized radiation, but are strong in case of TE polarization. Thus, HH-LH transitions can be employed for the detection and emission of radiation propagating normal to the sample surface. Additionally, in strained SiGe quantum well structures, the HH and LH bands couple for wave vectors further from the Γ-point. As a consequence, the second term in Eqn. 6.18 becomes finite in cases of high wave vectors, enabling optical transitions in TM polarization.

Therefore, in the p-type SiGe system surface emitting devices can be realized, making this material system a potential candidate for realizing surface emitting QCLs without any additional grating couplers.

6.1.4 Progress on the road to a SiGe QCL

Owing to the numerous potential advantages of a SiGe QCL over its III-V pedant, intensive research efforts have been devoted to the development of a group-IV source of coherent light. The major impulse given to this field was the first demonstration of electroluminescence in a SiGe quantum cascade structure by Dehlinger et al. in 2000 [12]. This section provides a short overview over the achievements in the field of SiGe quantum cascade emitters since this major breakthrough.

Electroluminescence

The first SiGe quantum cascade structure to exhibit electroluminescence utilized the spatially direct transition between the first excited HH state and the HH ground state in a p-type quantum well [12]. The luminescence peak was center around 130 meV, and the emitted light was TM polarized, as expected from the considerations given in chapter 6.1.3. The ground state of optical transition in the structure of [12] was at the same time the lowest state in each emitting quantum well period, thus the rapid depopulation of the ground state of the radiating transition required for a QCL was not included in the design concept. The ground state of the emitting transition could only be emptied via tunnelling into neighboring quantum wells. In the publications following, the structure was modified to reach emission energies between 125 and 154 meV [85]. Similar structures were used by Bormann et al., who demonstrated electroluminescence between 146 and 159 meV [86]. Diehl et al. demonstrated a structure grown on pseudosubstrate (see below) featuring an energy of emitted photons of 185 meV [87]. The above mentioned SiGe emitters showed proper electroluminescence for device temperatures up to 180 K [88].

Pseudosubstrate

Any structures mentioned above were grown on plain Si (100) wafers. As stated in [85], the design freedom as well as the number of layers, which can be grown pseudomorphically, is crucially restricted by the accumulation of strain for this kind of substrates. As a consequence, the structures in [12] and [14] featured the low number of 12 and 15 periods, respectively, while numbers of periods realized in III-V QCLs are usually beyond 50. In order to overcome this restriction, the employment of $Si_{1-x}Ge_x$ pseudosubstrates was suggested by several groups, where the germanium content x is matched to that of the quantum cascade structures. The use of pseudosubstrates was already discussed in detail in chapter 3. The performance of emitting devices grown on pseudosubstrates was studied in [87, 89] ($Si_{0.5}Ge_{0.5}$), in [79] ($Si_{0.8}Ge_{0.2}$), in [90] ($Si_{0.7}Ge_{0.3}$), in [84] ($Si_{0.77}Ge_{0.23}$) and [91] ($Si_{0.8}Ge_{0.2}$). In [91], Kelsall et al. reported on the growth of a cascaded structure with 600 periods featuring one quantum well each, an impressive number of layers comparable to those featured by III-V QCLs.

Even though SiGe emitter structures exhibit electroluminescence intensities competitive with those of equivalent III-V emitters [79], SiGe heterostructures have failed to show optical gain or stimulated emission so far. This is the case despite the fact that SiGe emitter structures exhibit dipole matrix elements sufficiently high for strong spontaneous emission. The reason for this setback for SiGe emitters in the mid-infrared has been generally attributed to the ultrafast intersubband relaxation processes for transition energies above the optical phonon energy, which seem to prevent such structures from reaching population inversion. The relaxation processes for transition energies above and below the optical phonon energies and their timescales will be highlighted in chapter 6.2.

Terahertz frequencies

The strong general interest in emitter devices in the terahertz region (see chapter 4.1) as well as the by more than one order of magnitude longer intersubband relaxation times have motivated extensive research on SiGe emitter devices operating in this spectral range. The SiGe cascade emitters demonstrated by Lynch et al. in [13] exhibited surface emission of photons at an energy of 12 meV. Bates et al. reported electroluminescence from SiGe cascade structures at 8 meV and 5 meV, both in TE polarization [79]. The devices presented by Kelsall et al. in [91] operate at 15 meV. All of the SiGe emitter structures operating in the THz regime as mentioned above are based on the optical transition between LH and HH states, which is allowed in TE polarization, as stated in section 6.1.3. However, in addition to the emission lines originating from intersubband transitions, the device presented in [13] exhibited purely TM polarized electroluminescence at photon energies of 37 meV. The latter results from intraband transitions between HH states, which are both confined within the cascade quantum wells. It has thus been demonstrated, that THz emitters can be realized in SiGe heterostructures, and that radiation of frequencies in the so-called terahertz gap between 1 and 10 THz can be reached in this material system. Additionally, the emission of terahertz radiation is not bound to TM polarization by selection rules. By employing intersubband transitions in the valence band, surface-normal emission has been demonstrated in the SiGe system [79].

Diagonal transitions

In addition to the concept of manipulating non-radiative relaxation times by modifying the energetic spacing between the involved states, the change in the spatial overlap between these states poses a possibility to tune intersubband lifetimes. Most of the SiGe emitter structures mentioned above [13, 84–89] are based on spatially direct transitions, they feature a vertical-transition active region, as Gmachl called it in [65]. So-called diagonal transition active regions form a means of increasing the non-radiative relaxation times in a QCL [65, 92], and therefore pose a promising concept for reaching population inversion in SiGe emitter structures. Diagonal radiating transitions in the SiGe system have been demonstrated in [79] and [86], where the structure presented by Bormann et al. featured an emitting transition between the HH states of adjacent quantum wells [86], while the device by Bates et al. used a transition between the HH state confined in one well and the LH state located in a neighboring well [79]. Due to the different nature of the transitions employed in the two afore mentioned publications, the emitted radiation differed both in polarization and wavelength (TM above 155 meV in [86], TE at 8 meV in [79]). Even though Bates et al. realized emitters based on diagonal transitions with transition energies far below the optical phonon energies, population inversion could not be achieved. The concept of diagonal transitions in QCLs and the associated prolongation of relaxation times is one of the key topics of this work and will be discussed in detail in the following chapters.

Simulation

Besides the molecular beam epitaxial fabrication of SiGe heterostructure emitters based on different design concepts and their thorough electronic and optical characterization, considerable efforts have been invested into the simulation of the carrier dynamics in SiGe quantum cascade structures [93–95]. As far as the development of laser resonators in the SiGe system is concerned, the use of highly doped semiconductor confinement layers is suggested for TM polarized modes by theoretical considerations in [96]. In the latter publication, Ikonic et al. propose the use of a novel metal\metal silicide waveguide for TE modes. This proposal was realized in [91], where a buried tungsten-silicide layer was used to confine the terahertz mode emitted by a SiGe quantum cascade structure.

Proposals for n-type devices

Very recently, proposals for an n-type SiGe quantum cascade emitter in the terahertz region have been published [97, 98]. These concepts are based on the strain-induced band-offset in the conduction band of SiGe layers of different Ge concentrations grown on a pseudosubstrate. In [98], Driscoll et al. suggest to employ L-valley electrons in Ge-rich systems (Ge content above 85%) for realizing terahertz emitters in order to circumvent the troubles accompanying the high longitudinal mass of the Δ-valley. Even more recent publications propose design concepts employing Δ-electrons in n-type SiGe structures grown on (111) substrate [99, 100]. However, neither of the mentioned concepts has up to now lead to the realization of *n-type* SiGe quantum cascade emitters.

Despite all the efforts put into the realization of a SiGe QCL and in spite of the numerous achievements on the road to success, structures in this material system up to now have ultimately failed by lacking population inversion. As highlighted several times throughout this chapter, the most fundamental issue concerning the implementation of population inversion is the intersubband relaxation time. The theoretical difference between the intersubband lifetimes in the conduction band of III-V QCLs and those in the valence band of SiGe heterostructures will be discussed in the next section.

6.2 Intersubband relaxation processes in SiGe

The means, by which the concept of QCLs could be successfully realized in III-V material systems despite the fast depopulation of the excited emitter state present there, is the even faster depopulation of the ground state of the lasing transition by the emission of polar optical phonons. This concept for reaching population inversion was already highlighted in chapter 6.1.2. As mentioned there, the efficient emission of polar optical phonons by the relaxation from the ground state of the lasing transition (2 in Fig. 6.1) into a lower state (1 in Fig. 6.1) is enabled by designing the spacing between states 1 and 2 in resonance with the LO phonon energy, thus exploiting the resonant behavior of the Fröhlich interaction. This resonant depopulation mechanism, which is essential for the design of III-V QCLs, lacks a counterpart in the SiGe system, as will be pointed out in the following section. The lack of an efficient depopulation process for the ground state of the lasing transition tremendously increases the importance of knowing the precise figure for the excited state lifetime. This makes the exact determination and understanding of intersubband relaxation times in the SiGe system one of the key issues for the realization of a QCL in this material system.

This chapter presents theoretical considerations as well as experimental results on the nature and quantitative timescale of intersubband relaxation processes in the SiGe system. As far as the interaction between charge carriers and phonons is concerned, the fundamental difference between the SiGe and III-V systems is that the SiGe material exhibits a non-polar, centrosymmetric crystalline structure. Thus, SiGe structures do not feature polar optical phonons. Additionally, no piezoelectric interaction between charge carriers and acoustic phonons exists in this system. The first-order interaction between charge carriers and phonons in the SiGe material is exclusively based on the energetic changes induced by the vibrational deformation of the crystal and the associated change in symmetry and periodicity.

6.2.1 Non-polar optical phonon scattering

Figures 6.4 and 6.5 present the calculated and measured phonon dispersion in bulk silicon [101,102] and germanium [103], respectively. Note the degeneracy of the TO and LO branches in the vicinity of the zone center, which is a result of the absence of polar phonons in the Si and Ge systems. The first-order interaction between charge carriers and phonons involves the absorption or emission of a phonon via a transition between two electronic states, whose energetic spacing matches the phonon energy. In a quantum cascade structure under ideal conditions, the lattice temperature is too low to allow a significant occupation of the optical phonon branch in Figs. 6.4 and 6.5. The charge carrier-LO phonon interaction in such a structure is thus limited to the *emission* of optical phonons. An illustration of this process

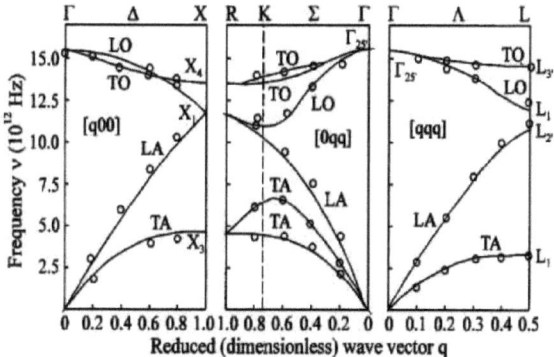

Figure 6.4: Phonon dispersion in bulk silicon, published in [101, 102].

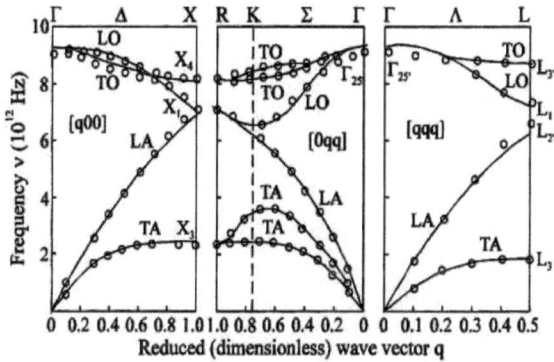

Figure 6.5: Phonon dispersion in bulk germanium, published in [103].

is given in Fig. 6.2.

Interaction potential

In contrast to acoustic phonons, which induce a volume dilation in the crystal due to a shift in the grid positions, long-wavelength optical phonons do not distort the lattice macroscopically ([73], p. 124). They rather lead to a displacement of the atoms within an unit cell in respect to each other. As the periodicity of the crystalic structure remains unchanged under the influence of an optical phonon, its perturbation Hamilonian does not depend on the volumetric change of the unit cell. Further, as due to the composition of an unit cell by two group-IV atoms optical phonons do not induce a polarization in the cell and are therefore non-polar, there is no Fröhlich interaction between charge carriers and optical phonons in the SiGe system, which dominates the phonon scattering in III-V structures (see chapter 6.1.2).

The interaction between charge carriers and non-polar optical phonons in the SiGe system is solely induced by the change in the bonding lengths between the atoms of the unit cell. The interaction via a change in the bonding length can be described in terms of deformation potentials. In [104], the deformation potential of non-polar optical phonons is treated in approximation of isotropic optical phonon dispersion. Within the frame of this approximation, the symmetry of the interaction Hamiltonian associated with non-polar optical phonons is not considered, the applied model thus does not account for the selection rules of the induced transitions. The isotropic phonon approximation delivers results of satisfying accuracy in the simulations reported in [104], and is sufficient for studying the dependence of scattering processes on the energetic spacing between the involved levels. Even though the numerical results obtained by Ikonic in [104] reflect the relations between HH2-HH1 and LH1-HH1 transitions well, it is incapable of providing analytical insight into the selection rules associated with deformation potential scattering. This topic is discussed in further detail in section 6.2.7. In the approximation of isotropic dispersion, the Hamiltonian for the coupling between electrons and non-polar optical phonons can be expressed in terms of a phenomenological optical phonon deformation potential D_O, as shown in Eqn. 6.20 [73].

$$\begin{aligned} H_{e-OP} &= D_O \sum_q u_{LO}^q(\boldsymbol{r},t) = \\ &= D_O \sum_q \sqrt{\frac{\hbar}{2\,N\,M\,\omega_{LO}^q}} (c_q^+ \, e^{\mathrm{i}(\boldsymbol{q}\,\boldsymbol{r}-\omega_{LO}^q t)} + \\ &+ c_q \, e^{-\mathrm{i}(\boldsymbol{q}\,\boldsymbol{r}-\omega_{LO}^q t)}) \end{aligned} \qquad (6.20)$$

In this equation, $u_{LO}^q(\boldsymbol{r},t)$ is defined as the absolute displacement between the atoms of an unit cell induced by a non-polar optical phonon of wave vector $\boldsymbol{q}$, which is expressed in terms of the phonon creation and annihilation operators analogous to chapter 6.1.2. Again, M is the reduced mass of the unit cell defined by Eqn. 6.7, and ω_{LO}^q represents the frequency of the optical phonon of wave vector $\boldsymbol{q}$ according to the dispersion relation, as illustrated in Figs. 6.4 and 6.5. Now, the fundamental difference in the deformation potential scattering between acoustic and optical phonons is that the Hamiltonian for the latter directly depends on the displacement $u_{LO}^q(\boldsymbol{r},t)$ itself instead of its derivative, which is in form of the strain tensor (see Eqn. 6.36) the quantity on which the Hamiltonian for acoustic phonons depends.

Scattering rate

In order to answer the question, whether the efficiency of the processes of emitting or absorbing non-polar optical phonons increases or decreases with raising energetic spacing between the initial and final states, the scattering rate between the initial state $|\widetilde{\phi}_i>$ and a final state $|\widetilde{\phi}_f>$ has to be calculated. For gaining the scattering rate, the perturbation matrix element $<\widetilde{\phi}_f|H_{e-OP}|\widetilde{\phi}_i>$ has to be determined. The initial and final states are elements of the direct product between the electron and phonon vector spaces with $|\widetilde{\phi}_i>=|\phi_i>|N_{ph}^q>$ and $|\widetilde{\phi}_f>=|\phi_f>|N_{ph}^q\pm 1>$, where N_{ph}^q is the number of phonons with wave vector $\boldsymbol{q}$. $|\phi_i>$ and $|\phi_f>$ are a generalized form of the initial and final states of the scattering process in the envelope function approach, including the coupling between the different hole bands at finite wave vectors $\boldsymbol{k}$. They are given by Eqn. 6.21.

$$\begin{aligned}
\phi_i &= S^{-\frac{1}{2}} \sum_n u_n(\boldsymbol{r}) e^{i\boldsymbol{k}_\parallel \cdot \boldsymbol{r}} \Phi_{i,n}(z) \\
&= S^{-\frac{1}{2}} \sum_n u_n(\boldsymbol{r}) \chi_{i,n}(\boldsymbol{r}) \\
\phi_f &= S^{-\frac{1}{2}} \sum_m u_m(\boldsymbol{r}) e^{i\boldsymbol{k}'_\parallel \cdot \boldsymbol{r}} \Phi_{f,m}(z) \\
&= S^{-\frac{1}{2}} \sum_m u_m(\boldsymbol{r}) \chi_{f,m}(\boldsymbol{r})
\end{aligned} \quad (6.21)$$

In this equation, the u_n are the known Bloch functions close to the zone center, where $\Phi_{i,n}(z)$ and $\Phi_{f,n}(z)$ are the associated envelope functions of the initial and final states for a heterostructure with charge carrier confinement in z-direction. Using the functions as defined in Eqn. 6.21 leads to the expression for the perturbation element $<\widetilde{\phi}_f|H_{e-OP}|\widetilde{\phi}_i>$ shown in Eqn. 6.22, where ρ is the density of the crystal.

$$\begin{aligned}
<\widetilde{\phi}_f|H_{e-OP}|\widetilde{\phi}_i> &= D_O \sqrt{\frac{\hbar}{2NV\rho}} S^{-1} \sum_{m,n} <u_m|u_n>_{cell} \cdot \\
&\quad \cdot \frac{1}{\sqrt{\omega_{LO}^q}} \sqrt{N_{ph}^q + \frac{1}{2} \pm \frac{1}{2}} \cdot <\chi_{f,m}|e^{\pm i\boldsymbol{q}\cdot\boldsymbol{r}}|\chi_{i,n}>= \\
&= D_O \sqrt{\frac{\hbar}{2NV\rho}} \sum_n \frac{1}{\sqrt{\omega_{LO}^q}} \sqrt{N_{ph}^q + \frac{1}{2} \pm \frac{1}{2}} \cdot <\chi_{f,n}|e^{\pm i\boldsymbol{q}\cdot\boldsymbol{r}}|\chi_{i,n}>
\end{aligned} \quad (6.22)$$

Equation 6.22 was obtained by considering the fact that $\chi_{i,n}(z)$, $\chi_{f,n}(z)$ and $e^{\pm iq_z z}$ vary slowly over one unit cell and therefore on the scale of the Bloch functions u_n. The integral thus can be split into one carried out over the unit cell volume to obtain $<u_m|u_n>_{cell}$, and a second one performed over the total volume in order to gain $<\chi_{f,m}|e^{\pm i\boldsymbol{q}\cdot\boldsymbol{r}}|\chi_{i,n}>$. Performing the integration over x and y gives a delta-function, ensuring the conservation of the total in-plane momentum and leading to Eqn. 6.23.

$$<\tilde{\phi}_f|H_{e-OP}|\tilde{\phi}_i> = D_O \sqrt{\frac{\hbar}{2NV\rho}} \sum_n \frac{1}{\sqrt{\omega_{LO}^q}} \sqrt{N_{ph}^q + \frac{1}{2} \pm \frac{1}{2}} \cdot$$
$$\cdot \; \delta(\boldsymbol{k}_\parallel - \boldsymbol{k}'_\parallel \pm \boldsymbol{q}_\parallel) \int_{-\infty}^{\infty} \Phi^\dagger_{f,n} \, e^{\pm iq_z z} \, \Phi_{i,n} dz \qquad (6.23)$$

For gaining the total non-polar optical phonon scattering rate between the electronic states $|\phi_i>$ and $|\phi_f>$ by employing Fermi's golden rule, the scattering rate contributions by all possible phononic initial and final states $|N_{ph}^q>$ and $|N_{ph}^q \pm 1 >$ have to be summed up, considering the volume in $\boldsymbol{q}$-space attributed to each phonon wave vector. This leads to Eqn. 6.24.

$$W_{if}^{OP} = \frac{D_O^2 \hbar}{2\rho(2\pi)^2} \int_{(q)} \frac{1}{\omega_{LO}^q \hbar} \left(N_{ph}^q + \frac{1}{2} \pm \frac{1}{2}\right) |G_{if}|^2 \cdot$$
$$\cdot \; \delta(\boldsymbol{k}_\parallel - \boldsymbol{k}'_\parallel \pm \boldsymbol{q}_\parallel) \, \delta(E_i - E_f \mp \hbar\omega_{LO}^q) \, d\boldsymbol{q}$$
$$N_{ph}^q = \frac{1}{e^{\frac{\hbar\omega_{LO}^q}{k_B T}} - 1}$$
$$G_{if} = \sum_n \int_{(z)} \Phi^\dagger_{f,n} \, e^{\pm iq_z z} \, \Phi_{i,n} \, dz \qquad (6.24)$$

The generalized overlap integral G_{if} between the final and initial states is a figure with crucial influence on the efficiency of the scattering process. G_{if} includes both the spatial overlap of the respective envelope functions and the amount of compositional overlap with respect to the Bloch functions u_n. G_{if} is thus high if the envelope functions of the initial and final states associated with the same Bloch functions overlap strongly, and is weak, if the states are spatially separated.

The predicted characteristic, that optical phonon scattering induced by the deformation potential increases in efficiency with the spatial overlap between the involved electronic states, is shared by a number of other scattering processes. Based on this quite intuitive effect is the concept of spatially indirect transitions in QCL structures. By spatially separating the excited state from the ground state of the lasing transition in a QCL by means of bandstructure design, the lifetime of the excited state is increased due to the resulting decrease in phonon scattering efficiency. The concept of a so-called diagonal transition active region [65, 92] has already been mentioned in chapter 6.1.4 and will remain a major topic in the following chapters.

Depopulation rate

In the previous section, an analytical expression for the non-polar optical phonon scattering rate between two distinct states, $|\phi_i>$ and $|\phi_f>$, was derived. These two states are, to be precisely, unambiguously characterized by a quantum number accounting for the quantization in growth direction of the heterostructure, and the in-plane momentum $k_\parallel$ of the state. A precise declaration of the final and initial states involved in the scattering event is given in Eqn. 6.25.

$$\begin{aligned} |\phi_i> &= |\phi_{i,k_\parallel}> \\ |\phi_f> &= |\phi_{f,k_\parallel'}> \end{aligned} \qquad (6.25)$$

Now, what is of practical interest for the operation of a quantum well device, is not the scattering rate between two distinct quantum mechanical states as those in Eqn. 6.25, but rather the relaxation rate of the total of charge carriers occupying a quantum well state i at various positions in $k_\parallel$-space into the zoo of $k_\parallel'$ states of a different level f. In order to gain this total relaxation rate $D_{i,k_\parallel}^{f\ OP}$ from an initial state $|\phi_{i,k_\parallel}>$ into the various $k_\parallel'$-states of a lower quantum number f, the scattering rate in Eqn. 6.24 is summed up over all possible final states. This summation leads to Eqn. 6.26.

$$\begin{aligned} D_{i,k_\parallel}^{f\ OP} &= \frac{D_O^2 \hbar}{2\rho(2\pi)^2} \int_{(q)} \frac{1}{\omega_{LO}^q \hbar} \left(N_{ph}^q + \frac{1}{2} \pm \frac{1}{2}\right) \cdot \\ &\quad \cdot |G_{i,k_\parallel}^{f,k_\parallel \pm q_\parallel}|^2 \, \delta(E_{i,k_\parallel} - E_{f,k_\parallel \pm q_\parallel} \mp \hbar\omega_{LO}^q) \, d\mathbf{q} \\ N_{ph}^q &= \frac{1}{e^{\frac{\hbar\omega_{LO}^q}{k_B T}} - 1} \\ G_{i,k_\parallel}^{f,k_\parallel'} &= \sum_n \int_{(z)} \Phi_{f,k_\parallel'}^{n\ \dagger} \, e^{\pm i q_z z} \, \Phi_{i,k_\parallel}^n \, dz \end{aligned} \qquad (6.26)$$

The determination of this relaxation rate of the occupation of one-dimensionally quantized states generally requires a numerical approach, like that demonstrated in [104]. However, in order to point out the fundamental difference between the Fröhlich interaction and the various scattering mechanisms in the SiGe system with respect to their depopulation efficiency and its dependence on transition energies, a rather rough analytic treatment is presented in this section. A number of approximations, which crucially restrict the validity of the given equations, will be applied. The conclusions drawn from the following derivations are thus merely qualitative ones, whose sole purpose is the comparison between the different scattering mechanisms. In order to simplify the analytical treatment, the depopulation rate of a charge carrier with zero in-plane momentum ($k_\parallel = 0$) in an excited level of quantum number i

is considered. Further, the band structure of the final state is approximated parabolically, leading to Eqn. 6.27, where E_f^0 represents the energy of level f at $\boldsymbol{k}_\parallel = 0$.

$$E_{f,\boldsymbol{k}_\parallel} = E_f^0 + \frac{(\hbar\,\boldsymbol{k}_\parallel)^2}{2\,m_{eff}} \tag{6.27}$$

Additionally, the in-plane phononic and electronic bandstructures are approximated in in-plane radial symmetry, thus neglecting the differing dispersions in (100) and (110) directions. However, this approximation will be sufficient for the demonstration of the difference in the dependence of the depopulation rate on the energetic spacing of the involved levels. In this approximation, both ω_{LO}^q and $G_{i,0}^{f,q_\parallel}$ depend exclusively on the absolute of $\boldsymbol{q}_\parallel$. As the depopulation rate of an excited state is discussed, the absorption of phonons represented by the plus-sign in the delta-function of Eqn. 6.26 is not considered. As a result, the integral in Eqn. 6.26 can be transformed into cylindric coordinates. Applying these various simplifications leads to Eqn. 6.28.

$$\widetilde{D}_{i,0}^{f\ OP} = \frac{D_O^2 \hbar}{2\rho 2\pi} \int_{(q)} \frac{1}{\omega_{LO}^{q_z,\,q_\parallel} \hbar} (N_{ph}^{q_z,\,q_\parallel} + 1) \cdot$$
$$\cdot\ |G_{i,0}^{f,q_z,\,q_\parallel}|^2\ \delta(E_i^0 - E_{f,q_\parallel} - \hbar\omega_{LO}^{q_z,\,q_\parallel})\ q_\parallel\ dq_\parallel\ dq_z \tag{6.28}$$

Carrying out the substitution $E = \frac{(\hbar\,q_\parallel)^2}{2\,m_{eff}}$ leads to Eqn. 6.29.

$$\widetilde{D}_{i,0}^{f\ OP} = \frac{D_O^2\,m_{eff}}{2\hbar^2 \rho 2\pi} \int_{(q_z,E)} \frac{1}{\omega_{LO}^{q_z,\,E}} (N_{ph}^{q_z,\,E} + 1) \cdot$$
$$\cdot\ |G_{i,0}^{f,q_z,E}|^2\ \delta(E_i^0 - E_f^0 - E - \hbar\omega_{LO}^{q_z,\,E})\ dE\ dq_z \tag{6.29}$$

The the delta-function in Eqn. 6.29 selects the value of the integrand at the position E' as given in Eqn. 6.30.

$$E' = (E_i^0 - E_f^0 - \hbar\omega_{LO}^{q_z,E'}) \tag{6.30}$$

As seen in Fig. 6.4, the optical phonons in Si exhibit a very flat dispersion, where the phonon energy varies by less than 20 % over the total Brillouin zone. For the simulations reported by Ikonic et al. in [104], the optical phonons were approximated as dispersionless. According to Ikonic, this simplification is justified due to the fact that phonons exhibiting a k_z exceeding 20% of the maximal Brillouin zone value do not contribute to the scattering rate due to the strong oscillations of the exponential in the definition of $G_{i,k_\parallel}^{f,k_\parallel}$ in Eqn. 6.25 and the resulting insignificance of the generalized overlap. Thus, even though $\omega_{LO}^{q_z,E'}$ depends implicitly on E', it can in good approximation be replaced by its value at the zone center, ω_{LO}, leading to

Eqn. 6.31.

$$\widetilde{D}_{i,0}^{f\ OP} = \frac{D_O^2\,m_{eff}}{2\hbar^2\rho 2\pi}\int_{(q_z)}\frac{1}{\omega_{LO}}\,(N_{ph}^{q_z,\,E_d}+1)\cdot|G_{i,0}^{f,q_z,E_d}|^2\,dq_z$$

$$E_d = (E_i^0 - E_f^0 - \hbar\omega_{LO})$$

$$G_{i,0}^{f,q_z,E_d} = 0 \quad \text{for } E_d < 0 \tag{6.31}$$

In this equation, E_d is equivalent to the detuning of the energetic spacing between the quantized levels i and f at $\boldsymbol{k}_\parallel = 0$, which is referred to as *transition energy* throughout this work, from the optical phonon energy. Equation 6.31 demonstrates, that within the boundaries of the applied approximation, the depopulation rate $\widetilde{D}_{i,0}^{f\ OP}$ does not exhibit any significant dependence on the detuning E_d. This is due to the independence of the Hamiltonian in Eqn. 6.20 of the phonon wave vector, combined with the flat dispersion of optical phonons and the constant two-dimensional density of states in quantum wells.

Thus, when increasing the spacing between the initial and final states beyond the optical phonon energy, the relaxation rate induced by non-polar optical phonon scattering does not decrease. Put differently, deformation potential scattering by non-polar optical phonons does not exhibit resonant behavior. Once the energetic spacing between two quantum well levels is higher than the optical phonon energy, charge carriers are expected to relax at high efficiency relatively independent of the precise value of the transition energy at $\boldsymbol{k}_\parallel = 0$ of these states.

Comparison to Fröhlich interaction

The crucial difference between perturbation via the deformation potential of non-polar optical phonons and that by Fröhlich interaction with polar optical phonons (see chapter 6.1.2) becomes evident when comparing Eqns. 6.20 and 6.10. The respective interaction Hamiltonians exhibit a dependence on the wave vector as given in Eqn. 6.32.

$$H_{e-OP}^q \sim \frac{1}{\sqrt{\omega_{LO}^q}}$$

$$H_{Fr}^q \sim \frac{\sqrt{\omega_{LO}^q}}{q} \tag{6.32}$$

The Hamiltonian of the Fröhlich interaction explicitly depends on the absolute wave vector of the scattering polar optical phonon. The derivation of the depopulation rate by polar optical phonon emission, which is carried out in complete analogy to that for non-polar counterparts including numerous approximations as described in the previous section, results in Eqn. 6.33.

$$\widetilde{D}_{i,0}^{f}{}^{Fr} = m_{eff}(\epsilon_\infty^{-1} - \epsilon_0^{-1}) \int_{(q_z)} \frac{\omega_{LO}^{q_z, E_d}}{2m_{eff}E_d + \hbar^2 q_z^2} (N_{ph}^{q_z, E_d} + 1) \cdot |G_{i,0}^{f,q_z,E_d}|^2 \, dq_z$$

$$E_d = (E_i^0 - E_f^0 - \hbar\omega_{LO}) \tag{6.33}$$

$$G_{i,0}^{f,q_z,E_d} = 0 \quad \text{for } E_d < 0 \tag{6.34}$$

The decrease of the Fröhlich interaction potential with the phonon wave vector leads to a decrease in the relaxation rate $\widetilde{D}_{i,0}^{f}{}^{Fr}$ with increasing energetic spacing between the initial and final levels and the associated increase in the detuning E_d. This behavior is reflected by the factor $\frac{1}{2m_{eff}E_d + \hbar^2 q_z^2}$ in Eqn. 6.33. Therefore, the relaxation rate by polar optical phonon scattering shows a maximum for transition energies between the initial and final states near the optical phonon energy and thus behaves resonantly. The scattering by non-polar optical phonons is non-resonant, preventing the use of conventional III-V QCL design for similar SiGe structures, as discussed in chapter 6.2.6.

It shall once more be pointed out that the equations given in this and the previous subsection have been derived by a number of approximations and are far from generally valid. They do not include the dependence of the in-plane dispersion on the crystal direction, nor the intrinsic dependence of the phonon energy fulfilling energy conservation on the phonon wave vector. In addition, the considerations in the previous chapters exclusively dealt with initial states at zero in-plane wave vector.

Therefore, this work does not claim that the relaxation rate induced by non-polar optical phonons is strictly independent of the transition energy. The theoretical considerations at hand merely serve the purpose of demonstrating, why the deformation potential scattering mechanism in SiGe structures is not expected to feature the resonant behavior shown by the Fröhlich interaction and exploited in III-V QCL design.

In a realistic system, a finite occupation of states with a significant $\boldsymbol{k}_\parallel$ occurs, enabling the emission of optical phonons even for an energetic spacing between the levels lower than the optical phonon energy. As a result, the scattering rate by optical phonons slowly increases with the transition energy of the two involved levels, until the emission of a optical phonon is energetically allowed from an initial state with $\boldsymbol{k}_\parallel = 0$, which induces a strong, jump-like increase in the scattering rate. This behavior has been numerically predicted in [106] and experimentally observed for LH-HH transitions in [107].

When drawing conclusions on the dependence of the relaxation rate on the transition energy, the dependence of the overlap integral $G_{i,0}^{f,q_z,E_d}$ on the energetic spacing was neglected, which is feasible for III-V devices operating in the conduction band, but not trivially acceptable for p-type SiGe structures, where mixing between the different hole bands occurs for higher values of $k_\parallel$ that are required as the detuning E_d increases.

What has been left unconsidered so far is the fact, that in an alloy of Si and Ge three different optical phonon modes exist, namely one for each combination of atoms in the unit cell. Associated with these modes are different phonon energies, which are 64 meV for the Si-Si mode, 38 meV for the Ge-Ge mode and 51 meV for the Si-Ge mode [108]. Now, as the energetic spacing between two quantum well states increases beyond one of the phonon energies, the relaxation channel associated with the respective phonon mode switches on, and the depopulation rate increases. This behavior is predicted by the simulations in [104] and [106], where the HH2-LHSO1 relaxation rate increases step-like with increasing energetic spacing between the states. The latter publication additionally predicts the HH2-HH1 relaxation rate to increase gradually with the transition energy. However, this increase is attributed to the stronger mixing of HH and LH/SO states for higher detuning E_d and the associated need for high phonon wave vectors (see Fig. 6.2). As deformation potential scattering is strong between different subbands, which will be discussed in more detail in section 6.2.7, the induced transition rate increases while the symmetry of the final state gains more and more LH/SO characteristics.

However, to the author's knowledge there is no experimental data nor theoretical predictions, that the relaxation rates between HH and LH states dominated by non-polar optical phonon scattering show any significant dependence on the quantitative transition energy, once the respective energetic spacing is high enough to enable the emission of optical phonons of any of the three modes. In [105], Woerner et al. conclude from simulations that "[...] the HH2→LHSO1 scattering is nearly constant if it is energetically possible."

6.2.2 Acoustic phonon scattering

In case of a transition energy below the optical phonon energy, the emission of optical phonons is forbidden due to energy conservation, and the variety of allowed phonon interactions reduces to the emission and absorption of acoustic phonons. The following considerations are based on [73], pp. 114.

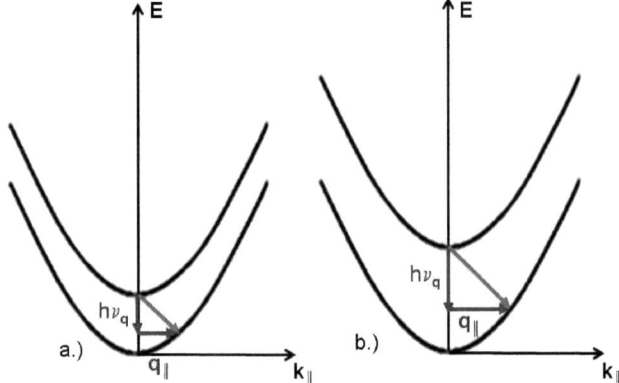

Figure 6.6: Position of acoustic phonon scattering processes in $k_\parallel$-space for a fixed q_z. Graphs (a) and (b) show the in-plane dispersion of quantum states confined in z-direction for different transition energies $E_2^0 - E_1^0$ (defined as the energetic spacing between the states at zero in-plane wave vector). The size of $h\nu_q$ is determined by the acoustic phonon dispersion $\nu_q = \nu(q_\parallel, q_z)$. Note that for acoustic phonon scattering direct transitions in k-space are not allowed for $q_z = 0$, since $\nu_{q=0} = 0$.

Acoustic phonon-induced transitions in k-space

As seen in Figs. 6.5 and 6.4, the fundamental difference between acoustic and optical phonons is that acoustic phonons do not carry a significant energy at zero wave vector, while optical phonons do. For acoustic phonons, interband transitions involving phonons of zero wave vector are therefore prohibited. However, for an arbitrary but fixed value of the phonon wave vector component q_z, larger in-plane phonon wave vectors are required for the scattering between states with increased transition energy, equivalent to the case of optical phonons. This circumstance is illustrated in Fig. 6.6.

Interaction potential

Acoustic phonons can be interpreted as sound waves in the crystal, as they result in a displacement $\delta \mathbf{R}$ of the lattice sites of the crystal. For an acoustic phonon of frequency ω and wave vector $\mathbf{q}$, this displacement follows Eqn. 6.35

$$\delta \mathbf{R} = \delta \mathbf{R}_0 \sin(\mathbf{r} \cdot \mathbf{q} - \omega t) \qquad (6.35)$$

For materials with a centrosymmetric crystalline structure like SiGe, electronic states are exclusively perturbed by the local deformation of the unit cell. This kind of phonon-electron interaction is referred to as deformation potential scattering. For long wave acoustic phonons, the local deformation of the unit cell can be described by the strain tensor e_{ij}, which can be calculated from the displacement δR according to Eqn. 6.36.

$$e_{ij} = \frac{1}{2}\left(\frac{\partial \delta R_i}{\partial x_j} + \frac{\partial \delta R_j}{\partial x_i}\right) \quad (6.36)$$

Acoustic phonons can induce a volumetric dilation $\frac{\delta V}{V}$ as well as a shear, where transverse acoustic phonons generate mostly shear. In the following considerations, the influence of the shear component on the electron-phonon interaction will be neglected. This approximation is particularly valid for longitudinal acoustic (LA) phonons and for scattering between non-degenerate bands. However, the lift of the band degeneracy, which is induced by the change in symmetry caused by the shear component of transverse acoustic (TA) phonons, is commonly not included in the numerical treatment of phonon-induced intersubband scattering. The dilation $\frac{\delta V}{V}$ induced by a LA phonon is equal to the trace of the strain tensor in 6.36, thus obeying Eqn. 6.37.

$$\frac{\delta V}{V} = \boldsymbol{q} \cdot \delta \boldsymbol{R} \quad (6.37)$$

The Hamiltonian of the electron-phonon interaction induced by the volume dilation can be written as in Eqn. 6.38, where a_{nk} represents the so-called volume deformation potential, which is, strictly speaking, dependent on n and $\boldsymbol{k}$ of the state, whose change in energy it describes.

$$H_{e-AP} = a_{nk} \sum_{q} \boldsymbol{q} \cdot \delta \boldsymbol{R}_q = \quad (6.38)$$

$$= a_{nk} \sum_{q} \boldsymbol{q} \cdot \boldsymbol{e}_q \sqrt{\frac{\hbar}{2NV\rho\, \omega_{LA}^q}} (c_q^+ \, e^{i(\boldsymbol{q}\,\boldsymbol{r}-\omega_{LA}^q t)}) +$$
$$+ \; c_q \, e^{-i(\boldsymbol{q}\,\boldsymbol{r}-\omega_{LA}^q t)}) \quad (6.39)$$

In order to obtain Eqn. 6.39, $\delta \boldsymbol{R}_q$ was expressed in terms of the annihilation and creation operators, c_q and c_q^+. In this equation, N represents the number of unit cells in the crystal, V and ρ the volume and density of the unit cell, and $\boldsymbol{e}_q$ the polarization of the acoustic phonon.

Scattering rate

The acoustic phonon scattering rate between subbands in a SiGe heterostructure was determined completely analogous to that for optical phonons. The result for the scattering rate between an initial electronic state ϕ_i and a final state ϕ_f, both of which are defined as in Eqn. 6.21, by acoustic phonons is given in Eqn. 6.40 [104].

$$\begin{aligned}
D_{i,k_\parallel}^{f\ AP} &= \frac{D_A^2 \hbar}{2\rho(2\pi)^2} \int_{(q)} \frac{(\boldsymbol{q}\cdot\boldsymbol{e}_q)^2}{\omega_{LA}^q \hbar} \left(N_{ph}^q + \frac{1}{2} \pm \frac{1}{2}\right) |G_{i,k_\parallel}^{f,k_\parallel \pm q_\parallel}|^2 \\
&\quad \cdot \delta(E_{i,k_\parallel} - E_{f,k_\parallel \pm q_\parallel} \mp \hbar\omega_{LA}^q)\, d\boldsymbol{q} \\
N_{ph}^q &= \frac{1}{e^{\frac{\hbar\omega_{LA}^q}{k_B T}} - 1} \\
G_{i,k_\parallel}^{f,k_\parallel'} &= \sum_n \int_{(z)} \Phi_{f,k_\parallel'}^{n\ \dagger} e^{\pm i q_z z}\, \Phi_{i,k_\parallel}^n\, dz
\end{aligned} \qquad (6.40)$$

In this equation, the state dependent deformation potential a_{nk} was replaced by an average acoustic deformation potential D_A, which is commonly used in literature (e.g. [95]). The rate of deformation potential scattering by acoustic phonons exhibits the same dependence on the generalized overlap integral G_{if} as that induced by non-polar optical phonons. It thus rises with increasing spatial overlap between the involved initial and final state's envelope functions. Therefore, decreasing this overlap by employing diagonal transitions, as already mentioned in chapter 6.2.2, forms a means of increasing the non-radiative relaxation time induced by acoustic phonons.

Comparison to Fröhlich interaction

As well as the electron interaction with optical phonons, that induced by acoustic phonons differs fundamentally from the Fröhlich interaction in its dependence on the phonon wave vector $\boldsymbol{q}$, which becomes evident when comparing Eqns. 6.39 and 6.10. This dependence of the contribution of a phonon of wave vector $\boldsymbol{q}$ to the interaction Hamiltonian is highlighted in Eqn. 6.41.

$$\begin{aligned}
H_{e-AP}^q &\sim \frac{\boldsymbol{q}\cdot\boldsymbol{e}_q}{\sqrt{\omega_{LA}^q}} \\
H_{Fr}^q &\sim \frac{\sqrt{\omega_{LO}^q}}{q}
\end{aligned} \qquad (6.41)$$

The Fröhlich interaction potential *decreases* with increasing q, resulting in the afore mentioned resonant behavior, which is used for the depopulation of the ground state in III-V QCLs. On the other hand, in case of deformation potential scattering by LA phonons, the strength of the interaction *increases* with increasing q. Therefore, at a first glance the relaxation based on the deformation potential scattering by acoustic phonons seems to increase in efficiency with the transition energy. However, thoroughly accounting for the relaxation rate of excited states requires the consideration of the full depopulation rate $\widetilde{D}_{i,0}^{f\ AP}$, which is derived analogous to that for optical phonons in chapter 6.2.1. For reasons of simplification, longitudinal phonons are considered, ($q \parallel e_q$). Due to the strong dispersion of acoustic phonons as compared to optical ones, $\omega_{q_z,E}$ cannot be approximated by its value at zero-wave vector during the evaluation of the delta-function, leading to Eqn. 6.42.

$$\widetilde{D}_{i,0}^{f\ AP} = \frac{D_A^2}{2\hbar^4\rho 2\pi} m_{eff} \int_{(E,q_z)} \frac{2m_{eff}E + \hbar^2 q_z^2}{\omega_{q_z,E}} (N_{ph}^{q_z,\ E} + 1) \cdot$$

$$\cdot |G_{i,0}^{f,q_z,E}|^2 \, \delta(E_i^0 - E_f^0 - E - \hbar\omega_{LA}^{q_z,\ E}) \, dE \, dq_z \qquad (6.42)$$

This equation demonstrates, that due to their significant dispersion the depopulation rate by acoustic phonons does not depend trivially on the energetic spacing between final and initial states. For a given q_z with $2m_{eff}E \gg \hbar^2 q_z^2$, the factor $\frac{2m_{eff}E+\hbar^2 q_z^2}{\omega_{LA}^{q_z,\ E}}$ increases with E due to the sublinear dispersion of $\omega_{q_z,E}$. However, $(N_{ph}^{q_z,E} + 1)$ significantly influences the contribution of a specific phonon to the depopulation rate, and the dependence of the phonon occupation on E strongly depends on the temperature of the crystal lattice. As a consequence, no general statement on the dependence of the relaxation rate by acoustic phonons on the transition energy can be given. Case studies and numerical considerations are required, some of which are given in [104, 109].

6.2.3 Alloy scattering

Alloy scattering is a process observed in compound semiconductors due to the fluctuating composition of the material. As a consequence of this inhomogeneous formation of a compound bulk, the band edges exhibit spatial fluctuations, which can be treated as a perturbation of the average Hamiltonian derived for the average composition of the material. The mechanism for alloy scattering is thus fundamentally different from that for phonon scattering, where the charge carrier interacts with a dynamic distortion of the periodic grid of the crystal by absorbing or emitting a vibrational mode.

Alloy induced transitions in k-space

While the electron-phonon scattering process is inelastic due to the associated change in the occupation number of the phonon modes, alloy scattering does not change the lattice energy and is therefore elastic. However, scattering between subbands is possible, as long as the energies of the initial and final states are equal, which can be fulfilled by transferring a significant amount of momentum. Fig. 6.7 illustrates alloy scattering induced transitions between excited states of low $k_\parallel$ and states of high in-plane wave vector in the lower level. From this plot it becomes evident that an increasing energetic spacing between the excited

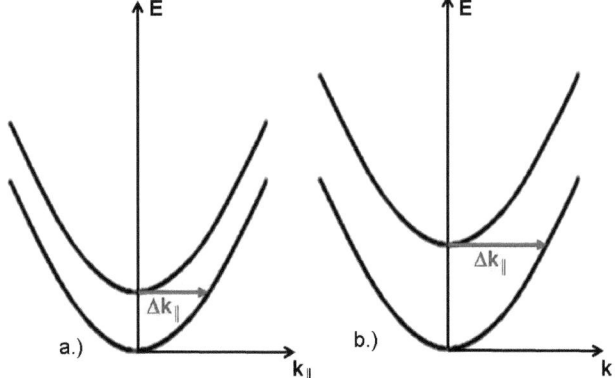

Figure 6.7: Position of alloy scattering processes in **k**-space. Graphs (a) and (b) show the in-plane dispersion of quantum states confined in z-direction for different transition energies $E_2^0 - E_1^0$ (defined as the energetic spacing between the states at zero in-plane wave vector). Note that alloy scattering is strictly elastic.

and the ground level demands an increase in the size of the transferred momentum in order to fulfill energy conservation, a behavior analogue to that for phonon scattering.

Interaction potential

In case of low temperatures and a low energetic intersubband spacing alloy scattering can grow the dominant process for intersubband relaxation in a heterostructure. Charge carriers of low in-plane momentum occupying excited subbands can relax into lower subband states with a high in-plane wave vector by alloy scattering, and further relax to low $k_\parallel$ states within the lower subband by the stepwise emission of phonons [104].

The reason for the ability of alloy scattering to efficiently inflict a significant change of the in-plane momentum on charge carriers lies in the narrow spatial confinement of its perturbation potential and the correlated wide spreading in **k**-space. In contrast, phonons and their associated perturbation potential spatially spread over the whole crystal volume and are therefore strongly confined in **k**-space.

In order to simplify the following derivation, it is assumed that the SiGe crystal is composed of two types of unit cells, those composed of two Si atoms and those including two Ge atoms. However, the obtained results can be easily generalized. In case of a $Si_{1-x}Ge_x$ layer, the average Ge concentration is x, and the average potential is given by Eqn. 6.43, if U_A and U_B are the potentials associated with a Ge-Ge and a Si-Si sites, respectively [110].

$$U_{av} = xU_A + (1-x)U_B \tag{6.43}$$

The perturbation potential associated with a single Ge or Si site in respect to the average Hamiltonian is commonly approximated by a spherical well potential with radius r_0, where the energetic depth is given by $(U_{av} - U_A)$ and $(U_{av} - U_B)$, respectively. The potential radius r_0 is chosen in a way assigning the volume occupied by each scattering site to the spherical potential, thus $\frac{4\pi r_0^3}{3} = \frac{a_0^3}{4}$, where a_0 is the lattice constant of the crystal. For a Ge site located at $\mathbf{r}_A$, the perturbation Hamiltonian therefore reads as in Eqn. 6.44.

$$H_A^{\mathbf{r}_A} = (U_{av} - U_A) \cdot \Theta(|\mathbf{r} - \mathbf{r}_A| - r_0) \tag{6.44}$$

Scattering rate

The perturbation matrix element of a Ge-Ge site located at $\mathbf{r}_A$ between two eigenstates $|\phi_i>$ and $|\phi_f>$ of the average Hamiltonian as defined in Eqn. 6.21 is presented in Eqn. 6.45.

$$\begin{aligned}
<\phi_f|H_A^{\mathbf{r}_A}|\phi_i> &= S^{-1}(U_{av} - U_A) \sum_{m,n} <u_m \mid u_n>_{cell} \cdot \\
&\quad \cdot <\chi_{f,m}|\Theta(|\mathbf{r}-\mathbf{r}_A|-r_0)|\chi_{i,n}> \\
&= S^{-1}(U_{av} - U_A) \sum_n \int_V \Phi_{f,n}^\dagger(z)\, e^{i\Delta\mathbf{k}_\parallel \mathbf{r}_\parallel}\, \Phi_{i,n}(z)\, \Theta(|\mathbf{r}-\mathbf{r}_A|-r_0)\, d^3\mathbf{r}
\end{aligned} \tag{6.45}$$

Once again, for evaluating Eqn. 6.45 the circumstance that the perturbation potential does not vary over one unit cell was considered, which is the case as the perturbation spreads exactly over one unit cell and is assumed constant within this volume. $\Delta\mathbf{k}_\parallel$ represents the difference in in-plane wave vector between the final and initial states. It shall further be assumed that the variation of the envelope functions $\Phi_{f,n}(z)$ and $\Phi_{i,n}(z)$ is very weak on the scale of r_0, which is reasonable for most heterostructures. This simplification leads to Eqn. 6.46.

$$\begin{aligned}
<\phi_f|H_A^{\mathbf{r}_A}|\phi_i> &= S^{-1}(U_{av} - U_A) \sum_{m,n} <u_m \mid u_n>_{cell} \cdot \\
&\quad \cdot \Phi_{f,n}^\dagger(z_A)\, \Phi_{i,n}(z_A) \int_V e^{i\Delta\mathbf{k}_\parallel \mathbf{r}_\parallel}\, \Theta(|\mathbf{r}-\mathbf{r}_A|-r_0)\, d^3\mathbf{r}
\end{aligned} \tag{6.46}$$

The evaluation of the integral in Eqn. 6.46 leads to Eqn. 6.47 with the form factor $F(\Delta \mathbf{k}_\parallel)$, which is depending on the momentum transfer.

$$
\begin{aligned}
<\phi_f|H_A^{r_A}|\phi_i> &= S^{-1}\,(U_{av}-U_A)\,\frac{4\pi r_0^3}{3}\,F(\Delta \mathbf{k}_\parallel)\cdot \\
& \quad \cdot \sum_n \Phi_{f,n}^\dagger(z_A)\,\Phi_{i,n}(z_A) \\
F(\Delta \mathbf{k}_\parallel) &= \frac{3\sin(\Delta k_\parallel\,r_0)-3\,(\Delta k_\parallel\,r_0)\cos(\Delta k_\parallel\,r_0)}{(\Delta k_\parallel\,r_0)^3}
\end{aligned}
\tag{6.47}
$$

A completely analogous equation can be derived for the perturbation matrix element of a Si site. Note that Eqn. 6.47 is exclusively dependent on the z-component of $\mathbf{r}_A$. In order to obtain the total scattering rate between $|\phi_i>$ and $|\phi_f>$, the rate contributions of of both several Si and Ge perturbation sites have to be summed up according to Fermi's golden rule. Furthermore, all possible wave vector transfer values $\Delta \mathbf{k}_\parallel$ have to be considered, leading to an additional integral in Eqn. 6.48.

$$
W_{if}^{Alloy} = \frac{2\pi}{\hbar}\int_{(\Delta \mathbf{k}_\parallel)}\sum_{r_A,r_B}[\,|<\phi_f|H_A^{r_A}|\phi_i>|^2 + |<\phi_f|H_B^{r_B}|\phi_i>|^2\,]\,\delta(E_f-E_i)\frac{A}{(2\pi)^2}\,d\Delta\mathbf{k}_\parallel
\tag{6.48}
$$

The summation is carried out over all lattice points with an unit cell of type Si-Si or Ge-Ge, respectively. Applying the following considerations, this sum can be converted into an integral. The number of scattering sites within a partial volume $d^3\mathbf{r}$ is $d^3\mathbf{r}\frac{N_{sc}}{V}$, where $N_{sc}=\frac{4V}{a_0^3}$ is the total number of scatterers and V the crystal volume. Of these scattering sites, a fraction x is of Ge type, and a fraction $1-x$ of type Si. In order to further generalize the model, it is allowed that *the average* Ge content varies over z, exhibiting a composition profile $x(z)$ [104]. The alloy scattering rate is in this case given by Eqn. 6.49. As already mentioned, H_B and H_A are only dependent of the z coordinate.

$$
\begin{aligned}
W_{if}^{Alloy} = \frac{2\pi}{\hbar}\frac{A}{(2\pi)^2}\int_{(\Delta \mathbf{k}_\parallel)}\int_V \frac{N_{sc}}{V}\,[\,x(z)\cdot|<\phi_f|H_A(z)|\phi_i>|^2 + \\
+ (1-x(z))\cdot|<\phi_f|H_B(z)|\phi_i>|^2\,]\,\delta(E_f-E_i)\,d^3\mathbf{r}\,d\Delta\mathbf{k}_\parallel
\end{aligned}
\tag{6.49}
$$

Equation 6.47 and the respective, analogous equation for $<\phi_f|H_B^{r_B}|\phi_i>$ are inserted into Eqn. 6.49.

$$
x\cdot(U_{av}-U_A)^2 + (1-x)\cdot(U_{av}-U_B)^2 = x\cdot(1-x)\cdot(U_A-U_B)^2
\tag{6.50}
$$

By further considering Eqn. 6.50, which is a direct consequence of Eqn. 6.43, the final expression for the alloy scattering rate as given in Eqn. 6.23 is obtained [104].

$$W_{if}^{Alloy} = \frac{1}{(2\pi)^2 \hbar} [\frac{a_0^3}{4} (U_A - U_B)]^2 \frac{N_{sc}}{V} \int_{(\Delta \mathbf{k}_\parallel)} |F(\Delta \mathbf{k}_\parallel)|^2 \cdot$$
$$\cdot \int x(z) \cdot (1 - x(z)) \; | \sum_n \Phi_{f,n}^\dagger(z) \, \Phi_{i,n}(z) |^2 \; \delta(E_f - E_i) \; dz \; d\Delta \mathbf{k}_\parallel$$

(6.51)

As is the case for both the deformation potential scattering by non-polar optical and acoustic phonons, the rate for alloy scattering strongly depends on the spatial overlap between the initial and final states, as becomes evident from Eqn. 6.51. However, there is a fundamental difference between phonon scattering and alloy scattering, concerning the form of the overlap integrals, as seen from comparing Eqn. 6.24 to the integral in Eqn. 6.51. This difference will be addressed in chapter 6.2.7. It shall be further mentioned, that due to its static nature, alloy scattering does not exhibit any dependence on the lattice temperature. The constant scattering rate by alloy disordering for any device temperature might prove a significant advantage for designing SiGe QCLs, as already mentioned in chapter 6.1.3.

Comparison to Fröhlich interaction

One important issue, which is still open for clarification, is that of the dependence of the alloy scattering rate on the energetic spacing between the final and initial states involved. This dependence is exclusively determined by the form factor $F(\Delta \mathbf{k}_\parallel)$, as indicated in the comparison between the Hamiltonians of the Fröhlich interaction and the alloy perturbation, which enter the expression for the form factor.

$$H_A^{r_A} \sim \frac{3 \sin(\Delta k_\parallel \; r_0) - 3 \, (\Delta k_\parallel \; r_0) \cos(\Delta k_\parallel \; r_0)}{(\Delta k_\parallel \; r_0)^3}$$

$$H_{Fr}^q \sim \frac{\sqrt{\omega_{LO}^q}}{q}$$

(6.52)

$F(\Delta \mathbf{k}_\parallel)$ is monotonically decreasing with $\Delta \mathbf{k}_\parallel$. However, once more the expression for the total relaxation rate of charge carriers with zero in-plane momentum into the various $\mathbf{k}_\parallel$-states of a lower level has to be considered in order to relate the relaxation rate to the energetic spacing between the respective levels. The depopulation rate by alloy scattering is given in Eqn. 6.53, as derived from Eqn. 6.51 analogous to that for optical phonon scattering in chapter 6.2.1.

$$\widetilde{D}_{i,0}^{f\ Alloy} = \frac{m_{eff}}{(2\pi)\hbar^3}[\frac{a_0^3}{4}(U_A - U_B)]^2 \frac{N}{V}|F(k_{tr})|^2 \cdot$$

$$\cdot \int x(z)\cdot(1-x(z))\ |\sum_n \Phi_{f,k_{tr}}^{\dagger\ n}(z)\ \Phi_{i,0}^n(z)|^2\ dz$$

$$E_{tr} = (E_i^0 - E_f^0)$$

$$k_{tr} = \frac{\sqrt{2m_{eff}E_{tr}}}{\hbar} \tag{6.53}$$

As already stated above, $F(k_{tr})$ is monotonically decreasing with the transferred in-plane wave vector k_{tr} and thus with the energetic spacing between the initial and final levels, E_{tr}. Although this implies a rather simple relation between the alloy-induced depopulation rate and the transition energy at zero wave vector, the following additional influence has to be taken into account for a thorough consideration of this relation.

Adjusting the transition energy of a semiconductor quantum well is usually achieved by varying its composition or its width, and thus by a change in the compositional profile $x(z)$. $\widetilde{D}_{i,0}^{f\ Alloy}$ not only varies via the dependence of the form factor on the transition energy, but also due to the terms explicitly containing $x(z)$ in Eqn. 6.53. It is therefore non-trivial to compare the alloy scattering of two different structures and requires a numerical case-by-case analysis. However, the change in the alloy relaxation rate induced by the variation in the width of a quantum well was theoretically reported to be rather weak in comparison to the difference in the phonon based depopulation rates [104]. In conclusion, there is no general indication that alloy scattering exhibits any resonant behavior similar to that of polar optical phonon scattering, and is therefore not expected to contribute to the functionality of conventional III-V inspired QCL design in the SiGe material system.

6.2.4 Interface roughness scattering

The scattering processes discussed so far shared the potential for crucially influencing intersubband relaxation rates, either by the direct emission of high-energetic phonons, or by the elastic scattering between subbands initializing further relaxation within this band by means of acoustic phonons, as is characteristic for alloy scattering. The above mentioned processes are able to transfer a significant amount of energy and/or momentum. Now, interface roughness scattering is capable of neither, and is therefore unable to exert a significant influence on intersubband lifetimes in the transition energy range of the structures presented in this work (≥ 30 meV). In case of significantly lower energies around 13 meV associated with the radiative transition of a quantum cascade structure, however, interface roughness scattering is theoretically predicted to significantly influence the conditions for population inversion [111]. Even though this scattering process is not expected to influence the intersubband relaxation times studied in this work due to its elastic nature, a short overview of this process is given in the section at hand.

Interaction potential

Typical quantum cascade structures contain hundreds of interfaces between material layers of different composition. Due to growth imperfections, these interfaces are not perfectly, mono-atomically flat, but exhibit a certain roughness. This roughness leads to scattering between the theoretical eigenstates of the perfect system. The scattering induced by the lateral variation of the actual vertical position of a given interface between two differently composed layers of a heterostructure can be understood by the following considerations.

For a given lateral position on a heterostructure surface, $r_\parallel$, the compositional profile along the growth direction at this respective lateral position is well defined. In an envelope function approach for gaining the eigenstates of a quantum well structure, this compositional profile in turn determines the one-dimensional potential, which is used for calculating the envelope functions $\Phi_{k,n}$ as given in Eqn. 6.21. Now, due to interface roughness, the precise shape of the compositional profile depends on the lateral position on the surface of a grown heterostructure. In the envelope function approach, this implies a dependence of the actually gained envelope function on the lateral position, at which the compositional profile is taken for the calculations. The physical reality of a varying compositional profile can be considered by defining an average compositional profile, and by treating the lateral deviations in this profile as a perturbation inducing scattering between the respective eigenstates of the ideal system.

Given an interface at an average z-position z_i, the perturbation potential induced by roughness can be approximated by Eqn. 6.54, roughly following [112, 113].

$$H_{int}^i \;=\; \delta U^i \; h_i(r_\parallel) \; \delta(z - z_i) \qquad (6.54)$$

In this equation, δU^i is the ideal potential step at the interface, and $h_i(r_\parallel)$ represents the roughness of the interface. The considerations above imply, that the z-position attributed to the interface has to be well defined on the scale of the envelope functions $\Phi_{k,n}(z)$ in Eqn. 6.21, but is arbitrary on the scale of the Bloch functions $u_n(r)$. This is considered by choosing a representation of $\delta(z - z_i)$, which is sharp on the scale of $\Phi_{k,n}(z)$, but broad on the scale of $u_n(r)$.

Now, interface roughness is not defined by an absolute profile, but rather by its autocorrelation, which is presented in Eqn. 6.55.

$$< h_i(\boldsymbol{r}_\|)h_i(\boldsymbol{r}_\| + \Delta \boldsymbol{r}_\|) > \;=\; \Delta^2\, e^{-\frac{|\Delta \boldsymbol{r}_\||^2}{\Lambda^2}} \tag{6.55}$$

Δ is equivalent to the roughness amplitude, while Λ gives its correlation length [113].

Scattering rate

The perturbation matrix element of an interface located at z_i between two eigenstates $|\phi_i>$ and $|\phi_f>$ of the average Hamiltonian as defined in Eqn. 6.21 is presented in Eqn. 6.56.

$$< \phi_f|H_{int}^i|\phi_i > \;=\; S^{-1}\,\delta U^i \sum_{m,n} < u_m \mid u_n >_{cell} \cdot$$
$$\cdot\; < \chi_{f,m} \mid h_i(\boldsymbol{r}_\|)\, \delta(z - z_i) \mid \chi_{i,n} > \tag{6.56}$$

Once more, for evaluating Eqn. 6.56 the circumstance that the perturbation potential does not vary over one unit cell was considered, which is ensured by using a representation of the delta-function $\delta(z-z_i)$, which is sharp on the scale of $\mid \chi_{i,n} >$ and $\mid \chi_{f,m} >$, but broad on the scale of one unit cell, as discussed above. Note that the delta-function *does not* describe the interface roughness itself, but exclusively defines the *average* position of the interface. The absolute square of the perturbation matrix element, which determines the scattering rate, is given by Eqn. 6.57, where the orthonormality of the Bloch functions was used.

$$|< \phi_f|H_{int}^i|\phi_i >|^2 \;=\; S^{-2}\,\delta U^{i\,2} \sum_{n \leq m} < \chi_{f,n} \mid h_i(\boldsymbol{r}_\|)\, \delta(z - z_i) \mid \chi_{i,n} > \cdot$$
$$\cdot\; < \chi_{f,m} \mid h_i(\boldsymbol{r}_\|)\, \delta(z - z_i) \mid \chi_{i,m} >^\dagger$$
$$=\; S^{-2}\,\delta U^{i\,2} \sum_{n \leq m} \int \int e^{-i\boldsymbol{k}_\|\cdot\boldsymbol{r}_\|}\, \Phi_{f,n}^\dagger(z)\, h_i(\boldsymbol{r}_\|)\, \delta(z-z_i) e^{i\boldsymbol{k}_\|\cdot\boldsymbol{r}_\|}\, \Phi_{i,n}(z) \cdot$$
$$\cdot\; e^{i\boldsymbol{k}_\|\cdot\boldsymbol{r}_\|'}\, \Phi_{f,m}(z')\, h_i(\boldsymbol{r}_\|')\, \delta(z'-z_i) e^{-i\boldsymbol{k}_\|\cdot\boldsymbol{r}_\|'}\, \Phi_{i,m}^\dagger(z')\, d^3\boldsymbol{r}'\, d^3\boldsymbol{r} \tag{6.57}$$

The explicit evaluation of the integrals in Eqn. 6.57 shows, that the z-integral selects the value of the envelope functions at position z_i, and leaves an in-plane integral over exponentials modifying the autocorrelation of the roughness $h_i(\boldsymbol{r}_\|)$ at the respective positions $\boldsymbol{r}_\|$ and $\boldsymbol{r}_\|'$, as seen in Eqn. 6.58.

$$
\begin{aligned}
|<\phi_f|H_{int}^i|\phi_i>|^2 &= S^{-2}\,\delta U^{i\,2} \sum_{n\leq m} \Phi_{f,n}^\dagger(z_i)\,\Phi_{i,m}^\dagger(z_i)\,\Phi_{f,m}(z_i)\,\Phi_{i,n}(z_i) \\
&\quad \cdot \iint e^{i(k_\parallel - k_\parallel')\cdot(r_\parallel - r_\parallel')}\,h_i(r_\parallel)\cdot h_i(r_\parallel')\,d^2 r_\parallel'\,d^2 r_\parallel \\
&= S^{-2}\,\delta U^{i\,2} \sum_{n\leq m} \Phi_{f,n}^\dagger(z_i)\,\Phi_{i,m}^\dagger(z_i)\,\Phi_{f,m}(z_i)\,\Phi_{i,n}(z_i) \\
&\quad \cdot \iint e^{-i\Delta k_\parallel \cdot \Delta r_\parallel}\,h_i(r_\parallel)\cdot h_i(r_\parallel + \Delta r_\parallel)\,d^2\Delta r_\parallel\,d^2 r_\parallel \\
&= S^{-2}\,\delta U^{i\,2} \sum_{n\leq m} \Phi_{f,n}^\dagger(z_i)\,\Phi_{i,m}^\dagger(z_i)\,\Phi_{f,m}(z_i)\,\Phi_{i,n}(z_i) \\
&\quad \cdot \int e^{-i\Delta k_\parallel \cdot \Delta r_\parallel}\,\Delta^2\,e^{-\frac{|\Delta r_\parallel|^2}{\Lambda^2}}\,d^2\Delta r_\parallel \\
&= S^{-2}\,\delta U^{i\,2}\,\Delta^2 \Lambda^2\,\pi \cdot e^{\frac{-|\Delta k_\parallel|^2 \Lambda^2}{4}} \\
&\quad \cdot \sum_{n\leq m} \Phi_{f,n}^\dagger(z_i)\,\Phi_{i,m}^\dagger(z_i)\,\Phi_{f,m}(z_i)\,\Phi_{i,n}(z_i)
\end{aligned} \qquad (6.58)
$$

As the autocorrelation of $h_i(r_\parallel)$ follows Eqn. 6.55 and the exponential function is exclusively dependent on $r_\parallel - r_\parallel'$, the substitution $\Delta r_\parallel = r_\parallel - r_\parallel'$ can be applied. The integration over $\Delta r_\parallel$ is equivalent to the Fourier transform of a Gaussian and gives a Gaussian again, as seen in Eqn. 6.59, which presents the final result for the interface roughness scattering rate W_{if}^{rough} after integrating over the momentum transfer $\Delta k_\parallel$.

$$
\begin{aligned}
W_{if}^{rough} &= \frac{V}{(2\pi)^3}\frac{2\pi}{\hbar}\,S^{-2}\,\delta U^{i\,2}\,\Delta^2\,\Lambda^2 \pi \int_{\Delta k_\parallel} e^{\frac{-|\Delta k_\parallel|^2 \Lambda^2}{4}} \\
&\quad \cdot \sum_{n\leq m} \Phi_{f,n}^\dagger(z_i)\,\Phi_{i,m}^\dagger(z_i)\,\Phi_{f,m}(z_i)\,\Phi_{i,n}(z_i) \cdot \delta(E_f - E_i)\,d\Delta k_\parallel
\end{aligned} \qquad (6.59)
$$

As seen from Eqn. 6.59, the interface roughness scattering rate exhibits a Gaussian dependence on the electronic wave vector $|\Delta k_\parallel|$ transferred during the process. The smaller the correlation length of the roughness, the more momentum can be transferred by the scattering event. However, as the correlation length in a high-quality structure spans at least several unit cells, the momentum transfer by roughness scattering is too low to enable elastic scattering between different subbands. Simulations by Valavanis et al. [112] predict that even for a low intersubband spacing of 10 meV and a low correlation length of $\Lambda = 5$ nm, the intersubband scattering rates by interface roughness are one order of magnitude below the alloy scattering rates in a SiGe single quantum well.

Further, without the possibility of intersubband scattering, the strict energy conservation in Eqn. 6.59 cannot be fulfilled for a finite momentum transfer. Thus, according to this equation, interface roughness is free of influence on charge carriers relaxation. However, in a system with finite state lifetimes, the energetic states are homogeneously broadened, and the energy conservation relaxes. In order to account for finite state lifetimes, the delta-function in Eqn. 6.59 is replaced by a Lorentzian characterized by the broadening factor Γ, as shown in Eqn. 6.60.

$$\widetilde{W}_{if}^{rough} = \frac{V}{(2\pi)^3} \frac{2\pi}{\hbar} S^{-2} \delta U^{i\,2} A \Delta^2 \Lambda^2 \pi \int_{\Delta k_\parallel} e^{\frac{-|\Delta k_\parallel|^2 \Lambda^2}{4}} \cdot$$

$$\cdot \sum_{n \leq m} \Phi_{f,n}^\dagger(z_i)\, \Phi_{i,m}^\dagger(z_i)\, \Phi_{f,m}(z_i)\, \Phi_{i,n}(z_i) \cdot \frac{\frac{\Gamma}{\pi}}{(E_f - E_i)^2 + \Gamma^2}\, d\Delta k_\parallel \quad (6.60)$$

The relaxation of the energy conservation allows the interface roughness scattering process to induce a broadening of the distribution of charge carriers in k-space. And even though the roughness-induced redistribution of charge carriers within the individual subbands leads to a significant inhomogeneous broadening of the electroluminescence signals from p-SiGe and n-GaInAs/AlInAs quantum cascade structures exceeding the homogeneous linewidth, as experimentally reported by Tsujino et al. in [113], the change of the distribution in k-space due to interface roughness is too insignificant to have any influence on intersubband relaxation times. This circumstance is predicted by the model discussed above, and no influence of interface roughness scattering on intersubband relaxation has been reported in literature.

In conclusion of this section it can be stated, that the process of interface roughness scattering is not expected, neither by theoretical considerations nor by experimental data, to have any influence on subband lifetimes and thus the formation of population inversion in a quantum cascade structure. Therefore it will not be included in the following considerations.

6.2.5 Lifetime limiting processes in SiGe structures

In the previous sections, a theoretical approach to the scattering processes in SiGe structures has been presented. However, one question has yet to be addressed:

Which processes dominate the intersubband lifetimes under which conditions?

In [104] and [109], Ikonic et al. calculated the intersubband relaxation times in SiGe quantum well structures for different temperatures, well widths and well compositions. In the course of these simulations, the band structure of the unperturbed Hamiltonian was calculated using a 6×6 $\boldsymbol{k.p}$ approach, and the scattering rates induced by the various processes between the respective bands were determined using the perturbation potentials presented in chapters 6.2.1, 6.2.2 and 6.2.3. This section summarizes some of the results given by Ikonic on the prediction of the dominant scattering processes for a given combination of temperature and transition energy.

Transition energies below the optical phonon energy at low temperature

In case of an energetic spacing between the excited state and the ground state of a quantum well below the LO phonon energy, an intersubband relaxation by the emission of an optical phonon is not possible due to reasons of energy conservation. The remaining feasible processes are alloy and acoustic phonon scattering. According to Ikonic, alloy scattering dominates the intersubband lifetimes at low temperatures. In case of a LH1-HH1 transition energy of 30 meV, alloy scattering forms the dominant process up to about 80 K, where the associated relaxation times are in the regime of 10 ps [109]. The same is the case for a HH2-HH1 transition with a transition energy of 25 meV. Around liquid helium temperature, the simulations of Ikonic predict intersubband relaxation times for transition energies below the optical phonon energy around 50 ps, which drop to 10 ps at liquid nitrogen temperature.

Transition energies below the optical phonon energy at high temperature

Due to the static nature of alloy perturbations, the associated scattering rate is independent of the lattice temperature of the structure. However, as temperature increases, the occupation of phonon modes (N_{ph}^q in the equations of the previous sections) rises. As a consequence, the increase in temperature generally leads to a domination of the acoustic phonon scattering over alloy scattering regarding intersubband relaxation times. For the examples given in [104], namely a LH1-HH1 transition with an energy of 31 meV and a HH2-LH1 transitions of 7 meV, the threshold temperature for the domination of phonons lies above 80 K. A further rise in the lattice temperature increases the influence of optical phonon scattering due to the occupation of electronic states up sufficiently high in $\boldsymbol{k}_\parallel$-space for the emission of optical phonons. A decrease of the intersubband relaxation times down to about 1 ps results. However, this growing influence of non-polar optical phonons is in the simulations of [104] only observed at high temperatures above 150 K, and for transition energies already close to the optical phonon energy.

Transition energies above the optical phonon energy at low temperature

As soon as the transition energy between two quantum well states is high enough to emit an optical phonon, intersubband relaxation times for holes in SiGe heterostructures at low temperatures are determined by the deformation potential scattering induced by these phonons. In [104], Ikonic presents examples for a HH2-HH1 and a LH1-HH1 transition at 79 meV and 72 meV, respectively. The intersubband relaxation times between those states are dominated by non-polar optical phonon scattering up to about 60 K. The simulations presented by Ikonic predict non-polar optical phonon scattering to shorten intersubband relaxation times down to the ps-regime already at liquid helium temperatures.

Transition energies above the optical phonon energy at high temperature

As the temperature rises above 60 K, scattering by acoustic phonons gains significance for the relaxation rate between two states energetically further apart than the optical phonon energy. The reason for the change in the relative dominance of the respective phonon scattering processes with temperature is related to the occupation of energetically higher acoustic phonon states exhibiting long wave vectors, and the occupation of electronic states with high $k_\parallel$ values. According to [104], the latter results in a compensation of the increase in the number of optical phonons with temperature by changing the mixed composition of the electronic states, which leads to a reduction of the overlap $G_{i,k_\parallel}^{f,k_\parallel}$ in Eqn. 6.25. As a consequence, the simulations in [104] predict the non-polar optical phonon scattering to exhibit only a slight dependence on the lattice temperature. This theoretical conclusion is consistent with the calculations reported in [106], where the deformation potential scattering rate is predicted to remain constant up to 150 K, and to be only slightly increasing with the lattice temperature beyond 150 K. Therefore, non-polar optical phonon limited intersubband relaxation times drop slowly with the lattice temperature below 1 ps. From a temperature of about 200 K on, both phonon types contribute equally to the depopulation of excited states.

Table 6.1 gives a rough summary of the type of scattering mechanisms, which are according to the simulations in [104] expected to dominate intersubband relaxation times between two quantum well states energetically separated by $E_f^0 - E_i^0$ in the respective temperature range. In this table, the dominant mechanism is given by bold characters, while processes with significant but not dominating influence are given by plain text. Further, the order of intersubband relaxation times at the respective parameters is given. As the temperature range of interest for this work is close to liquid helium temperature, the intersubband relaxation processes of relevance are alloy scattering for transition energies below the optical phonon energy, and deformation potential scattering by non-polar optical phonons for an energetic spacing between the states exceeding the optical phonon energy.

Table 6.1: Scattering processes dominating the intersubband relaxation times in SiGe heterostructures for transition energies $E_f^0 - E_i^0$ and different temperatures [104]:

	Low temperature (liquid helium)	Medium temperature (liquid nitrogen)	High temperature (room)
$E_f^0 - E_i^0 < \hbar\omega_{LO}$	alloy ~ 50ps	alloy acoustic phonons ~ 10ps	**acoustic phonons** ~ 1ps
$E_f^0 - E_i^0 > \hbar\omega_{LO}$	**optical phonons** ~ 1ps	**optical phonons** acoustic phonons < 1ps	**optical phonons** **acoustic phonons** < 1ps

6.2.6 Disadvantage of non-polar scattering for QCL development

One of the main difficulties for the implementation of the concept of QCLs in the SiGe system are the fundamentally different properties of the non-polar optical phonons in this system in comparison to the polar optical phonons in the III-V material. As concluded in chapter 6.2.1, the rate of non-polar optical phonon scattering does not decrease with increasing energetic spacing between the final and initial states. As a consequence, the design concept commonly used to rapidly depopulate the ground state of the lasing transition in III-V QCLs cannot be applied to SiGe emitter structures.

This concept is based on introducing a quantum state, for which the energetic spacing to the ground state of the lasing transition is in resonance with the LO phonon energy. In Fig. 6.1, this depopulation state is labelled 1, while 2 is the ground state of the lasing transition. In addition, as the transition energy between the excited state (3) of the optical transition and state 1 is intrinsically higher than that between states 2 and 1, the non-radiative relaxation process between 3 and 1 is the slowest of the individual relaxation mechanisms, which is strongly in favor of the total lifetime of excited state carriers.

The common QCL design rule to implement a 1-2 spacing energetically smaller than the energy of the lasing transition $3 \to 2$ cannot be applied to the SiGe system. In a SiGe structure, such a configuration would lead to a non-radiative relaxation time between state 2 and 1 similar to that between state 3 and 2, and reaching population inversion would be principally hindered. Moreover, theoretical approaches like that of Ikonic in [104,109] predict intersubband relaxation times by non-polar optical phonons in SiGe structures to be of the order of that induced by polar-optical phonons in the III-V material. Thus, without the resonant behavior of the scattering process, the ultra-fast phonon relaxation processes present in the SiGe system pose the same barrier as they did in the beginning of QCL development in the III-V materials.

Therefore, novel concepts are required if a SiGe QCL is to be successfully implemented. One of the concepts to achieve a ratio between the lifetimes favorable for population inversion is that of diagonal radiative transitions, which are in III-V QCLs employed for increasing the total lifetime of the lasing transition's excited state. QCL designs with a diagonal active region are based on the structural manipulation of the spatial overlap between initial and final states. This spatial overlap is discussed in detail in the next subsection. Another concept for circumventing the cumbersome properties of non-polar optical phonons is to shift the operation wavelength of the emitting device into a regime, where optical phonon scattering does not have any influence, namely the terahertz region.

6.2.7 Overlap integrals and selection rules

Overlap

Section 6.2.5 concluded, that alloy and non-polar optical phonon scattering determine the intersubband lifetimes in SiGe heterostructures at operation temperatures relevant for emitter structures. As further stated in section 6.2.6, the energetic spacing between states in a SiGe structure cannot be employed as a parameter to efficiently manipulate depopulation rates. Therefore, another parameter for the design-induced change in intersubband relaxation times is required. One of the parameters crucially influencing the efficiency of the scattering process by both phonons and alloy fluctuations is the quantitative overlap of the envelope functions of the initial and final states.

Even though both scattering processes increase with rising spatial overlap between the wave functions, the detailed terms including a generalized overlap integral differ, as seen by comparing Eqns. 6.51 and 6.24. This difference shall be illustrated in the following by a simple example. If a constant composition profile $x(z)$ is assumed, and if further both the initial and final states are purely composed of the same Bloch function, the overlap term in Eqn. 6.51 reduces to Eqn. 6.61

$$W_{if}^{Alloy} \sim \int |\Phi_{f,n}(z)|^2 \, |\Phi_{i,n}(z)|^2 \, dz \quad \textbf{Alloy scattering} \quad (6.61)$$

On the other hand, the overlap term of Eqn. 6.24 simplifies into Eqn. 6.62.

$$W_{if}^{OP} \sim |\int \Phi_{f,n}^{\dagger} \, e^{\pm iq_z z} \, \Phi_{i,n} \, dz|^2 \quad \textbf{Phonon scattering} \quad (6.62)$$

Equation 6.61 shows that alloy scattering between states composed of the same Bloch function is dependent on the overlap between the *charge carrier densities* of the final and initial states. For the overlap integral in Eqn. 6.61, the phase of the respective wavefunctions is irrelevant, and strong scattering between states of orthogonal envelope functions (e.g. between two HH states) is possible.

In contrast, phonon scattering is influenced by the overlap between the *wave functions* of the involved states, modulated by an exponential function depending on the wave vector of the involved phonon. According to reference [104], phonons of a q_z component exceeding 20% of the maximal Brillouin zone wave vector do not contribute to scattering, as in this case the oscillations induced by the exponential in Eqn. 6.62 decrease the value of the overlap integral. Further, due to the wavefunction overlap integral in this equation, the scattering between two states of orthogonal envelope functions is weaker than for two functions of the same symmetry.

Selection rules

It has to be kept in mind, that $G_{i,k_\parallel}^{f,k_\parallel'}$ in Eqn. 6.25 includes the overlap of the basic Bloch functions as well as that of the envelope functions. As already mentioned in section 6.2.1, assuming a simplified, isotropic dispersion of the phonons for the analytical consideration leads to incorrect predictions of the selection rules. While the approximation in [104] assumes the deformation potential to be linearly dependent on the product of the scalar quantities D_O and u_{LO}^q, it is in the more accurate harmonic approximation following Eqn. 6.63 [106].

$$H_{e-OP} = \sum_q D_O u_{LO}^q(r,t) \qquad (6.63)$$

In this equation, $\boldsymbol{u}_{LO}^q(\boldsymbol{r},t)$ is the displacement *vector* of the respective phonon, and $\boldsymbol{D}_O$ the deformation potential *tensor* in the basis of the Bloch functions. In order to determine the form of $\boldsymbol{D}_O$, the symmetry of the initial and final basis functions as well as that of the respective phonon have to be taken into account. According to reference [106], $\boldsymbol{D}_O$ is exclusively built up by non-diagonal elements in respect to the Bloch basis functions. At zero wave vector, the deformation potential scattering by non-polar optical phonons is therefore only allowed between hole states with wave function components in different subspaces (HH, LH, SO).

Non-polar optical phonon scattering is thus expected to be extremely efficient for LH-HH transitions, but weaker for HH-HH and LH-LH transitions. This theoretical finding was reported in [85], [114] and [106].

Apart from the mentioned differences in the detailed form of their overlap integrals, the efficiency of both the phonon and alloy scattering processes reduces for spatially separating initial and final states, a feature exploited for increasing intersubband relaxation times in diagonal transition active regions of a QCL (see chapter 6.1.4). Therefore, for both relaxation mechanisms, the relative spatial position of the respective quantum well states forms one of the few remaining parameters allowing the manipulation of relaxation times by means of band structure design. Within this work, the possibility to apply the concept of increasing intersubband relaxation times by spatially diagonal transitions to SiGe heterostructures is studied experimentally.

6.3 Measurement of intersubband relaxation times in SiGe

Due to its crucial role for the achievement of population inversion, the experimental access to intersubband relaxation times in the SiGe material system has attracted considerable research interest since the first demonstration of quantum cascade electroluminescence in a SiGe structure [12]. Proceeding on the road to a SiGe QCL beyond the stage of electroluminescence by reaching population inversion and finally gain requires a detailed study of intersubband scattering rates, in order to exploit the underlying processes by smart and careful band structure design.

The previous section presented a theoretical introduction into the lifetime-limiting processes in the SiGe system, along with some examples of numerical results obtained for the scattering rates by several groups. However, owing to the high complexity of the valence band of the SiGe system and the additional challenges of numerically treating scattering processes as cumbersome as alloy scattering with scattering parameters of strong uncertainty, the experimental determination of relaxation times in this material system is even more crucial than it was for the III-V systems. Still another urge for experimental access is formed by the non-resonant nature of scattering processes in SiGe, as discussed in chapter 6.2, which deems the simple design rules for energetic spacings in III-V systems worthless and makes careful design on the basis of experimentally well founded parameters necessary.

The time scales of intersubband relaxation times in the SiGe material range from hundreds of femtoseconds into the picosecond regime and can therefore be classified as ultrashort. Among the possibilities of extracting these lifetimes from experimentally obtained data, so-called pump-probe techniques involving ultra-short laser pulses are the most common ([126], pp. 13). In addition, there are intrinsically indirect methods for extracting the intersubband relaxation times from experiments not involving time-resolved techniques, such as treating the relaxation time as a parameter in the simulation of electroluminescence intensity and subsequently correlating the simulation results with the measured integral electroluminescence [116]. These methods suffer from the complex influence of numerous simulation parameters and lead to results of reduced reliability. The following subsection gives a short introduction into the pump-probe method of accessing ultrafast processes.

6.3.1 Pump-probe technique

The pump-probe measurement technique is based on establishing a non-equilibrium occupation of an excited state by illuminating the sample under investigation with one laser beam, and probing the change in the sample's optical properties by a second beam. By synchronizing both beams and adjusting a well-defined delay between them, the recovery of the optical properties of the sample can be studied as a function of the delay.

Pump-probe experiments are extensively employed to study a variety of ultrafast process from chemical reactions in molecular beams to intersubband relaxation in solid state devices. In order to investigate a specific process, laser pulses significantly shorter than the expected timescale of the mechanism are required. For processes, whose transition energies are in the range of far- to mid-infrared radiation, free electron lasers (FELs) are a preferable source of radiation, as both their wavelength and pulse duration can be adjusted over a wide region in parameter space. In the terahertz regime, the FEL is the *only* source providing pulses short enough at the required intensity to perform pump-probe experiments [117]. The simplest way to gain two synchronized laser pulses is to use one source and split the beam into two. In this setup configuration, which is called degenerate pump-probe technique, pump and probe beam are of the same wavelength. After being routed through two different optical paths, both beams are focused onto the sample under investigation. In order to prevent coherent effects between the two beams and to be able to distinguish between them in respect to detection, the polarization of one of the two pulses can be rotated by 90°. In order to apply an adjustable temporal delay between the two pulses, a movable mirror is employed to vary the optical path length of one of the pulses on a micrometer scale. As a result, the delay between the two beams can be tuned in the range of femtoseconds.

The simplest form of pump-probe experiments employs the absorption properties of the studied transition as its characteristic quantity. These transmission pump-probe experiments are, among other applications, employed for the measurement of intersubband relaxation times in semiconductor heterostructures. Figure 6.8 illustrates the flow chart of a degenerate pump-probe experiment aimed at determining the relaxation rate between two quantum mechanical states 2 and 1. The system under investigation is cooled down to a temperature at which a large majority of charge carriers is occupying the ground state of the system, as shown in Fig. 6.8 (a). The wavelength of the source of ultrashort laser pulses is chosen in resonance with the probed transition. This tuning of the laser source into resonance with the stud-

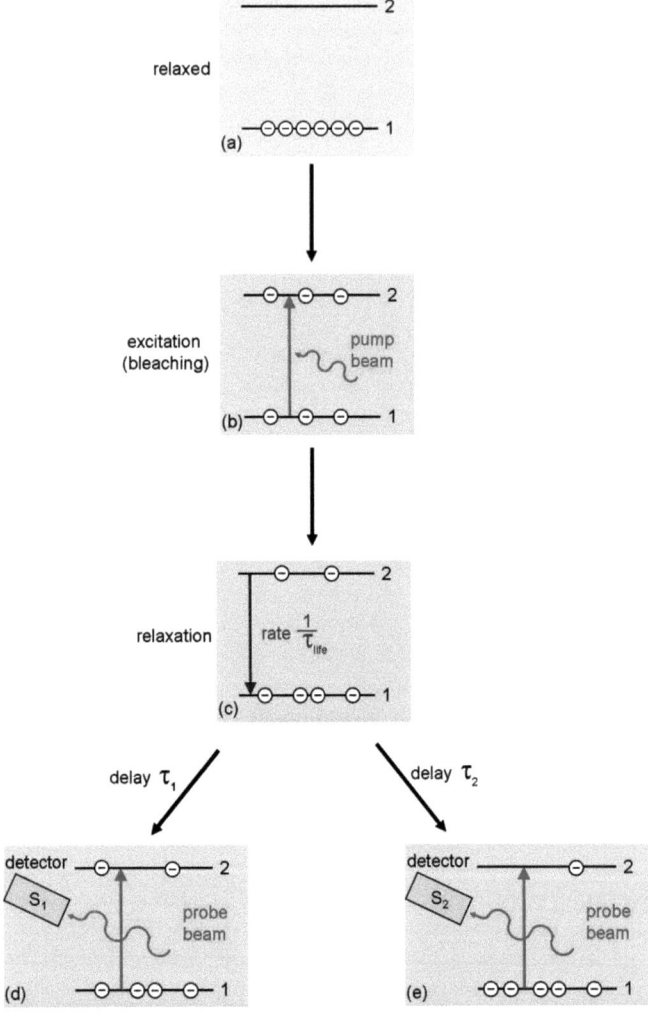

Figure 6.8: Basic scheme of the pump-probe technique. Details are given in the text.

ied system can be conveniently performed when using a FEL source. The first of two laser pulses reaching the quantum mechanical system, the so-called pump-beam, excites a non-equilibrium occupation of state 2 (see Fig. 6.8 (b)), where the polarization of the incident beam has to be chosen in accordance with the selection rules of the transition. The induced change in the relative occupation between the two states alters the associated absorption coefficient, as seen in Eqn. 6.64.

$$\alpha(\nu,t) = \frac{e^2}{m_{effav}^2 \, \nu \, n_r \, c \, \epsilon_0 \, V} \sum_{k_\parallel} | <\Phi_{2,k_\parallel} |\pi| \Phi_{1,k_\parallel}> \cdot \epsilon_\lambda |^2 \cdot \frac{\frac{\Gamma}{\pi}}{(E_{1,k_\parallel} + h\nu - E_{2,k_\parallel})^2 + \Gamma^2} \cdot$$
$$\cdot (n_{1,k_\parallel}(t) - n_{2,k_\parallel}(t)) \tag{6.64}$$

This equation has been derived from Eqn. 2.3 by choosing fixed initial and final subbands and explicitly carrying out the summation over all in-plane wave vectors $k_\parallel$. The equilibrium occupation functions of the involved states in Eqn. 2.3 have been replaced by their non-equilibrium, time dependent counterparts $n_{1,k_\parallel}(t)$ and $n_{2,k_\parallel}(t)$. Under the influence of the intense pump beam, the absorption coefficient induced by the two states 1 and 2 reduces due to the increase in the stimulated emission by the transition $2 \to 1$ while reducing the absorption by the process $1 \to 2$. Under ideal experimental condition, the two processes completely balance each other. This situation is called bleaching, as the system is then absolutely transparent ($\alpha = 0$), and involves an occupation ratio between the two states of approximately 50:50.

As soon as the ultrashort pump-pulse has passed the studied sample and its intensity therefore reduced to insignificance, the system starts relaxing towards its equilibrium condition by a rate $\frac{1}{\tau_{life}}$, where τ_{life} is the lifetime of the excited state and the figure under investigation (Fig. 6.8 (c)). Alongside with the relative occupation, the absorption coefficient in Eqn. 6.65 relaxes towards its equilibrium with $n_{2,k_\parallel}(t \to \infty) = 0$ for all in-plane wave vectors.

The second pulse irradiating the investigated system, which arrives delayed in respect to the pump-pulse by a time span determined by the difference in the optical paths, is employed to determine the actual absorption coefficient as a function of the adjusted delay. For this purpose, the intensity of the second pulse transmitted through the bleached system is measured for several delay times, as illustrated in Fig. 6.8 (d) and (e) for two different delay values τ_1 and τ_2. As the occupation of the excited state is at τ_2 reduced due to relaxation in comparison to the value at τ_1 with $\tau_1 < \tau_2$, the transmission observed with the help of the second, so-called probe pulse, decreases from τ_1 to τ_2. The probe pulse is usually of weaker intensity than the pump-pulse, so that it does not change the occupation significantly by itself.

The transmission through the investigated system is now determined for a row of delays by measuring the transmitted intensity of the probe beam using a detector, where the pump-pulse can be blocked from the detecting device by an analyzing polarizer, as far as orthogonally polarized beams are employed. The detector is usually not able to temporally resolve the shape of the ultrashort probe pulse, but rather delivers an electric signal stretching over a certain time span determined by the detector characteristics and the processing electronics. However, the integral electric signal is proportional to the integral intensity of either one pulse or a series of pulses bundled in a so-called macropulse, as is common for FEL sources (see chapter 7.5.1 and Fig. 7.9). In the latter case, it is essential that the repetition rate of the micropulses is significantly lower than the studied relaxation rate. The integrated detector signal is then proportional to the quantity given in Eqn. 6.65.

$$\Lambda(\tau) = \int_{-\infty}^{\infty} I(t,\tau) \, e^{-\alpha(\nu,\ t)d} \, dt \tag{6.65}$$

In this equation $I(t,\tau)$ is the temporally evolving intensity of the probe pulse at a given delay τ. As long as the time resolution of the measurement setup is much higher than the quantity under investigation, meaning that the laser pulse width is negligible in comparison to the decay time, $\alpha(\nu,\ \tau)$ stays constant for the time span attributed to the probing pulse. Further, as most intersubband absorption processes exhibit only weak absorption coefficients, the exponential can be approximated by its taylor series up to first order, resulting in Eqn. 6.66, where $\widetilde{I}_0$ represents the total integral intensity of the probe pulse.

$$\begin{aligned}\Lambda(\tau) &\simeq (1 - \alpha(\nu,\ \tau)d) \int_{-\infty}^{\infty} I(t,\tau) \, dt \\ &\simeq (1 - \alpha(\nu,\ \tau)d)\widetilde{I}_0\end{aligned} \tag{6.66}$$

In order to enable the accurate extraction of relaxation times from data gained in transmission pump-probe experiments, the electronic dispersion as well as the dependence of the matrix element $|<\Phi_{2,k_\parallel}|\boldsymbol{\pi}|\Phi_{1,k_\parallel}>\cdot\boldsymbol{\epsilon}_\lambda|$ in Eqn. 6.64 on the in-plane wave vector $\boldsymbol{k}_\parallel$ has to be taken into account. In general, pump-probe experiments monitor the relaxation of the *optical properties* rather than the *overall relaxation* of charge carriers from an excited level into the ground level. The latter is related to the quantity in Eqns. 6.31, 6.42 and 6.53.

Under ideal experimental conditions and thus in absence of hot carriers occupying high $\boldsymbol{k}_\parallel$ states, the complications emerging from including in-plane dispersion can be avoided by applying a one-dimensional two-level model like that illustrated in Fig. 6.8 in order to interpret the experimental pump-probe data. Equation 6.64 then simplifies into Eqn. 6.67

$$\alpha(\nu, t) = \frac{e^2}{m_{effav}^2 \, \nu \, n_r \, c \, \epsilon_0 \, V} \quad |\widetilde{\boldsymbol{\pi}}_{1,2} \cdot \boldsymbol{\epsilon}_\lambda|^2 \cdot \frac{\frac{\Gamma}{\pi}}{(E_{1,0} + h\nu - E_{2,0})^2 + \Gamma^2} \cdot$$

$$\cdot (\widetilde{n}_1(t) - \widetilde{n}_2(t)) \tag{6.67}$$

In this equation, $\tilde{n}_1(t)$ and $\tilde{n}_2(t)$ resemble the time devolution of the *total* population of the ground and excited level, respectively. $\tilde{\pi}_{1,2}$ represents the effective dipole matrix element between the two quantized levels, averaged over the respective $k_\parallel$-values weighed by their significance. Within this simple two-level model, a total depopulation lifetime τ_{life} between the two levels can be defined, and the rate equations for non-radiative decay read as in Eqn. 6.68.

$$\frac{d}{dt}\tilde{n}_1(t) = -\frac{d}{dt}\tilde{n}_2(t) = \frac{\tilde{n}_2(t)}{\tau_{life}} = \frac{n_{tot} - \tilde{n}_1(t)}{\tau_{life}} \quad (6.68)$$

Here, n_{tot} is the total number of charge carriers in the system. As stated in Eqn. 6.69, the lifetime directly relates to the depopulation rate for the respective scattering mechanisms as defined in Eqns. 6.53, 6.42 and 6.31.

$$\frac{1}{\tau_{life}} = \tilde{D}^1_{2,0} \quad (6.69)$$

Inserting Eqn. 6.68 into Eqn. 6.67 leads to the simple differential equation for the time dependence of the absorption coefficient as given in Eqn. 6.70, where $\alpha_0(\nu)$ is the equilibrium absorption coefficient of the two-level system.

$$\frac{d}{dt}\alpha(\nu,t) = -\frac{\alpha(\nu,t) - \alpha_0(\nu)}{\tau_{life}}$$

$$\alpha_0(\nu) = \frac{e^2}{m^2_{effav}\,\nu\,n_r\,c\epsilon_0\,V}\,|\tilde{\pi}_{1,2}\cdot\epsilon_\lambda|^2 \cdot \frac{\frac{\Gamma}{\pi}}{(E_{1,0} + h\nu - E_{2,0})^2 + \Gamma^2} \cdot n_{tot} \quad (6.70)$$

The solution of this differential equation is given in Eqn. 6.71, where $\Delta\alpha(\nu)$ represents the change in transmission directly after the excitation of the system by the pump-pulse.

$$\alpha(\nu,\tau) \simeq \alpha_0(\nu) - \Delta\alpha(\nu)e^{-\frac{\tau}{\tau_{life}}} \quad (6.71)$$

The transient integral signal recorded by the detector of the measurement setup and its read-out electronics consequently reads as in Eqn. 6.72.

$$\Lambda(\tau) \simeq (1 - \alpha_0\,d)\,\tilde{I}_0 + \Delta\alpha\,d\,\tilde{I}_0\,e^{-\frac{\tau}{\tau_{life}}} \quad (6.72)$$

Therefore, under ideal experimental conditions and in absence of any heating effects resulting in the occupation of high $k_\parallel$ states, the integral transmission signal recorded in dependence of the temporal delay between the pump and probe pulses exhibits a peak at zero-delay and then decays exponentially with the total relaxation rate between the two involved levels. The relaxation time can in such a case be directly extracted from the experimental data by fitting this exponential decay. The highest accuracy in fitting the relaxation time τ_{life} is achieved by completely bleaching the optical transition, which results in a maximal change in the absorption coefficient with $\Delta\alpha = \alpha_0$. In case of heating effects contributing to the outcome of the pump-probe experiment, models of higher degrees of sophistication are required to extract the studied lifetimes from the experimental data, resulting in a degrading accuracy in the determination of the individual fit parameters.

It shall be once more pointed out, that pump-probe experiments monitor the relaxation of the *optical properties* induced by the population of the respective states rather than the relaxation of the level's *population* itself. This circumstance forms one of the major shortcomings of the transmission pump-probe technique. Another deficiency of this experimental method is that its successful application is limited to systems with reasonably strong absorption, in order to reach a signal-to-noise ratio of the increased transmission sufficient for the extraction of the lifetime parameters. Both shortcomings are not shared by the photocurrent pump-pump technique, which will form the experimental basis of this work and will be discussed in detail throughout the following chapters.

6.3.2 Below the optical phonon energy: Experimental results so far

Intersubband relaxation times between confined states in p-type SiGe heterostructures energetically separated by less than the LO phonon energy are relatively well experimentally accessible through transmission pump-probe experiments, as long as a FEL source is available. There has been a number of reports on the successful experimental determination of intersubband relaxation times below the optical phonon energy in the SiGe system [80, 107, 117, 118, 120]. This subsection gives an overview over the experimental results reported by a number of groups up to now.

One of the first measurements in this field was reported by Heiss in [118] even before reliable theoretical predictions on the involved scattering rates were available. In this publication, a value of 30 ± 10 ps was found for the intersubband relaxation time between two levels energetically separated by 33.5 meV by performing transmission pump-probe experiments. The relaxation time value was extracted from the experimental data by fitting a single exponential, as described in the previous subsection.

In [80] and [120], Murzyn and Pidgeon reported on pump-probe experiments on SiGe quantum well structures grown on $Si_{0.78}Ge_{0.22}$ pseudosubstrates for a series of temperatures. For a transition energy of 27 meV between the HH ground state and the lowest LH state, where the first excited HH and the lowest LH state were degenerate, a constant value of about 10 ps was reported for sample temperatures between 4 and 100 K. This value is in good agreement with the theoretically predicted intersubband relaxation times dominated by alloy scattering as reported in [104]. The transmission pump-probe characteristics presented in [80, 120] exhibit a complex shape with a fast decay for low delay times followed by a slow rise, and a subsequent long relaxation to equilibrium values. Murzyn et al. attribute this behavior to charge carriers escaping into slowly decaying spacer layer states. The temperature independence experimentally observed for intersubband relaxation times between states of a transition energy below the optical phonon energy within a temperature range between 4 and 100 K is in accordance with the theoretical predictions based on alloy scattering.

In [107], Kelsall et al. extended the studies on intersubband relaxation times of transitions with energies below the LO phonon energy from quantum wells separated by spacer layers to structures with coupled quantum wells. For one of these structures diagonal intersubband electroluminescence was reported in [79]. Both samples were grown on a $Si_{0.7}Ge_{0.3}$ pseudosubstrate. Due to the lack of spacer layers in these systems, the transmission transient characteristics obtained in pump-probe experiments were less complex than that reported in [80], as no long-living spacer states existed. Therefore a simple, one-dimensional two level model as described in chapter 6.3.1 was sufficient to extract the intersubband lifetimes from the pump-probe data. One of the two coupled well structures exhibited a transition energy of 13 meV between the HH1 and the LH1 subbands. For this transition, a relaxation time between 20 and 25 ps could be extracted from the measurement data by performing a single-exponential fit. The energetic HH1-LH1 spacing of the second structure in [107] was 32.7 meV, for which a relaxation time of 2 ps was measured. Kelsall attributed the reason for this short lifetime to the fact that 32.7 meV are much closer to the LO phonon energy than 13 meV and therefore optical phonons contribute stronger to the total scattering rate. According to Kelsall the phenomenon occurs, "[...]that the LH1-HH1 lifetime increases progressively as the subband spacing is reduced[...]". Once more, the experimentally determined intersubband relaxation times did not exhibit any dependence on the sample temperature, which is in agreement with theory.

Motivated by the potential of diagonal transitions as a means of increasing intersubband relaxation times, a study on the change in the recovery time of the absorption properties of a heterostructure with decreasing confinement of the excited state in a quantum well was presented in [117]. The samples investigated in this publication were grown on $Si_{0.8}Ge_{0.2}$ pseudosubstrates, where each structural period contained a single quantum well separated from the spacer region by pure Si barriers. Each sample featured a different barrier and well width. These parameter were chosen in a way that the excited LH1 state of the well was energetically located far below the spacer band edge for sample 1, slightly below for sample 2, and in resonance with the spacer edge for sample 3, resulting in a decreasing confinement of the excited state of the studied transition when going from sample 1 to sample 3. Due to the decrease of the overlap integral between the HH ground state and the excited LH state with the reduction of the confinement of the latter, the scattering rates between those two states are expected to decrease from sample 1 to 3. The experimental confirmation of this theoretical prediction was reported in [117], utilizing pump-probe transmission measurements at a FEL facility, where LH1-HH1 relaxation times of approximately 20 ps, 40 ps and 40 ps were determined for sample 1, 2 and 3, respectively. Similar to the experiments reported in [80], the influence of long-living spacer states on the pump-probe characteristics was observed in [117].

Even though reference [117] reported on the correlation between the confinement of the excited state and its relaxation time into the ground state, it is not trivial to generalize this experimental finding and draw conclusions on diagonal transitions. In order to do so, aspects of the structural difference between the investigated samples in addition to the confinement of the LH1 state have to be taken into account. One of these aspects is the difference in the LH1-HH1 transition energy between the samples, where the calculated energetic spacing for sample 1 is 35 meV and that for sample 3 values 24 meV. As reported in [107], for transition energies below the LO phonon energy the total scattering rate between two states is expected to increase with increasing transition energy. This difficulty in applying the conclusions drawn from the experimental results in [117] to diagonal transitions in general is inherent in the measurement method itself, which requires the comparison of differently designed and grown heterostructures. In order to draw direct general conclusions on the behavior of diagonal transitions in the SiGe system from time-resolved experiments, the comparison of measurements on one and the same structure is required. This demands means of controlling the overlap between the initial and final state of an optical transitions other than the change in growth parameters. One promising means of tuning this overlap is applying a controllable voltage to the structure. This approach is pursued in the course of this work, where the details are given in chapter 8.

Table 6.2 summarizes the values for the LH-HH intersubband relaxation times reported in literature for transition energies below the LO phonon energy. They are in the range of several 10 ps, where the discrepancies between the individual values can be explained by either structural differences of the investigated samples or by measurement and interpretation uncertainties. The observance of relatively long intersubband relaxation times for low transition energies in SiGe heterostructures seems highly promising for the realization of a SiGe QCL device. Nevertheless, it is to be kept in mind that in addition to a long-living excited state, a fast depopulation of the ground state of the emitting transition has to be realized in order to achieve population inversion. Therefore, even though the typical relaxation times between LH and HH states seem satisfyingly high, their precise measurement and manipulation remain crucial topics for the development of a QCL design which employs *both* long-living excited and short-living ground states.

Table 6.2: Reported LH-HH intersubband relaxation times in the SiGe system determined by pump-probe experiments for transition energies below the LO phonon energy:

LH-HH transition energy	Reported relaxation time	Type	Reference
13 meV	20 - 25 ps	miniband	[107]
24 meV	~ 40 ps	weakly confined excited state	[117]
27 meV	10 ps	separated QW	[80, 120]
32.7 meV	< 2 ps	miniband	[107]
33.5 meV	30 ± 10 ps	separated QW	[118]
35 meV	~ 20 ps	weakly confined excited state	[117]

6.3.3 Above the optical phonon energy: Experimental results so far

In case of transitions with an energetic spacing above the optical phonon energy, the experimental determination of intersubband relaxation times is substantially more cumbersome than for intersubband spacings below the LO phonon energy. This is due to the ultra-short time scale of relaxation processes based on the emission of optical phonons, which lies in the sub-picosecond range. Achieving stable FEL pulses of sufficiently short duration to resolve these processes is highly demanding. Furthermore, the intensity of the pump beam required in order to reach a degree of bleaching in the studied system sufficient for a tolerable signal-to-noise ratio in pump-probe experiments increases with decreasing relaxation time. This is due to the fact that at too low intensities the decay characteristics of the monitored tranmission drop below the noise level of the measurement already within the time span occupied by the pump pulse. Under such conditions, the data relevant for the relaxation time extraction is dominated by the FEL pulse characteristics, and the extraction of a reliable lifetime value is impossible.

On the other hand, large pump beam intensities lead to a heating of the charge carrier ensemble and result in a change in relaxation properties, and are thus highly unfavorable. As a consequence of these experimental difficulties, reports on the measurement of intersubband relaxation times in the SiGe system for transition energies above the LO phonon energy suffer from either a poor signal-to-noise ratio of the pump-probe characteristics or from heavy heating contributions, which in turn have to be deconvoluted from the obtained data. Each of the two effects leads to a significant uncertainty in the determined lifetime values.

The first report on the experimental determination of intersubband relaxation times induced by LO phonon scattering was given in [121]. For an HH2-HH1 subband spacing of 130 meV, an associated relaxation time between the states of approximately 400 fs was reported. However, as the duration of the FEL micropulses employed for the pump-probe measurements, on which the publication is based, was above 500 fs, the time resolution of the experimental setup given in [121] was simply insufficient for determining the exact value of the intersubband lifetime. This circumstance becomes evident from the plot of the pump-probe characteristics presented in the publication, which in addition suffered heavily from coherent effects leading to a complete modulation of the transmission signal. As Kaindl puts it in [114], "However, first time-resolved experiments for subbands spaced more than the energy of an optical phonon [...] lack sufficient time resolution [...] and thus give no information on this important issue."

In [114], Kaindl et al. reported on the determination of an intersubband relaxation time of 250 ± 100 fs for a HH2-HH1 transition spanning 167 meV. The experimental data presented in this publication had been gathered during pump-probe experiments employing FEL pulses of 150 fs length. However, the reported experiments suffered from a strong contribution of hole heating processes, as a result of which no clear exponential decay could be observed in the transmission transients. Consequently a complex procedure had to be developed in order to extract the transmission change induced by the intersubband relaxation process based on LO phonons, rendering the measurement rather indirect and imposing a significant uncertainty on the extracted lifetime value. The heating of the hole ensemble induced by the pump-pulse resulted in a redistribution in $k_\parallel$ space. As the dipole matrix element between HH1 and HH2 states exhibiting a finite $k_\parallel$ value is smaller than that for states at $k_\parallel = 0$, and in addition the wavelength of the radiation absorbed by finite wave vector states shifts towards longer values, this redistribution of HH1 holes within the band leads to an increase in transmission for radiation in resonance with the subband spacing at $k_\parallel = 0$. This heating-induced enhanced transmission, which exerted a strong influence on the measurements presented in [114] throughout the whole range of used pump intensities, decays slowly on a time scale of 25 ps [122] based on intraband relaxation. In order to extract the desired intersubband relaxation time from the experimental data, Kaindl et al. measured the transmission transient for a range of pump intensities and a series of FEL wavelengths. For low pump intensities, the dependence of the transmission on the delay time was exclusively determined by the heating redistribution of holes within the HH1 band, as the population of the HH2 band was too weak to be observed. As a consequence, at an FEL photon energy equal to the intersubband spacing, an enhanced transition exclusively induced by hole heating was observed, while at lower photon energy the transmission was reduced due to the occupation of high-$k_\parallel$ states enabling the absorption of these wavelengths. The obtained transient transmission spectra were then used to subtract the contributions by heating at resonant wavelengths from the total excess transmission for high pump intensities, where the influence of the HH2 occupation grew relevant. This was done by scaling the spectra measured at low intensities to match the curves at high pump intensity in the spectral region of reduced transmission, and then subtracting the integral spectra measured for low intensities around the resonance from those gained at high intensities. The resulting transient curves were fitted exponentially, giving the reported value of 250 ± 100 fs. According to the simulations, to which the experimentally obtained results in [114] were compared, the dominant process determining the HH2-HH1 relaxation time is not the direct scattering between the two states, but the depopulation via the intermediary mixed LHSO level. Kaindl et al. reported a calculated relaxation time of 760 fs for the direct HH2-HH1 scattering process, of 319 fs for the HH2-LHSO1 scattering and of 170 fs for the scattering between LHSO1 and HH1. In [105], the conclusion was drawn from this theoretical finding that decreasing the energetic spacing between HH2 and HH1 below 100 meV might induce a significant increase in the total HH2-HH1 relaxation time, as the emission of a LO phonon during the HH2-LHSO1 transition would then be energetically impossible. This would switch off the respective relaxation channel.

At the beginning of the work on this thesis the measurements reported in [114] gave the most reliable experimental values for ultrafast intersubband relaxation times in the SiGe system for transition energies above the LO phonon energy up to date. However, the influence of heating contributions and the consequential complicated and indirect parameter extraction process raised the need for a different experimental method allowing a *direct* determination of this relaxation time without the need for complex interpretation of the obtained data. The establishment of such a method and its application in the SiGe system forms the topic of chapter 7 of this work.

In [116], Bormann et al. reported on electroluminescence experiments on a series of SiGe quantum cascade emitters in the mid-infrared. The emission of these devices is based on a diagonal transition between HH ground states of two neighboring quantum wells, where the respective structures differ in the width of the barrier separating the two wells. By comparing the integral intensity of the electroluminescence centered around a photon energy of about 180 meV, conclusions on the relaxation time between the two HH states are drawn in this publication. Bormann et al. state that even though the squared dipole matrix element between the two emitting heavy hole states is predicted to decrease by a factor 13 between the sample with the thinnest separating barrier and that with the thickest one, no significant decrease in the electroluminescence intensity could be observed. The authors concluded, that the non-radiative relaxation time of the emitting transition had to be by a factor 17 higher for the sample featuring the thick barrier layer in comparison to that with the thin separation.

The HH2-HH1 relaxation times extracted from the electroluminescence data were 0.7 ps and 12 ps, respectively.

Even though reference [116] reports on an increase in the non-radiative relaxation time of a diagonal transition with rising spatial separation between the involved states, the reliability of the conclusions drawn from electroluminescence data is limited. The major flaw of the reported findings lies in the complete absence of any time-resolved studies in the course of the experiments. The data obtained in [116] was exclusively gained under equilibrium conditions. Due to the consequential lack in transient information, the possible influence of heating effects is completely unaccessible and cannot be distinguished from the intersubband relaxation effects. Further, the experimental data was obtained for three samples grown and studied separately. The huge amount of possibly influential parameters ranging from structural deviations to inhomogeneous field distributions and coupling efficiencies therefore further decrease the reliability of the conclusions drawn in [116], particularly as they are based on the *non-observance* of a significant change in the electroluminescence intensity between the studied samples, whose number of *three* is extremely low for studying a trend.

6.3.4 Missing information on intersubband relaxation times

Despite the availability of experimentally determined values for intersubband relaxation times in the SiGe system for transition energies below the LO phonon energy (see chapter 6.3.2), there is a detrimental lack in and a consequential strong need for reliable and directly measured data on intersubband lifetimes dominated optical phonon scattering. As stated in chapter 6.3.3, the experiments aimed at determining these ultrafast scattering times in SiGe heterostructures suffered heavily from contributions of hole heating effects, requiring complex means for parameter extraction and thus drastically reducing the reliability of the reported results. Put differently, up to now no *direct* monitoring of intersubband relaxation processes for transition energies above the LO phonon energy has been reported for the SiGe system. Furthermore, the pump-probe experiments reported in [121] and [114] have been performed on un-biased structures, leaving the study of the voltage dependence of intersubband relaxation times and the possible changes in relaxation times due to band bending and carrier transport unstudied. Particularly in respect to the highly complex valence band structure in SiGe heterostructures and the associated problems with simulations, the experimental conditions for lifetime measurements should be chosen as similar to those of an operating emitter device as possible. This calls for the possibility of performing time-resolved measurements on a biased, current-carrying heterostructure.

Motivated by the crucial importance of experimental data on ultrafast intersubband relaxation processes in the SiGe system for the design and implementation of a QCL in this material system, chapter 7 covers an experimental alternative to transmission pump-probe experiments, which allows the *direct* determination of ultrafast lifetimes induced by optical phonon scattering in *biased* SiGe heterostructures.

Moreover, the concept of increasing intersubband relaxation times in the SiGe system by employing diagonal transitions has not been studied directly by time-resolved experiments so far. As diagonal transitions are among the most promising concepts for increasing intersubband relaxation times sufficiently for achieving population inversion in SiGe quantum cascades, time-resolved studies of this concept on one and the same sample are of high interest. Chapter 8 of this thesis deals with the experimental access to the change in intersubband scattering rates with the amount of spatial overlap between the involved states, where their spatial separation is controlled by an applied voltage. This enables the comparison of different overlap situations for one and the same sample and thus excludes the influence of a series of unknown parameters diminishing the reliability of experiments based on the comparison of data sets on different samples.

Chapter 7

Monitoring the intersubband relaxation by LO phonons in SiGe

Gaining direct and detailed insight into the ultrafast relaxation processes by optical phonon scattering in the SiGe system is a demanding experimental task, as discussed in chapters 6.3.3 and 6.3.4, and promises information of essential importance for the design and realization of a SiGe QCL. This chapter presents time-resolved experimental work on a SiGe quantum cascade structure, which enabled the *direct* determination of the up to now only indirectly accessible relaxation times for intersubband transitions between states with an energetic spacing above the LO phonon energy. By employing photocurrent pump-pump experiments involving a free electron laser, these ultrafast relaxation processes could be monitored. The results presented and discussed in this chapter were published in references [123] and [124].

7.1 Design and fabrication

7.1.1 Structure design and growth

In order to monitor optical phonon-induced relaxation processes in a SiGe heterostructure with a design as close to an operational emitter as possible, sample H019 was chosen as a subject for the time-resolved measurements in this work. The design of the layer sequence of sample H019 (see table 7.1) is based on the emitter structure, for which quantum cascade electroluminescence was demonstrated for the first time in the SiGe system, as reported in [12]. Figure 7.1 (a) shows the calculated HH (blue), LH (green) and SO (red) band edges of two subsequent periods of the structure, on which reference [12] reported, at an applied field of 50 kV/cm. Along with the band edges, the energetic positions of the structure's eigenstates are shown together with the shape of the absolute squared envelope functions of the respective

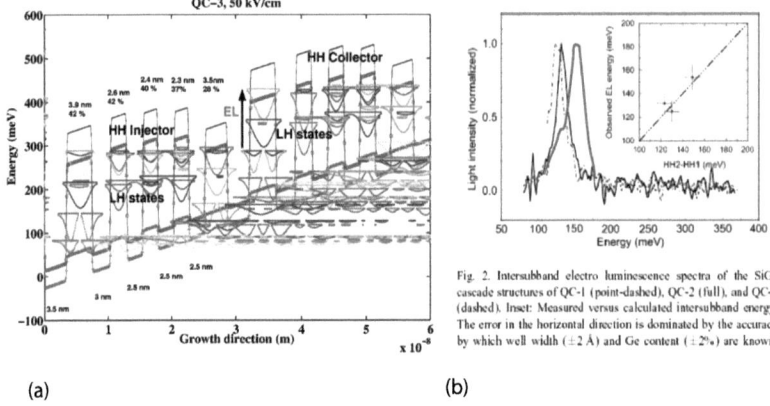

Figure 7.1: Band structure and electroluminescence spectrum of the SiGe emitter structure reported on in reference [12]. The design of sample H019 is based on the structure shown in (a), where the Si separation layer between neighboring periods has been increased in thickness for H019 in order to prevent tunneling transport between the periods. In the chapter at hand the non-radiative relaxation time between the HH2 and HH1 states, which form the basis of the electroluminescence peak around 160 meV in (b), is to be determined on the altered structure H019.

levels. At the presented bias, the HH ground state of the shallow well of each period (labelled 'HH Injector') is aligned with the first excited HH state of the deep well of the neighboring period. This enables the transfer of holes from the 'HH Injector' into the excited HH2 state, which can subsequently emit a photon by undergoing a transition from the HH2 to the HH1 state in the deep well. After this optical transition, they are extracted from the ground state via the 'HH collector'. In [12], the observation of an electroluminescence peak around 160 meV was reported for this structure at an applied field of 50 kV/cm. Figure 7.1 (b) shows the electroluminescence characteristics published in [12].

The non-radiative lifetime of the emitting transition of the structure demonstrated in [12] is of high interest, as it ultimately determines the relative amount of charge carriers, which decay from HH2 to HH1 by emitting an unwanted phonon instead of a desired photon. The exact knowledge of the value for the relaxation time induced by LO phonons is particularly crucial for further proceeding on the road to a QCL by establishing population inversion.

The experimental method for monitoring ultrafast processes employed in this work is based on

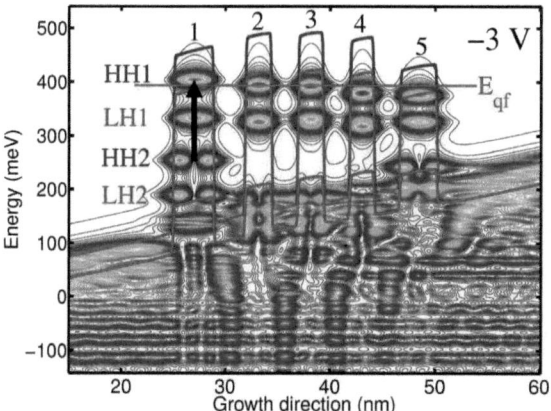

Figure 7.2: Band structure of H019. The plot shows the calculated heavy-, light- and split-off- (blue, green, red) hole band edges for an applied voltage (field) of -3 V (39 kV/cm) and a contour plot of the absolute squared wave functions. The spreading of the eigenstates along the energy-axis accounts for their homogenous broadening. For the chosen applied voltage, the charge carriers are located in W1. The labels indicate the character of the excited states (HH, LH). The straight red line indicates the quasi-Fermi-level as calculated in the simulation, the black arrow marks the transition under investigation in this work.

the measurement of a photocurrent induced by a two-photon process under non-equilibrium conditions. In order to eliminate the dark-current observed in the structure demonstrated in [12] due to tunnelling transport, the quantum well regions of subsequent periods of H019 were decoupled from each other via the introduction of a 50 nm wide silicon barrier between them. At the same time the quantum well region of each of H019's periods was kept equal to that of the sample in [12]. The band structure of H019 is shown in Fig. 7.2.

Structure H019 was grown pseudomorphically on a Si substrate by G. Dehlinger, H. Sigg and D. Grützmacher at the Paul Scherrer Institut in Villigen, Switzerland, using low temperature ($\sim 350^\circ$C) molecular beam epitaxy. Directly on the (100) substrate a highly p-type (boron) doped contact layer (doping concentration $2 \cdot 10^{18}$cm^{-3}) of 300 nm thickness was grown. This bottom contact layer was followed by ten periods of the sequence given in table 7.1, where the values last in the table are those closest to the substrate. The ten periods of optically active material were completed by a 100 nm thick top contact layer of a boron doping concentration of $2 \cdot 10^{18}$cm^{-3}. X-ray reflectivity measurements on sample H019

Table 7.1: MBE growth sequence and doping concentration of sample H019:

Thickness	Ge concentration	Doping concentration	Band edge profile
25 nm	0	0	separation
3.5 nm	28 %	$5 \cdot 10^{17} \mathrm{cm}^{-3}$	well 5
2.5 nm	0	$5 \cdot 10^{17} \mathrm{cm}^{-3}$	barrier
2.3 nm	37 %	$5 \cdot 10^{17} \mathrm{cm}^{-3}$	well 4
2.5 nm	0	$5 \cdot 10^{17} \mathrm{cm}^{-3}$	barrier
2.4 nm	40 %	$5 \cdot 10^{17} \mathrm{cm}^{-3}$	well 3
2.5 nm	0	$5 \cdot 10^{17} \mathrm{cm}^{-3}$	barrier
2.6 nm	42 %	$5 \cdot 10^{17} \mathrm{cm}^{-3}$	well 2
3 nm	0	$5 \cdot 10^{17} \mathrm{cm}^{-3}$	barrier
3.9 nm	42 %	$5 \cdot 10^{17} \mathrm{cm}^{-3}$	well 1
25 nm	0	0	separation

show that the actual structural parameters are typically within 1% of the design parameters.

The band edge profile induced by this layer structure is presented in Fig. 2.2. As already discussed in chapter 2.1.1 and in more detail in reference [20], the grown heterostructure induces five valence band quantum wells per period. The transition under investigation in this work is located in the deepest quantum well W1, which is structurally equivalent to the emitting transition in reference [12].

7.1.2 Sample processing

In order to be able to apply a voltage to the quantum well structure for inducing a field distribution over the quantum well region analogous to that of the operating quantum cascade emitter device in reference [12], contacted mesas were processed into sample H019. The process steps were in complete analogy to that for sample K091 in chapter 2.1.2. Nevertheless, they are listed in table 7.1.2 for the sake of completeness.

The mesa, on which the time-resolved experiments of this chapter were performed on, featured an area of 7×0.5 mm. Its surface was completely covered by the top contact, leading to a field enhancement for TM polarized radiation at the position of the optically active structure. The experiments devoted to precharacterizing the HH1-HH2 transition energy, on which chapters 7.3.1 and 7.3.2 report, were also carried out on this mesa.

7.2 Selection rules

In the course of studying the ultrafast relaxation times between subbands, the respective optical selection rules formed a crucial element for successfully distinguishing between the monitored decay process and effects caused by the time resolution of the experimental setup, as is discussed later on in this work.

In order to predict the dependence of transitions in structure H019 on the polarization of the involved radiation, band structure calculations were performed. The possible final states for an optical transition from the HH1 ground state were mapped in respect to their relevance,

Table 7.2: Sample processing steps for H019

Step No.	Process	Equipment	Parameters
1	cleaning	acetone, methanol	
2	resist deposition	resist 1818, spinner	40 s, 4000/s
3	softbake	oven	90°C, 15 min
4	photolithography mesa	mask aligner EV420	4.5 s exposition
5	developing	developer	1 min
6	mesa etching	reactive-ion-etcher	100% SF_6, 50% O_2, 40 mT, 15% RF, 950 nm
7	cleaning	acetone, methanol	
8	resist deposition	resist 1818, spinner	40 s, 4000/s
9	softbake	oven	90°C, 15 min
10	photolithography bottom contact	mask aligner EV420	4.5 s exposition
11	developing	developer	1 min
12	native oxide removal	hydrofluoric acid	
13	vapor deposition	evaporation chamber, Al target	200 nm Al
14	lift-off	acetone	
15	cleaning	acetone, methanol	
16	resist deposition	resist 1818, spinner	40 s, 4000/s
17	softbake	oven	90°C, 15 min
18	photolithography top contact	mask aligner EV420	4.5 s exposition
19	developing	developer	1 min

Step No.	Process	Equipment	Parameters
20	native oxide removal	hydrofluoric acid	
21	vapor deposition	evaporation chamber, Al, Si targets	20 nm Al, 2nm Si, 20 nm Al, 2nm Si, 20 nm Al, 2nm Si, 40 nm Al
22	lift-off	acetone	
23	contact alloying	rapid thermal annealing oven	380°C, 20 sec $N_2 + H_2$
24	cleaning	acetone, methanol	
25	resist deposition	resist 1818, spinner	40 s, 4000/s
26	softbake	oven	90°C, 15 min
27	photolithography gold contact	mask aligner EV420	4.5 s exposition
28	developing	developer	1 min
29	vapor deposition	evaporation chamber, Ti, Au targets	10 nm Ti 100 nm Au
30	lift-off	acetone	
31	backside polishing	polishing wheel paper: 4000, 1000 diamond spray: 1 μm, 0.25 μm	
32	facette polishing	polishing wheel paper: 4000, 1000 diamond spray: 1 μm, 0.25 μm	30° to (100) surface
33	bonding	bonder, Au wire	70°C

as described in chapter 2.2.2. The results of these theoretical considerations are presented in Fig. 7.3. For comparison, Fig. 7.2 presents the total, un-weighted band structure of sample H019, where the arrow indicates the transition to be investigated in this work. Figure 7.3 implies that the transition between the HH1 and HH2 states is strong for TM polarized radiation, while it is very weak in TE polarization. In other words, the transition, whose non-radiative relaxation time is to be determined in this work, can only interact with TM radiation.

The simulation results were further used to predict the strength of an optical transition between the excited HH2 state of well 1 and the unconfined continuum states reached for a

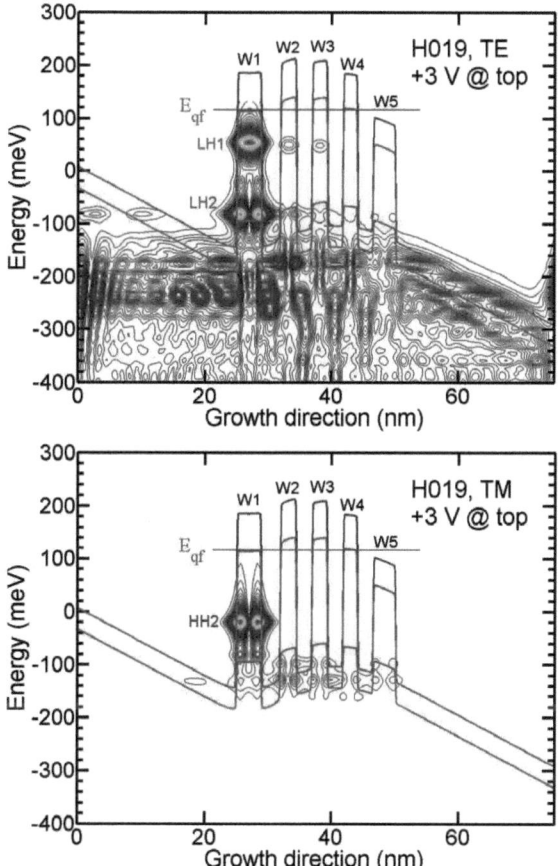

Figure 7.3: Final states relevant for absorption of TE and TM polarized radiation from the ground state of structure H019 at +3 V bias. At this applied voltage, charge carriers are confined in well 1. The plot is realized by weighting the band structure according to the states' relevance for the absorption of radiation as described in the text. Note that in TM polarization the transition from the ground state of W1 to its HH2 state is strong. This transition is negligible for TE polarized light, where the strong transitions occur to the LH1 and LH2 states.

transition energy of 160 meV. The calculations show that this transition is allowed for both TE and TM polarized radiation, as the respective final continuum states are strongly mixed. Table 7.3 gives a summary of the theoretically predicted polarization dependence of the most important transitions in the deepest well 1 of structure H019.

7.3 Spectral characterization of the HH2-HH1 transition

Previous photocurrent studies on structure H019 in the mid-infrared, published in [16, 20], have led to a detailed understanding of optical transitions from the bound HH1 state into continuum states, including their selection rules as well as their voltage dependence.

In order to be able to tune the FEL wavelength into resonance with the investigated HH2-HH1 transition of structure H019, it was necessary to spectrally characterize this transition, preferentially for exactly the same mesa, on which the time-resolved experiments were to be performed. Photocurrent experiments such as those reported on in [16, 20] do not give insight into this transition, as the excited HH2 state is well confined and does not contribute to a current. This circumstance is illustrated in Fig. 7.2, which presents the simulated band structure of H019 calculated as described in [20] and [22].

7.3.1 Voltage modulated waveguide transmission

One possibility to characterize the HH2-HH1 transition of the processed sample H019 exploits the known fact [16] that the amount of charge carriers located in the deep quantum well 1 of each period can be varied by changing the voltage applied to the photocurrent sample. For positive applied biases, a majority of holes occupies the HH1 state of well 1 at liquid helium temperature. By changing the sign of the applied bias, these holes can be transferred to the shallower wells 2-5, a fact which has been employed for tuning the spectral onset of the photocurrent through the sample in reference [16]. As only well 1 of the structure features a well-confined HH2 state due to its relatively large width, exclusively holes located in this well contribute to an absorption around 160 meV. By changing the sign of the bias applied to the

Table 7.3: Allowed transitions in quantum well 1 of H019:

Transition	Energy	TE polarization	TM polarization
HH1-LH1	72 meV	strong	negligible
HH1-HH2	160 meV	negligible	strong
HH1-LH2	215	strong	negligible
HH2-continuum	160 meV	moderate	moderate

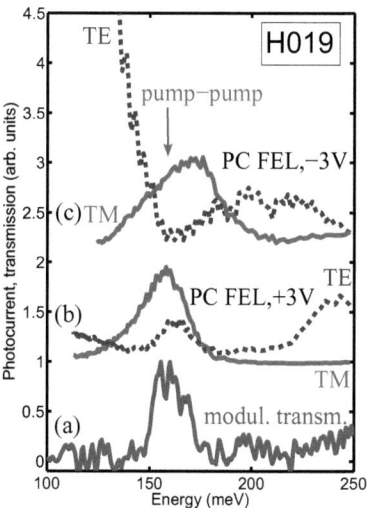

Figure 7.4: Characterization of the HH1-HH2 transition energy by modulated transmission and two-photon photocurrent experiments. The modulated waveguide transmission spectrum presented by the green curve in (a) exhibits a peak at 160 meV, as do the FEL photocurrent spectra given for a bias of +3 V in (b) and of -3 V in (c) in case of TM polarized radiation (solid red lines). Due to selection rules, the resonance in (b) and (c) is not observed for TE polarized FEL radiation (broken blue lines). The graphs were offset for clarity. The arrow marks the photon energy, at which the time-resolved measurements were performed. [124]

sample, the dip in the sample transmission around 160 meV induced by the HH1-HH2 transition can therefore be modulated in strength. By periodically varying the applied voltage and acquiring the correlated transmission spectrum in lock-in-technique, the spectral position of the HH1-HH2 transition could be obtained.

For the modulated transmission measurements, sample H019 was placed in a helium cryostat with ZnSe windows and cooled down to liquid helium temperature. The radiation from a globar source of the Bruker IFS 66 spectrometer was coupled into the sample in waveguide geometry via the polished side facette (see Fig. 2.5). The sample surface was positioned orthogonally to the incident radiation, where the coupling facette was located at the top edge of the sample. The infrared globar radiation passed the sample and exited at the backside via

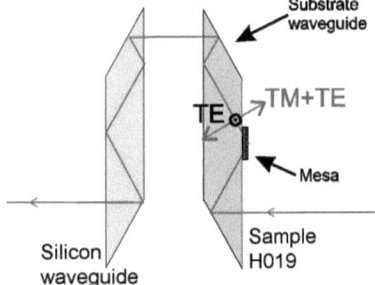

Figure 7.5: Double-periscope geometry for the voltage modulated waveguide transmission experiments. A waveguide composed of pure silicon compensates the offset in the optical path induced by the substrate waveguide of sample H019.

the polished facette located at its bottom edge (periscope geometry). In order to be able to detect the transmitted radiation by the MCT detector of the IFS 66, the optical path of the radiation exiting the sample was made collinear to that of the incoming beam by positioning a completely symmetric sample composed of undoped silicon back-to-back with H019. The second "periscope" therefore compensated the offset induced in the optical path by the first one. The double-periscope geometry of the setup is sketched in Fig. 7.5.

Using a pulse generator, an ac bias switching between +2 V and -2 V at a frequency of 50 kHz was applied to the top contact of the investigated sample mesa in respect to its bottom contact. The sample was illuminated by TM polarized radiation in the configuration described above. The radiation transmitted through the modulation-biased sample was measured by the MCT detector of the interferometer in lock-in-technique, where the trigger signal of the pulse generator was used as reference signal for the Stanford Research Systems SR830 DSP lock-in-amplifier. The output of the SR830 was fed into the IFS 66 spectrometer, which recorded the spectrum presented in Fig. 7.4 (a) in step-scan mode (see [20], pp. 55).

Figure 7.4 (a) shows the region of the transmission spectrum for TM polarized light, which is affected by modulating the bias applied to the sample. As discussed above, the amount of radiation absorbed by the HH1-HH2 transition in well 1 is expected to vary with the electric field along the structure due to a change in the number of charge carriers located in this well. Absorption induced by this transition is strong in TM polarization only (see section 7.2). Thus the peak in the modulated transmission spectrum in Fig. 7.4 (a) is associated with the HH2-HH1 transition and allows the precise characterization of its resonance energy for the fully processed structure, which is to be studied by time-resolved experiments.

7.3.2 FEL photocurrent spectra based on two-photon-absorption

In previous photocurrent experiments reported on in [16,20], where a globar lightsource was used for exciting confined charge carriers into the continuum, no photocurrent was observed for an exciting phonon energy of 160 meV. However, the FEL used for the time-resolved experiments constitutes a tunable radiation source of extremely high intensity. Even though a charge carrier initially confined in the HH1 ground state of well 1 in Fig. 7.2 cannot be excited into the continuum by the absorption of a single photon with an energy of about 160 meV, the continuum can be reached by the absorption of two photons. The FEL intensity is high enough to generate a measurable photocurrent by this two-photon process. As according to simulations (see Fig. 7.2) two photons of 160 meV each exhibit a total energy sufficiently high for exciting a HH1 carrier in well 1 into continuum states capable of carrier transport, the photocurrent is expected to be resonantly enhanced at photon energies equal to the HH1-HH2 transition energy. Two-photon photocurrent spectra thus constitute, next to voltage modulated transmission experiments, the second means of characterizing the HH2-HH1 transition energy in the fully processed sample H019.

The FEL photocurrent spectra were recorded in basically the same setup, in which the time-resolved measurements were performed. This setup is in detail described in chapter 7.5. For the spectrally resolved photocurrent measurements one of the two beams in Fig. 7.10 was blocked, the position of the movable mirror was fixed, and the FEL peak wavelength was swept at a constant micropulse energy, while recording the photocurrent through the sample for each FEL wavelength configuration. The spectra were measured for TM (TE pulse blocked) and TE (TM pulse blocked) polarized FEL radiation, and for two different voltages of +3 V and -3 V applied to the top contact of the sample mesa in respect to its bottom contact. In order to compensate for eventual temporal fluctuations in the FEL intensity, the intensity of the beam leaving the beam splitter B2 normally to the sample direction was measured using an MCT detector. The photocurrent spectra measured for the sample were then normalized to the simultaneously recorded beam intensity. However, this normalization did not change the basic shape of the observed spectral dependence of the photocurrent. The results of this experiment are presented in Fig. 7.4 (b) and (c).

As seen in this figure, for both applied voltages the photocurrent caused by a two-photon absorption is resonantly enhanced at the HH2-HH1 transition energy of 160 meV in case of TM polarized radiation. The resonance is not observed for the TE polarized FEL pulse, as expected by the selection rules discussed in section 7.2. The experimentally confirmed theoretical expectation, that TM polarized radiation in resonance with the HH1-HH2 transition energy in well 1 is able to induce a photocurrent by a two-photon-absorption, forms the very base of the time-resolved experiments presented in this work.

7.4 Principle of photocurrent pump-pump experiments

As transmission pump-probe experiments studying intersubband relaxation times in the SiGe system suffered heavily from a low signal-to-noise ratio [121] and a domination by hole heating effects [114], a different approach was followed in this work by employing the photocurrent instead of the transmission through the studied sample as a measurable indicator for the non-equilibrium occupation of the excited state. Among the advantages of photocurrent experiments is their expected high sensitivity in comparison to conventional pump-probe measurements. In the experience of the author, the characterization of intersubband transitions by photocurrent spectrometry reaches a sensitivity superior to transmission experiments, given a proper sample design, processing and a final state in the continuum. Sample H019 exhibits excellent photoconducting qualities, as reported previously in [16], even qualifying as a competitive detector in the mid-infrared. Further advantages of using photocurrent as a measurable quantity in pump-pump experiments are discussed later on in this section.

Figure 7.6 illustrates the principle of photocurrent pump-pump experiments, which is slightly more complex than that of transmission pump-probe measurements as presented in chapter 6.3.1. A full discussion of the flowchart presented in this figure is given later on during the discussion of the experimental results in chapter 7.6.4. For the illustration of the basic idea behind the photocurrent pump-pump method, the left branch of the diagramm is sufficient. In a transmission pump-probe experiment studying the HH2-HH1 relaxation time, only these two states are relevant, where HH2 would equal state 2 in Fig. 6.8, and HH1 would be represented by state 1. As a consequence, the transition involved in probing is exactly the same as the pumped one. In photocurrent pump-pump experiments, this is not the case.

Analogous to measurements in transmission, the first step in photocurrent pump-pump experiments is to establish a non-equilibrium occupation of the excited state of the studied transition by pumping with a resonant laser beam (pump 1) of matching polarization, as shown in the left branch of Fig. 7.6 (b). After the pump beam 1 has passed the system, the excited state relaxes into the ground state at a rate $\frac{1}{\tau_{life}}$, where τ_{life} is the relaxation time to be determined. This process is still in analogy with transmission experiments and is illustrated in Fig. 7.6 (c).

The basic requirement for photocurrent pump-pump experiments is that the excited state

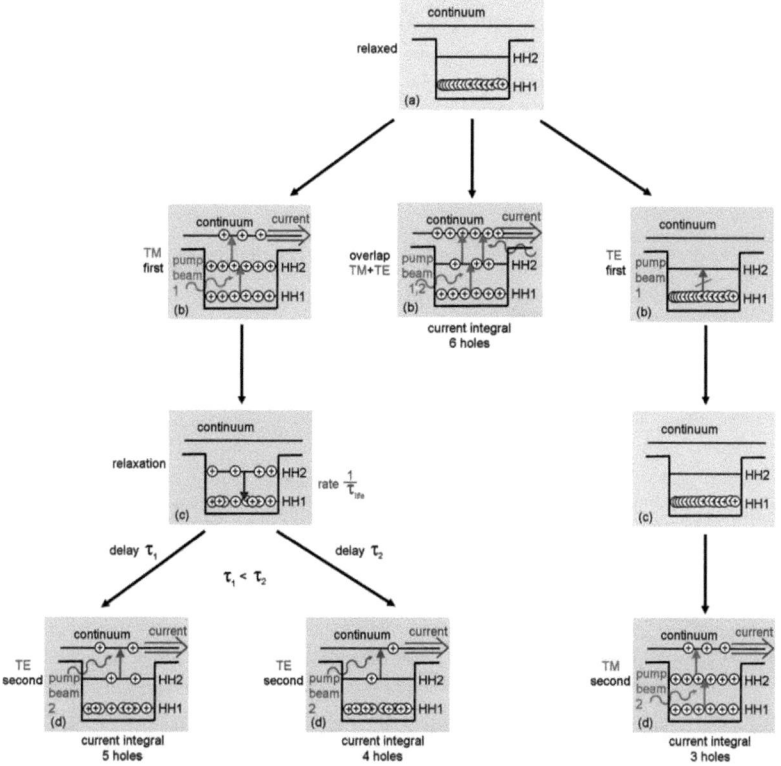

Figure 7.6: Basic scheme of photocurrent pump-pump experiments as employed in the course of this work. (a) shows the investigated system in equilibrium. (b) illustrates the excitation by the first of the two incident pulses (pump 1). In case of a pump 1 beam with matching polarization, a non-equilibrium occupation of the excited HH2 state is established, which decays with the HH2-HH1 relaxation time τ_{life}, as sketched in (c). In (d), the occupation of HH2 remaining after a delay time τ is probed into the continuum by the second beam (pump 2) and generates a photocurrent, which is measured as a function of the delay. The three main branches of the chartflow represent the individual temporal configurations of the TE and TM pulses. Details are given in the text of this section and in chapter 7.6.4.

of the investigated transition is well confined and occupying charge carriers do not contribute to a photocurrent. In other words, the basic experimental condition is that for a relaxed system as shown in Fig. 7.6 (a) two photons of the pump-beam are required to generate a photocurrent, as the absorption of only one photon can merely reach another bound state. This non-linear dependence of the measured quantity on the pump intensity is the fundamental basis of any pump-probe experiment, no matter what the exact nature of the monitored quantity is.

Degenerate photocurrent pump-pump experiments, that means experiments with pump 1 and pump 2 beams of equivalent wavelength, require the presence of continuum states at an energetic distance from the excited state, which is equal to the resonance energy of the investigated transition. In practise, this demands a continuum onset, which is separated from the excited state by the resonance energy of the studied transition or less.

The second beam, labelled pump beam 2, reaches the sample after a variable delay, and probes the *occupation* of the excited HH2 state *into the continuum*, that means it excites charge carriers left in the HH2 states after the particular delay time into continuum states, where they are capable of generating a photocurrent. The sooner the second pulse follows the first one, the more charge carriers are still left in the excited HH2 state, and the higher the photocurrent generated by the second pulse is. This is the case regardless whether selection rules allow the second pulse to also generate a photocurrent by two-photon absorption from the ground state or not, as a charge carrier in the HH2 state will always contribute stronger to the generation of photocurrent than one in the HH1 state. In the first case, an interaction with one phonon only is required, while in the second case a two-photon-process has to take place. However, in the experiments at hand the polarization of the second pulse in the left branch was chosen in a way that prevents this pulse from exciting ground state carriers into the continuum. It thus exclusively probes the occupation of the HH2 state by generating a photocurrent *proportional* to the number of carriers remaining in this state.

In the left branch of Fig. 7.6, which presents the flowchart of the experimental processes actually monitoring the HH2 occupation, the pump beam 2 does not interact with the HH1 ground state. Therefore, any heating of holes leading to a $k_\parallel$-space redistribution of charge carriers in the ground state *does not* effect the probing process, which is exclusively dependent on the interaction of the second pulse with the HH2 state. This fact poses the second fundamental advantage of photocurrent pump-pump experiments over transmission pump-probe measurements, which intrinsically involve the ground state in the probing process. Heating of ground state holes was the effect crucially degrading the reliability of the experiments presented in [114]. The photocurrent experiments presented in this chapter therefore show a

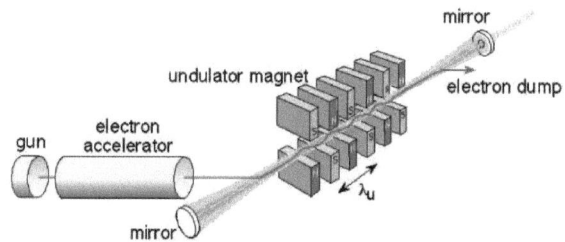

Figure 7.7: Schematic of a free electron laser. An electron beam, which is sped up to relativistic velocities in a linear accelerator, emits radiation by oscillating in a periodic magnetic field induced by undulator magnets. Details are given in the text. (Source: www.rijnhuizen.nl)

means of circumventing this detrimental difficulty by separating the transition involved in the probing process from the pumped one, and consequentially gaining independence of ground state effects.

In the course of the photocurrent pump-pump experiments, the *integral* photocurrent through the sample over a train of laser micropulse pairs is measured as a function of the delay between the two pump pulses. Trivially, it is important that the temporal spacing between subsequent pump-pump pulse pairs by far exceeds the scale of the delay time range of the experiment and thus the investigated relaxation time. The integral current through the sample induced by the uncorrelated pump beams for long delay times, which allow a complete relaxation of the pumped transition prior to the impact of the second pulse, gives a finite PC signal background. The HH2 population persistent at the time of incidence of the second pulse causes an additional integral photocurrent in case of short delays. This additional integral photocurrent decays exponentially with the delay by the HH2-HH1 intersubband relaxation time. Under ideal experimental conditions, the intersubband relaxation time thus can be directly extracted from the measurement data by fitting the decay of the integral excess photocurrent with the pulse delay by a single exponential function.

7.5 Experimental setup for pump-pump experiments

7.5.1 The free electron laser source

Favorable radiation sources for pump-probe experiments are free electron lasers, as they feature a wavelength tunability covering a huge spectral range from the mid-infrared to the terahertz, ultrashort pulses and high intensities.

Figure 7.8: The free electron lasers FELIX and FELICE at the FOM Institute for Plasma Physics. (Source: www.rijnhuizen.nl)

Free electron laser basics

FEL sources are based on the stimulated emission of radiation by a relativistic electron beam, which is forced on an oscillating path by a periodically changing magnetic field. The schematic sketch of an FEL setup is presented in Fig. 7.7. The employed electron beam is emitted by an electron gun and then increased in speed up to relativistic values in a linear accelerator. The interaction of this electron beam with radiation of the desired wavelength is achieved by a magnetic field periodically changing its orientation induced by so-called undulator magnets (or wiggler), placed at a period of λ_u. The magnetic field of the undulator is aligned normally to the path of the electrons.

As the electron beam travels through the periodic undulator field, it oscillates and spontaneously emits radiation in analogy to a dipole antenna. However, as the electron velocity is relativistically high, the wavelength of the emitted radiation is strongly Doppler blue shifted from the periodicity length of the undulator field. This allows the emission of radiation in the mid-infrared for undulator periods of e.g. 65 mm (FELIX). The mechanism for spontaneous emission is equivalent for synchrotron sources and FELs. However, in case of a randomly distributed phase of the electrons in a bunch, an equal fraction of electrons is slowed down by the emission of a photon as is sped up by its absorption, and no net energy transfer between the electron beam and the radiation field occurs. In order to establish an energy transfer to the radiation field, coherence between the electrons on the scale of the radiation wavelength is required.

The fundamental difference between a synchrotron light source and a FEL, which allows the latter to emit coherent radiation, lies in the formation of a so-called pondermotive wave [125] by the interference between the magnetic field components of the undulators and that of the emitted electromagnetic radiation of a frequency ω. The pondermotive wave is therefore basically a beat wave between the high-frequency electromagnetic wave and the low-frequency undulator wave, travelling at the velocity $v_p = \frac{\omega}{k+k_u}$, where k is the wavevector of the electromagnetic field, and k_u that of the undulator field. For FEL operation, the sufficiently low v_p is matched with the velocity of the electron beam, thus allowing a coherent propagation of the pondermotive wave and bunches of free electrons. This process is called bunching. For an FEL with a helical wiggler, the wavelength of the emitted radiation λ is given by Eqn. 7.1 [125], where B_u is the undulator amplitude and E_b the kinetic energy of an electron in the beam and m_e its rest mass.

$$\lambda = (1 + a_u^2)\frac{\lambda_u}{2\gamma_0^2}$$
$$a_w = \frac{eB_u}{m_e c^2 k_u^2}$$
$$\gamma_0 = 1 + \frac{E_b}{m_e c^2} \tag{7.1}$$

Now, the duration of the pulse emitted by a FEL is equal to the time span required for the electron beam to traverse the undulator. The number of oscillations in the electromagnetic wave package is equal to the number of oscillation undergone by an electron while passing the periodic magnetic field and thus to the number of undulator periods, N_u. A direct consequence of the short duration of the radiation pulse emitted by a FEL is its finite spectral width, where the two quantities are fundamentally related via the Fourier transform.

The basic parameters of the FEL setup, namely the velocity of the electron beam, the undulator period and its magnetic amplitude, can be used for selecting and tuning the FEL radiation wavelength and temporal pulse width. The wide range, over which those instrument parameters can be chosen, forms the base for the tremendous tunability of FELs. In contrast to gas and solid state laser sources, FELs are intrinsically free of limitations imposed by material properties. Both their wavelength tunability and ultrashort pulses make them an ideal tool for the study of intersubband transitions.

Felix

The time-resolved measurements forming the base of this work were carried out at the FOM (Stichting voor Fundamenteel Onderzoek der Materie) Institute for Plasma Physics Rijnhuizen, employing the FEL source FELIX, a photography of which is shown in Fig. 7.8. The wavelength of FELIX can be chosen between 4.5 μm and 35 μm using line FEL 2, and between 35 μm and 250 μm for FEL 1. Fast wavelength sweeps by a factor two or three can be performed by tuning the undulator field and therefore changing a_w in Eqn. 7.1. Such

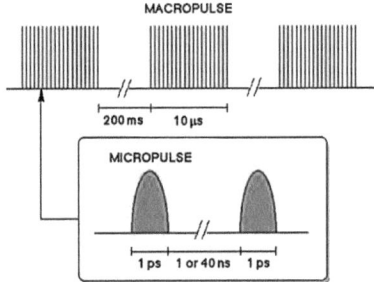

Figure 7.9: Pulse structure of FELIX. A macropulse is 10 μs long and consists of micropulses repeated at a rate of 25 MHz. The macropulse duty cycle was 10 Hz for the experiments of this work. (Source: www.rijnhuizen.nl)

wavelength scans were employed for gaining the data presented in chapter 7.3.2. The FELIX radiation is linearly polarized, where the beam enters the user station in horizontal polarization. The micropulse repetition rate used for the experiments of this work was 25 MHz, with a macropulse length of 10 μs and a macropulse repetition rate of 10 Hz. Figure 7.9 illustrates the temporal pulse structure of FELIX, where the value of 40 ns between two micropulses is valid for the employed repetition rate of 25 MHz. Micropulse energies of up to 50 μJ can be reached with FELIX. By adjusting the cavity length of the laser, the micropulse length of FELIX can be tuned down to 6 optical cycles, which equal 150 fs for a photon energy of 160 meV. FELIX thus offers a time resolution sufficiently high for experiments on intersubband relaxation times.

7.5.2 Pump-pump setup

In the course of this work, the intersubband relaxation of sample H019 was monitored by degenerate photocurrent pump-pump experiments in the setup configuration presented in Fig. 7.10. In order to use the radiation pulses generated by FELIX as both pump 1 and 2, the horizontally polarized incoming beam was split by a beam splitter B1. The adjustable temporal delay between the two resulting pulses was induced by routing the two beams along different optical paths. One beam passes mirror M2 and retro-reflector R2, while the other beam traverses a retro-reflector R1 mounted on a movable stage and mirror M1. The movable stage was used to define an adjustable difference in the length of the optical paths of the two beams on the scale of micrometers, leading to a temporal delay between the respective traversing times in the femtosecond range.

The measurements were carried out in cross polarization of the beams, which prevents interference of the two pump pulses. Even more important, the orthogonal polarization of pump 1 in respect to pump 2 enables the identification of relevant processes, even if they appear on a time scale close to the duration of the FEL pulse. This can be achieved by the thorough consideration of selection rules, as discussed in more detail in 7.6.4. Polarization rotators built up by three gold mirrors were employed to adjust the polarization of the pulses. The FELIX beam entered the optical table in horizontal polarization. Polarization rotator PR1 was adjusted in a way not affecting the polarization of the beam passing the movable rectifier, leaving it in horizontal polarization (TE when hitting the optically active region of the sample in waveguide geometry). PR1 was merely required to symmetrize the optical pathes of the two beams. Rotator PR2 was used to turn the polarization of the beam passing the fixed rectifier into vertical polarization (mixed TM+TE when hitting the optically active region of the sample in waveguide geometry).

Beam splitter B2 was employed to make the two beams collinear again. They were focused onto the location of the sample mesa under investigation by a parabolic mirror. In order to be able to monitor and record the intensity of the FEL, the radiation leaving the beam splitter B2 normally to the path leading to the sample was detected by an MCT detector. In this way, the FEL intensity could be recorded simultaneously with the sample signal, allowing to compensate the measured data for eventual fluctuations in the FEL power.

Note that the optical pathes guiding the FEL radiation to the sample are completely symmetric for the two beams in respect to the number of included transmissions and reflections, as seen in Fig. 7.10. The same holds true for the two pathes leading to the reference detector. This is important to ensure an equal intensity ratio between the two beams of approximately one at both the position of the sample and the reference detector.

The optical setup was placed in a metallic vacuum box, preventing the absorption of the FEL radiation by water. The optical path between the second beam splitter B2 and the sample cryostat was set up in a box purged by nitrogen, again to suppress water absorption. The FEL radiation was coupled into the substrate waveguide using the 30° facette cleaved into the sample. Thus the horizontally polarized beam hit the sample mesa in TE polarization, while the vertically polarized beam exhibited a TM:TE component ratio of 3:1, induced by the coupling geometry. However, as the investigated mesa was completely covered with a metallic contact, the TE component of the vertically polarized pulse was further suppressed at the position of the optically active structure. As a result, the vertically polarized pulse in Fig. 7.10 is in good approximation purely TM polarized at the location of the quantum well structure.

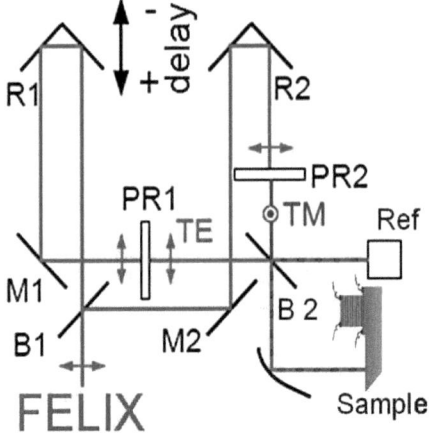

Figure 7.10: Pump-pump setup. The FEL beam enters the optical setup in horizontal (TE in respect to the sample) polarization and is splitted by a beam splitter (B1). The polarization of one of the beams is turned by 90° into vertical (TM+TE in respect to the sample) polarization (PR2). The beam remaining in horizontal polarization is reflected by a movable mirror (R1), allowing to adjust the delay between TE and TM pulse on a femtosecond scale. Before being coupled into the sample waveguide, the beams are made collinear again by the use of a second beam splitter (B2). The integral photocurrent through the variably biased sample originating from the delayed TE and TM micropulses is measured as a function of the delay in the electronic setup presented in Fig. 7.11.

The desired wavelength of the FEL radiation, its intensity and pulse duration were skillfully adjusted by the scientific staff of the FOM Rijnhuizen, Jonathan Phillips and Nguyen Q. Vinh. With a fixed output intensity of FELIX, the intensity actually hitting the sample could be reduced by using a set of attenuators, featuring 3 dB, 5 dB, and three times 10 dB. Due to the non-linear nature of the investigated processes, the careful choice of the radiation intensity was of high importance, as is discussed in section 7.6.1.

While the macropulse train of delayed pump 1 and 2 pulses exerts its impact on structure H019, the response of the sample to this incident radiation was determined. In order to be able to study the relaxation behavior of a biased structure, a variable voltage was applied

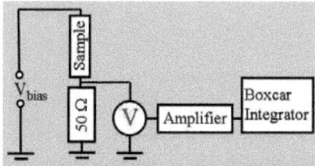

Figure 7.11: Electronic setup for the pump-pump experiments on sample H019. The photocurrent induced in the biased sample by the FEL radiation is measured as a voltage drop over a 50 Ω resistor via a voltage amplifier. Details are given in the text.

to the top contact of the sample in respect to ground. The bottom contact of the mesa was grounded via a 50 Ω resistor, as illustrated in Fig. 7.11. The incident FEL radiation pulses induced photocurrent pulses in the biased mesa. This current pulses were measured as a voltage drop over the 50 Ω resistor by a Princeton Applied Research S113 voltage amplifier connected in parallel to the resistor. As a resistance of 50 Ω is negligible in comparison to that of the sample, nearly the total applied bias drops over the sample, and the influence of the 50 Ω resistor on the field distribution is negligible.

The voltage response of the electrical circuit to the current pulse induced by the FEL radiation is expected to rise and decay with the time constant of the preamplifier's low-pass filter (1-10 μs), by far exceeding the micropulse separation of 40 ns (see Fig. 7.9). The change in the conductivity induced by the FEL radiation in the sample itself relaxes completely between two subsequent micropulses, as this relaxation occurs on the time scale of 10 ps, known from previous experiments reported in [16]. As further the external measurement circuit exhibits a linear system response, its response to the micropulse train that makes up a macropulse is proportional to the sum over N identically decaying exponentials excited at times $t_{0,...,N-1}$, at which the N micropulses hit the sample.

During the time-resolved experiment, the output of the voltage amplifier is integrated over a FEL macropulse by a box-car integrator. This box-car integrator electronically integrates the response of the measurement circuit to a FEL macropulse as given at the preamplifier output, and subtracts a dark-current offset obtained from the region of the electronic trace between two macropulses. Following linear signal theory, this integral is proportional to the integral over the initially induced change in the conductivity of the sample, independent of the detailed system response function, and, according to the discussion above, therefore proportional to the system response to a single micropulse. Furthermore, the change in conductivity of the sample is proportional to the number of charge carriers excited into the continuum, which is a general property of any QWIP operating in the linear regime.

To conclude these considerations, the integrated output of the voltage preamplifier is proportional to the total sum of charge carriers excited into the continuum of structure H019 by a single pair of pump 1 and 2 pulses. In order to study the time evolution of the quantum well system, the integral output of the preamplifier was recorded for a series of delays between the two FEL pump pulses. The results are presented in the following section.

After establishing the optical setup as sketched in Fig. 7.10, the position of the movable delay stage, for which the optical paths for the two pump beams are equivalent with an accuracy on a micrometer scale, had to be determined. For this purpose, a pinhole was placed at the position of the sample, and an MCT detector was set up directly behind the pinhole. Further, the polarization of both beams was turned into horizontal direction. Then the combined intensity of the beams was measured by the MCT detector while sweeping the position of the movable stage. The center peak of the resulting interferogram marked the position of zero-delay for the stage.

7.6 FEL experiments

7.6.1 Power dependence

Even though the experimental evidence strongly suggests, that the photocurrent peak in Fig. 7.4 stems from a two-photon absorption from the HH1 ground state into the continuum via the intermediate HH2 state, a final confirmation was sought by studying the dependence of the photocurrent peak around 160 meV on the intensity of the incoming FEL radiation.

For this purpose, the TE pulse of the pump-pump setup (see Fig. 7.10) was blocked, and the sample was illuminated exclusively by the beam in TM polarization. The FEL wavelength was tuned into resonance with the HH1-HH2 transition in the deepest well 1 of structure H019, equivalent to 7.9 μm according to the data presented in section 7.3. The FEL macropulse energy of the TM pulse at the position of the sample was about 200 μJ without any attunuation, equivalent to a micropulse energy of approximately 1 μJ. The top contact of the investigated mesa was biased at +3 V in respect to ground, leading to a confinement of the charge carriers in the HH1 state of well 1 according to [16]. The box-car output, which is proportional to the integral photocurrent induced by the TM pulse in the structure, was recorded for a series of different attenuations of the FEL beam. The obtained data are presented in Fig. 7.12.

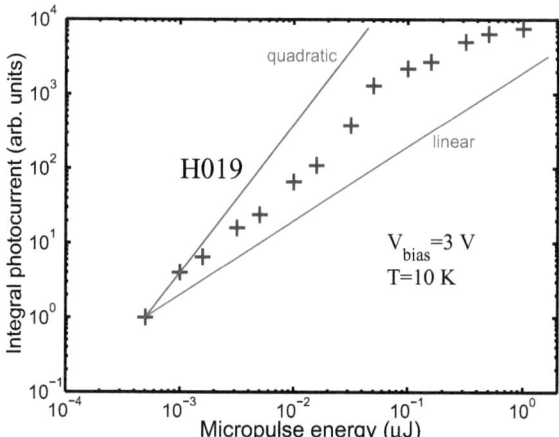

Figure 7.12: Dependence of the integral photocurrent through the sample on the micropulse energy of an FEL pulse in TM polarization. Sample H019 was biased at 3 V. The red lines indicate the slopes associated with a linear and a quadratic dependence on the micropulse energy. For low micropulse intensities, the photocurrent clearly exhibits a superlinear behavior, indicating its dependence on a two-photon-process. At high micropulse energies, the characteristics saturate due to bleaching effects of the HH1-HH2 transition.

This figure shows a double logarithmic plot of the integral photocurrent J_{int} generated by a TM micropulse in dependence on the micropulse energy L. For micropulse energies below 30 nJ, the slope of the graph, s, is clearly above 1, indicating a superlinear dependence of the integral current on the micropulse energy, as expressed by Eqn. 7.2.

$$J_{int} = \int j(t)dt \sim L^s$$
$$s > 1 \quad \text{for} \quad L < 30\text{nJ} \tag{7.2}$$

Naively, for a photocurrent originating from a two-photon absorption process a quadratic dependence on the intensity of the incoming radiation is expected. Now, even though the graph in Fig. 7.12 shows superlinear characteristics, it features a slope of less than two. The reason for this is found in the fact that the process for the generation of a photocurrent involves three quantum mechanical states and two optical transitions between them, both of which can be bleached. The precise slope in Fig. 7.12 therefore is determined by the population dynamics of the three states, and the detailed interpretation of the presented data is non-trivial and requires numerical simulations, which are given in chapter 7.7.3 of this work.

However, the superlinear dependence of the integral photocurrent through sample H019 on the intensity of the incoming radiation as shown by Fig. 7.12 leads to the definite conclusion, that the photocurrent is generated by a two-photon process. As further concluded from the data in this figure, the intensity dependence of the photocurrent saturates for micropulse energies above 30 nJ, which can be directly associated with a bleaching process in the quantum level system under investigation. Now, as for TM polarization the dipole matrix element between the HH1 and HH2 states by far exceeds that between the HH2 state and resonant continuum states, as seen from Fig. 7.3, and in addition continuum states decay rapidly, the observed bleaching is attributed to the HH1-HH2 transition. On first sight, this bleaching of the HH1-HH2 transition is favorable for the time-resolved experiments, as it ensures a maximal occupation of the HH2 state, which is then probed into the continuum by the second pulse. However, a more thorough consideration in the following paragraphs leads to the opposite conclusion that employing pump-pulses of a micropulse energy sufficiently high for bleaching is rather unfavorable for photocurrent pump-pump experiments, and that low pulse energies are preferable.

7.6.2 Role of the FEL intensity for the experimental time resolution

For time-resolved experiments on transitions with relaxation times close to the FEL pulse duration, the appropriate and educated choice of FEL parameters is of fundamental importance, especially since beam time is usually strictly limited. As seen in reference [114], the choice of high pump intensity can lead to carrier heating and associated effects, which prevent a direct measurement of the investigated relaxation times and degrade the reliability of the extracted lifetimes. On the other hand, insufficiently high laser intensities lead to a population change in the studied system too insignificant to result in a effect measurable at sufficient accuracy, as seen in reference [121].

Photocurrent pump-pump experiments are expected to be far less sensitive to carrier heating than transmission pump-probe measurements (see section 7.4). Nevertheless, the correct choice of the intensity of the employed FEL beam is of fundamental importance for the successful determination of ultrafast intersubband relaxation times in photocurrent experiments. This statement is based on the considerations given in this section and was confirmed by the experience collected in the course of the experimental work.

From theory (see reference [114] and chapter 6.2.5), the HH2-HH1 relaxation times are expected to be close to the minimum pulse length feasible for FELIX (6 optical cycles, equivalent to 160 fs at 7.9 μm). Therefore, the most cumbersome problem during the experiments carried out for this work was actually distinguishing between the exponential signal decay stemming from the intersubband relaxation and the additional photocurrent originating from the temporal overlap of the two pump pulses. The latter is simply a consequence of the non-linear dependence of the photocurrent through the sample on the radiation intensity for a wavelength of 7.9 μm, as discussed in chapters 7.3.2 and 7.6.1. If the two pump pulses overlap at low delay times (as illustrated by the central branch in Fig. 7.6), the integral photocurrent is higher than the sum of the two current pulses produced by the uncorrelated beams. The following considerations illustrate, how the time resolution of the pump-pump photocurrent measurements depend on the FEL intensity in absence of any heating effects.

In an ideal pump-pump setup, the pump pulse features a temporal width vanishing on the scale of the experiment, and the non-equilibrium occupation of the excited level would be built up instantly on the arrival of the pulse. In such an experiment, *any* finite decay characteristics in the obtained data could be associated with an actual relaxation process within the sample. However, as the FEL pulse employed for this work exhibited a non-negligible length on the scale of the relaxation times to investigate, the non-equilibrium HH2 occupation took a finite time to build up. Under such experimental conditions it occurs, that even though the occupation peak has been reached and is decaying, the flank of the pump 1 pulse still excites additional carriers and therefore broadens the signal characteristics. Thus, if the investigated system itself relaxes instantly on the time scale of the laser pulse, the measured signal exhibits a finite width, with exclusively originates from the temporal width of the involved pulse.

For the following considerations a two-level system shall be assumed, with a ground state A and an excited state B. It shall be assumed, that the B-A relaxation occurs on a time scale negligible on the scale of the FEL pulse width. This means, that the two-level system reaches equilibrium for a given pump intensity $I(T)$ instantly on the scale of the FEL pulse width. The temporal shape of the pulse intensity is assumed gaussian and follows Eqn. 7.3.

$$I(t) = I_0\, e^{-\frac{1}{2}(\frac{t}{\sigma_t})^2} \quad (7.3)$$

For these considerations, a realistic value of 200 fs is assumed for σ_t.

As can be derived from the density matrix equations of a two-level system in rotating wave approximation, the bleaching of the A-B transition under resonant excitation is described by Eqn. 7.4 [127].

$$\begin{aligned}\frac{n_{occ}}{N} &= \frac{1}{2} \cdot \frac{\frac{I}{I_{sat}}}{1 + \frac{I}{I_{sat}}} \\ I_{sat} &= \frac{c\, \Gamma\, \sqrt{\epsilon\epsilon_0}}{2\, \mu^2\, \tau_{life}}\end{aligned} \quad (7.4)$$

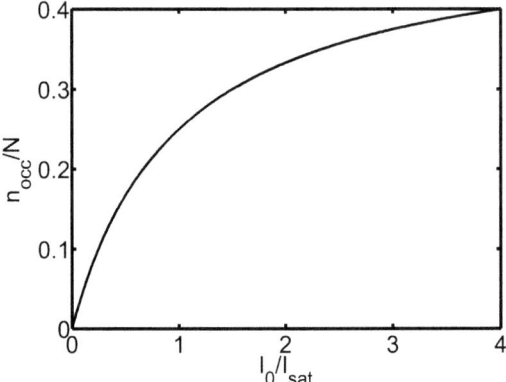

Figure 7.13: Saturation characteristics of the excited state occupation of a two level system under resonant excitation, as given by Eqn. 7.4.

In this equation, n_{occ} is the occupation of the excited level B of the system, N the total number of charge carriers, and I_{sat} the intensity, at which a relative excited state occupation of 0.25 is reached. Γ represents the phase relaxation rate associated with the two states, and μ the respective dipole matrix element. τ_{life} is the non-radiative carrier relaxation time between states B and A. Figure 7.13 presents a plot of the relative occupation characteristics of the excited state as given by Eqn. 7.4. Now, in case of a pump 1 pulse of low intensity hitting the studied system at $t = 0$, the excited state occupation is linearly dependent on the incoming intensity, and the temporal evolution of n_{occ} mimics the pulse shape. This is illustrated by the yellow curve in Fig. 7.14. This figure shows half of the occupation pulse originating from a pump pulse centered at $t = 0$ for a series of pulse intensities. The response characteristics are symmetric for negative times.

The FWHM of the temporal evolution of the occupation is in case of low intensities far from bleaching conditions equivalent to that of the FEL pulse. However, as the peak intensity I_0 rises, the high intensity components of the pulse contribute sub-linearly to the change in the occupation, and the shape of n_{occ} flattens. The blue line crossing the occupation characteristics in Fig. 7.14 marks their FWHM position and clearly indicates, that the FWHM rises with increasing pump pulse intensity.

In the experimental configuration employed for this work, the non-equilibrium occupation of the excited state is probed into the valence band continuum by the pump 2 pulse (see

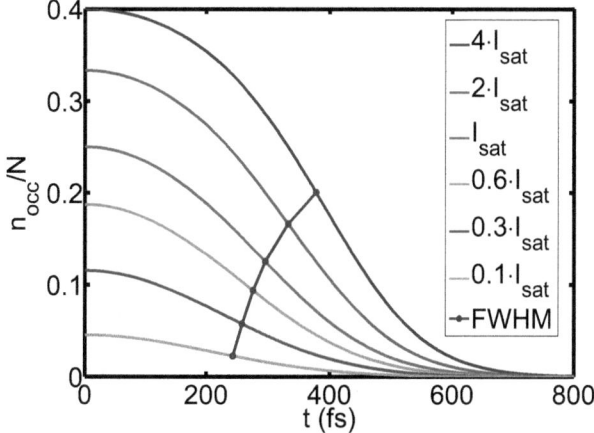

Figure 7.14: Broadening of the pulse-limited width of the temporal occupation characteristics for level B. As the intensity of the pump pulse increases, the FWHM of the pulse limited evolution of n_{occ} rises. Therefore, for the pump-pump experiments presented in this work, the FEL pulse energy has to be chosen carefully in order to achieve the required time resolution.

chapter 7.4). Now, in an experimental setup, in which the expected value for the relaxation time under investigation is close to the pulse duration, it is essential to keep the pulse limited temporal broadening of the excited state occupation as low as possible. In case of a pulse broadening of the measured signal and a relaxation time of similar values, a clear distinguishing between the quantity to measure and the laser pulse properties is not possible.

Thus, for the experiments at hand, the pulse intensity was extremely carefully chosen, where the data gained by the power dependent measurements in chapter 7.6.1 proved essential. Moreover, a clear distinction between effects caused by the time resolution of the experimental setup and the characteristics relevant for the extraction of the relaxation times was made possible by the thorough consideration of selection rules. A detailed discussion of these subtle experimental procedures is given in chapter 7.6.4.

7.6.3 Photocurrent pump-pump experiments: Results

The time-resolved experiments aiming at the determination of the HH2-HH1 relaxation time of structure H019 were performed in the setup described in chapter 7.5, where the FEL was

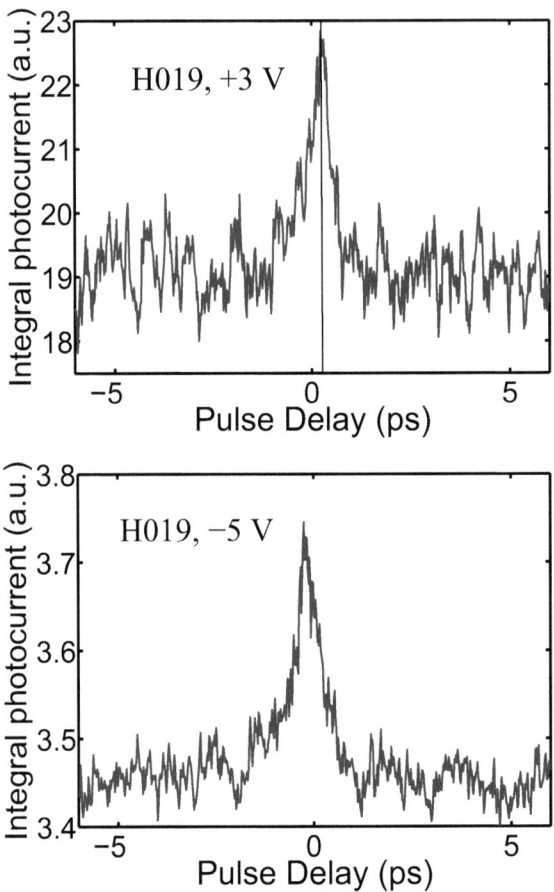

Figure 7.15: Photocurrent pump-pump results for H019 at a bias of +3 V and -5 V, respectively. For negative delays, the TM pulse hits the sample first, for positive delays the TE pulse is first. The integral photocurrent through the sample features a peak at zero-delay between the two FEL micropulses. This excess photocurrent exhibits a clear asymmetry in respect to the sign of the delay, where the decay with increasing delay is slower for negative delays (TM first).

adjusted to a peak wavelength of 7.9 µm, and its temporal micropulse length was kept as short as possible, giving durations of about 200 fs. The measurements were performed for two different biases applied to the sample contacts, one negative and one positive, in order to determine wether the voltage-tunability of the occupation of the different quantum wells 1 to 5 in structure H019 (see reference [16]), affects the monitored relaxation time. Based on the considerations in chapter 7.6.2 and the data obtained for the power dependence of the photocurrent through the sample as presented in chapter 7.6.1, the micropulse energy for the time-resolved measurements was chosen carefully. The precise value for this energy was determined using a pyrometer.

For the experiments carried out under a bias of +3 V to the mesa top contact, the TM pulse featured a macropulse energy of 120 µJ (micropulse 480 nJ), and the TE pulse of 160 µJ (micropulse 640 nJ). Both beams were attenuated by 20 dB, giving a micropulse energy of 4.8 nJ and 6.4 nJ for the TM and TE pulse, respectively. As seen in Fig. 7.12, these micropulse energy values lie well within the energy range of superlinear behavior. In this experimental configuration, the transitions in the quantum mechanical level system under investigation are thus far from bleaching, and the FEL pulse induced broadening of the excess photocurrent is expected to be minimal, as discussed in chapter 7.6.2.

For the experiments carried out under a bias of -5 V, the TM pulse exhibited a macropulse energy of 160 µJ (micropulse 640 nJ), and the TE pulse of 130 µJ (micropulse 520 nJ). Now, for the experiments carried out under an applied bias of -5 V, a beam attenuation of 5 dB could be used, giving a micropulse energy of 200 nJ and 160 nJ for the TM and TE pulse, respectively. These micropulse intensities would clearly result in bleaching effects under an applied bias of +3 V according to Fig. 7.12.

The reason for the possibility to employ these high micropulse intensities for a bias of -5 V is attributed to the difference in band structure between the two bias configurations. As observed during the photocurrent experiments published in [16], at an applied bias of +3 V the vast majority of holes is located in the deep well 1, and the average matrix element between the occupied ground states and the HH2 state in well 1 is high. Therefore, bleaching of the transition between the occupied ground states and the HH2 state can be reached at relatively low FEL intensities, as seen in Fig. 7.12. As the bias is changed to -5 V, charge carriers are redistributed and occupy the lowest HH states of all wells 1-5. Thus, the average dipole matrix element between the occupied ground states and the HH2 state in well 1 reduces, and higher micropulse energies are required for bleaching the transition. As a consequence, at a bias of -5 V higher micropulse energies can be employed without risking to enter the regime of saturation broadening of the pulse-limited excess photocurrent.

The photocurrent pump-pump experiments were carried out by sweeping the position of the movable rectifier mirror R1 (see Fig. 7.10) relative to its zero-delay point over a total length of 3.6 mm, thus covering delays between the TM and TE micropulses from -6 to +6 ps. For each mirror position, the integral photocurrent through the sample was recorded as described in chapter 7.5, where the average over three box-car values was taken. The mirror step width was 3 μm (20 fs). Simultaneously to the integral photocurrent through the sample, the output of a nitrogen-cooled MCT detector was recorded. As the response of the MCT detector is linearly dependent on the FEL pulse intensity, its output delivers a value for the total macropulse energy for each recorded photocurrent data point. This information was used to compensate for eventual temporal fluctuations in the FEL intensity. For this purpose, the obtained photocurrent pump-pump data was divided by the simultaneously recorded MCT data. In order to improve the signal-to-noise ratio of the measurement, the delay sweep was performed ten times, where each run took approximately five minutes.

The average over the integral photocurrent data recorded during these ten delay sweeps is displayed in Fig. 7.15 for both bias configurations. For large delays, values of about 19 (+3 V) and 3.45 (-5 V) are observed for the integral photocurrent baseline, originating from the excitation of ground state carriers into the valence band continuum by the uncorrelated pump 1 and 2 beams. For short delays between the two pulses, a clear excess photocurrent based on the non-linearity of the structure's photocurrent response to resonant radiation is observed. This excess photocurrent spans ten percent of the total photo-generated integral current. Furthermore, a clear asymmetry of this excess current in respect to the sign of the delay between the two pulses is observed, where negative delays are associated with the situation of the TM pulse reaching the sample first, and positive delays represent the case of a first impact by the TE pulse. The asymmetry is a direct result of the cross polarization between the two beams in combination with the selection rules of the transitions in the structure, and is a crucial indication for the relevance of the results extracted from the experimental data, as discussed in the following section.

7.6.4 Photocurrent pump-pump experiments: Interpretation

The interpretation of the experimentally obtained data given in the previous section is illustrated by the flowchart in Fig. 7.6. Selection rules of the transitions involved in the pump-pump experiment as presented in chapter 7.2 play an essential part in the interpretation of the experimentally observed behavior. The following discussions exclusively deal with the HH1-HH2 transition in the deep quantum well 1 of structure H019, as the FEL radiation is in resonance with the HH1-HH2 transition of this well only. The schematics in Fig. 7.6

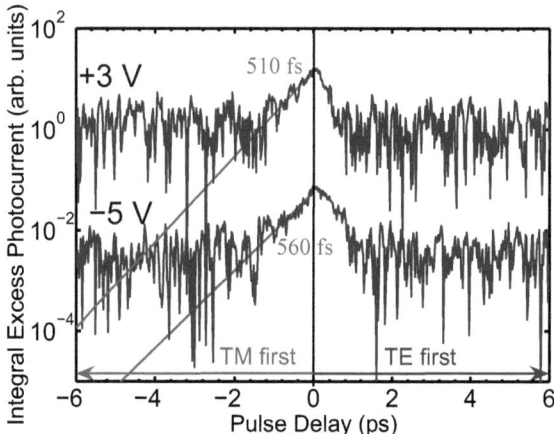

Figure 7.16: Logarithmic plot of the integral excess photocurrent through sample H019 as a function of the delay between the two FEL pulses. The blue lines show the data given in Fig. 7.15 after the subtraction of the integral current baseline and a shift of the zero-delay position to the peak value of the curve. A clear asymmetry of the data in respect to the sign of the delay is observed. The red lines indicate the fitted single-exponential function, as given in Eqn. 7.5, which reproduces the experimental data well over one order of magnitude. Thus, the extracted HH2-HH1 intersubband relaxation times of 510 fs and 560 fs are highly reliable. Note that no voltage dependence of the relaxation time could be observed. The curves have been offset vertically for clarity.

thus illustrate processes localized in well 1 of the structure.

The integral current baseline in Fig. 7.15 corresponds to the current generated by two uncorrelated pump pulses, at long delays. The current answer given by a system with a linear response to the FEL radiation would be exclusively reduced to this baseline. The deviation of the experimentally observed characteristics from the current baseline is a clear indication for non-linear processes. The information of relevance for the extraction of an intersubband relaxation time is the dependence of the excess current on the delay between the two pulses. Figure 7.16 presents the data shown in Fig. 7.15, where the current baseline has been subtracted, and the curves have been normalized to the maximum integral current values. Further, the delay time associated with the maximum excess photocurrent in Fig. 7.15 has been re-defined as zero-delay, the peak maxima have thus been shifted to $\tau = 0$. The characteristics for the two different biases, for which the experiment was carried out, have been vertically offset for clarity.

Negative delay: TM first

For negative delay values, the two excess photocurrent graphs presented in Fig. 7.16 reach a maximum for small delay times, and clearly decay exponentially with increasing delay (linear slope in the logarithmic plot) down to the noise level of the measured quantity. The origin of the excess current and its exponential decay can be illustrated by the flowchart diagram in Fig. 7.6, where the left branch represents negative delays. Prior to the impact of pump pulse 1, charge carriers in the system exclusively occupy the HH1 ground state at low temperatures (Fig. 7.6 (a)).

Now, the pump 1 beam is employed to excite a non-equilibrium occupation of the HH2 state, whose relaxation behavior shall then be monitored. According to the theoretical and experimental data presented in Figs. 7.3 and 7.4, the HH1-HH2 transition is strong in TM polarization. Thus, as the first (TM) pulse hits the sample, charge carriers are excited into the HH2 state and, subsequently, into the continuum (Fig. 7.6 (b)). As a simple example, which shall illustrate the principle of the actually observed processes, the integral current directly generated by the pump 1 beam via two-photon-absorption (shown as a blue arrow in Fig. 7.6 (b)) is assumed to amount 3 holes.

The pump 1 beam establishes an occupation of the HH2 state along with a significant current through the sample. The HH2 occupation now decays with the HH2-HH1 relaxation time τ_{life}, which is the figure to extract from the performed experiments (Fig. 7.6 (c)).

The pump 2 beam is used to monitor the decay of the HH2 occupation. It reaches the sample after a certain delay time τ, which is determined by the position of the movable rectifier mirror in Fig. 7.10. In the case discussed presently, the pump 2 pulse is TE polarized, which according to the selection rules is forbidden to induce transitions between the HH1 and HH2 states. It therefore *does not* excite additional carriers from the ground state into the HH2 state. However, it can excite the holes still present in the confined HH2 state into the continuum, where they contribute to a current (Fig. 7.6 (d)). The pump 2 beam in TE polarization is thus capable of directly probing the occupation of the HH2 state into the continuum, completely independent of the distribution of charge carriers in the ground state.

As the valence band continuum can be reached from the HH2 state by the absorption of one FEL photon, the amount of current generated by pump 2 is proportional to the numbers of charge carriers remaining in the HH2 state at the time of its impact. For small delay times

τ_1 more holes are left in the excited state than for large delays τ_2 (see Fig. 7.6 (d)). Pump 2 directly probes the occupation of the excited state, and in the simple example in this figure generates an integral photocurrent of two and one holes at a delay of τ_1 and τ_2, respectively. The total integral photocurrent generated by both beams thus amounts 5 holes for a delay of τ_1 and 4 holes for τ_2. For delay times long enough to allow a total relaxation of the system until the impact of pump 2, the total integral photocurrent in this example is equal to 3 holes.

Thus, as the non-equilibrium occupation of the HH2 state established by the pump 1 beam decays exponentially with the HH2-HH1 intersubband relaxation time, so does the photocurrent linearly generated by the pump 2 beam, and with it the integral photocurrent measured in the experiment. The exponential decay of the measured integral excess photocurrent for negative delay values in Fig. 7.16 can therefore directly be associated with the HH2-HH1 intersubband relaxation.

Positive delay: TE first

For positive delay, the asymmetric characteristics in Fig. 7.16 as well reach their maximum value for small delays. In comparison to negative delays, the integral excess photocurrent decays much more rapidly with increasing positive delay times. The reason for this is once again illustrated by the schematics in Fig. 7.6, where the right branch sketches the case of positive delays.

For positive delays, the first pulse to hit the sample is TE polarized. According to the selection rules, the transition from HH1 to HH2 is forbidden in this polarization. Therefore, the pump 1 beam does not induce any non-equilibrium population of the HH2 level (b), and no photocurrent is generated. The TM polarized pump 2 beam reaching the sample after a certain delay time finds the system in its relaxed state, and excites charge carriers in the same way as in the case of negative delays (Fig. 7.6 (d)). Therefore, in the simple example picture given in Fig. 7.6, the integral photocurrent for the TE-first-configuration is equal to the uncorrelated value of 3 holes.

Put differently, for positive delay values the pump 1 beam is not expected to excite the quantum level system under investigation into any non-equilibrium state. Nevertheless, the experimental data does show a clear, quickly decaying excess current for positive delays. This excess current does not stem from a decaying occupation of the excited HH2 state, but is a direct result of the finite duration of the FEL pulses. It is generated by the simultaneous presence of both pump pulses, a situation which is illustrated by the central branch in Fig. 7.6. Even though the peak intensity of the TE polarized pump 1 pulse hits the sample before the maximum of the TM polarized pump 2 pulse arrives, for very short delays between the pulses the flank of the temporal intensity distribution of pump 1 coincides with fractions of pump 2. Within a delay span dependent on the temporal width of the two pulses, which determines the amount of overlap for a certain delay, pump 1 is thus able to excite carriers brought into the HH2 state by pump 2.

The width of the integral excess photocurrent in Fig. 7.16 for positive delays thus gives a measure for the delay time dependence of the overlap between the two pulses. In a quantum level system with a non-linear photocurrent response to FEL radiation and relaxation times far *below* the FEL pulse width, photocurrent pump-pump curves are exclusively determined by this pulse overlap characteristics. The photocurrent pump-pump characteristics of a system with experimentally unaccessible relaxation times would thus be symmetric in respect to the sign of the delay time, and its width would be determined by the FEL pulse duration, as discussed in chapter 7.6.2.

To sum up, the temporal width of the integral excess photocurrent for positive delays in Fig. 7.16 indicates the fundamental time resolution of the experimental setup employed for the pump-pump experiments, which is determined by the FEL pulse duration. The clear asymmetry of the observed characteristics in this figure with its longer decay for negative delay values is a definite indication for the observance of an intersubband relaxation process, whose relaxation time is clearly above the time resolution of the experiment.

Relaxation time extraction

As discussed above, by a thorough consideration and exploitation of the selection rules of the quantum mechanical system under investigation and a careful choice of FEL pulse intensities, a clear distinguishing of the monitored HH2-HH1 relaxation process in structure H019 from the pulse-width-limited pump-pump characteristics was made possible, even though the studied relaxation times were close to the time resolution of the experimental setup. As already stated, in case of photocurrent pump-pump characteristics determined by the relaxation of the excess photocurrent originating from non-equilibrium HH2 carriers, the measured integral photocurrent decays exponentially with the HH2-HH1 relaxation time. The intersubband lifetime of the HH2 state can thus be extracted from the data for negative delays in Fig. 7.15 by a simple single-exponential fit. The used fit function is given by Eqn. 7.5.

$$\frac{I(\tau)}{I_{max}} = [c_1 \, e^{-\frac{|\tau_0 + \tau_m - \tau|}{\tau_{life}}} + c_2] \cdot \Theta(-\tau + \tau_0 + \tau_m) + \Theta(\tau - \tau_0 - \tau_m) \quad (7.5)$$

In this equation, $I(\tau)$ is the dependence of the integral excess current on the delay time in Fig. 7.15, I_{max} its maximum value and τ_m is the delay time position of the maximum. c_1, c_2,

τ_{life} and τ_0 are the parameters of the fit, where τ_{life} is the HH2-HH1 relaxation time under investigation, τ_0 allows a small offset between the precise zero-delay and the excess current maximum, c_2 represents the current baseline, c_1 the integral excess current amplitude, and Θ denotes the Heaviside step function. The HH2-HH1 relaxation time is numerically extracted from the experimental data shown in Fig. 7.15 by performing a least-squares-fit on the data points of delay values below τ_m. The fit results for the data sets obtained at the two voltages are presented in table 7.4.

The fit parameter values for c_1 and c_2 trivially reflect the contribution of the integral excess photocurrent to the total signal, while the figures gained for the zero-delay-shift τ_m are negligibly small on the scale of the extracted relaxation times.

The relaxation times of the integral excess photocurrent of sample H019, which are equivalent to the HH2-HH1 intersubband relaxation times, are 510 ± 60 fs and 560 ± 18 fs for applied biases of +3 V and -5 V, respectively. In Fig. 7.15, the fit functions as given by Eqn. 7.5 are plotted as red lines for the optimal parameter set given in table 7.4. As can be seen, the exponential functions decaying with the two relaxation times of 510 fs and 560 fs fit the experimentally obtained data well over one order of magnitude down to the noise level of the measurement.

Discussion of the extracted values and their significance

In order to quantify the asymmetry of the experimental curves presented in Fig. 7.15, and thus the observance that the monitored relaxation process occurs on a time scale well above the time resolution of the experimental setup, a fit analogous to that for extracting the HH2-HH1 relaxation time was performed for the data obtained at positive delays. The resulting decay times are 277 fs and 400 fs for an applied voltage of +3 V and -5 V, respectively. These times are equivalent to the time resolution of the pump-pump setup employed in the course of this work. If the investigated intersubband relaxation time would be too short to be resolved, the width of the excess photocurrent would be pulse limited, and its decay would imitate a relaxation time of 277 fs and 400 fs. However, the asymmetry in the acquired characteristics and the fitted long relaxation times for the data of negative delay clearly indicate,

Table 7.4: Optimal fit parameters determined for the data in Fig. 7.15:

Bias	τ_{life}	c_1	c_2	τ_0
+3 V	**510 fs**	0.15	0.84	1.5 fs
-5 V	**561 fs**	0.06	0.92	2 fs

that the intersubband relaxation time can be experimentally resolved by the photocurrent pump-pump experiments.

As Fig. 7.15 shows, no hole heating contributions to the relaxation characteristics were observed in the presented measurements. The intersubband relaxation times could be *directly* extracted from the experimental data by a simple single-exponential fit. The photocurrent pump-pump experiments carried out allowed the *direct* monitoring of the decaying HH2 occupation by probing exactly this occupation into the valence band continuum. The measurement was not influenced by the charge carrier distribution in the HH1 ground state, since it was not involved in the probing process, in contrast to conventional transmission pump-probe experiments. Therefore, as the relaxation characteristics of the measured quantity were exclusively determined by the HH2-HH1 intersubband relaxation process, no de-convolution of the effects of any other processes was necessary during data evaluation. This increases the reliability of the obtained lifetimes significantly over those given in [114], where hole heating effects dominantly influenced the relaxation characteristics and had to be de-convoluted numerically before extracting the reported relaxation time values. The integral excess current characteristics in Fig. 7.16 exhibit a good signal-to-noise ratio, allowing the extraction of the intersubband relaxation times with an unmatched accuracy (for comparison, see e.g. [121]).

Strictly speaking, the photocurrent pump-pump experiments of this work monitor the total depopulation of the HH2 state rather than the relaxation rate between the HH2 and HH1 state. For the LO phonon depopulation of the HH2 state, two channels are available, one into the LH1 level and one into the HH1 state. According to reference [114], the LH1-HH1 relaxation is with a predicted relaxation time of 170 fs theoretically expected to be nearly two times faster than the HH2-LH1 transition with 319 fs. Thus, any charge carrier relaxing from the HH2 state into the LH1 state is expected to be immediately transferred into the HH1 ground state due to the extremely efficient LH1-HH1 relaxation, and no significant LH1 population builds up. As a consequence, the total HH2 depopulation rate is in the system under investigation equivalent to the total HH2-HH1 relaxation rate via two channels, a direct one between HH2 and HH1, and one via the LH1 intermediate state.

The intersubband lifetime values extracted from the pump-pump data of about 550 fs are well within the theoretically predicted range, as discussed in chapter 6.2.5. However, they are a factor of two above the value of 250 ± 100 fs reported by Kaindl et al. The higher intersubband relaxation time values acquired in the course of this thesis, which are of higher reliability than those reported in [114], are in favor of the development of a SiGe QCL.

To summarize, the lifetime figures obtained in this work are the most accurate and reliable experimental values for intersubband relaxation times in the SiGe system for transition energies above the LO phonon energy, and the first *directly* determined values in this transition energy range. As discussed in chapter 6, a thorough knowledge of these values is essential for the conception and design of SiGe quantum cascade emitters on the road to a SiGe QCL, particularly in respect to the difficulties arising from the non-resonant behavior of intersubband scattering by LO phonons in this material system (see chapter 6.2.6). Furthermore, the determination of intersubband relaxation times in structure H019 was carried out on an electrically biased sample. The experimental conditions present during the measurements thus resemble the actual working conditions in an operating QCL to a higher degree than any previous time-resolved studies of intersubband relaxation times in SiGe heterostructures. Their determination under biasing conditions further increases the reliability of the HH2-HH1 relaxation times for the use in the design of quantum cascade emitter devices.

Voltage dependence of the extracted lifetimes

The time-resolved experiments on structure H019 were performed for two different biases applied to the top contact of the investigated mesa in respect to its bottom contact. As known from previous photocurrent experiments on H019 reported in [16, 20], the bias change from +3 V to -5 V redistributes charge carriers in the relaxed system from the HH1 ground state of well 1 into wells 1 to 5. The bias change is expected to result in a change in the amount of charge carriers excited into the HH2 level by the TM polarized pump 1 beam, as fewer charge carriers are located in well 1 at -5 V than for +3 V, and the FEL radiation is in resonance with the HH1-HH2 transition of well 1 only. This led to an observed change in the optimal pump intensities for the time-resolved measurements, as discussed in chapters 7.6.2 and 7.6.3.

However, despite the experimentally indicated changes in the band structure of sample H019 with the applied bias, no significant dependence of the HH2-HH1 relaxation time on the sample voltage could be observed, as shown by the similar values of 510 fs and 560 fs found for +3 V and -5 V, respectively. The explanation for this insensitivity of the intersubband relaxation time to the applied field can be found in the fact that the distribution of charge carriers in the ground state does not influence the outcome of photocurrent pump-pump experiments, as stated several times before. As discussed in chapter 6.2.7, the phonon scattering rate between two quantum states is essentially influenced by the wavefunction overlap associated with these states (see Eqn. 6.62). Further, as mentioned in the previous section, the overall depopulation rate of the monitored HH2 state in the structure under investigation is determined by two relaxation pathes, namely a direct HH2-HH1 relaxation process and a decay path via the intermediary LH1 state. Consequentially, the total depopulation rate of the excited HH2 state is determined by the wavefunction overlaps between the HH2 state in well 1 occupied by non-equilibrium charge carriers, and the LH1 intermediate state as well as the HH1 ground state in this well. These wavefunction overlaps *do not change* with biasing. As a consequence, the relaxation time stays the same when changing the applied bias from +3 V to -5 V.

Put differently, the observed insensitivity of the HH2-HH1 relaxation time of H019 in respect to the applied field is based on the circumstance, that a bias change does not alter the spatial overlap between the excited state and possible final states of a scattering process in well 1 of this structure. This is in strong contrast to the band configuration presented in chapter 8 of this work, where a tunability of the intersubband relaxation time with the applied field *is* observed.

7.7 Density matrix simulations

In chapter 7.6.4, an HH2-HH1 intersubband relaxation time of about 550 fs for a transition energy of 160 meV was extracted from experimentally acquired photocurrent pump-pump data by a simple single-exponential fit. In order to obtain a more detailed picture of the evolution of the individual states' occupation and further gain insight into the bleaching mechanisms leading to the intensity dependence of the photocurrent through the sample as presented in Fig. 7.12, density matrix simulations were performed. It shall be pointed out, that the density matrix simulations were not required to extract the intersubband relaxation times from the pump-pump measurement results, but rather confirmed the interpretation of the experimental data by showing that within the same theoretical model both the intensity dependence in Fig. 7.12 and the pump-pump data in Fig. 7.15 can be simulated with high accuracy.

7.7.1 Density matrix basics

The dipole coupling of a quantum mechanical system to a radiation field is commonly treated in the basis of the uncoupled light and matter states, with the interaction Hamiltonian given in Eqn. 7.6, where x is the location operator, $E(x,t)$ the electric field component of the radiation, and μ the dipole operator (see [23], pp. 298 and [126], pp. 28).

$$\begin{aligned} H_d &= e\, x \cdot E(x,t) \\ &= -\mu \cdot E(x,t) \end{aligned} \quad (7.6)$$

This interaction operator can be used to determine the coherent temporal evolution of a quantum mechanical system under the influence of a radiation field.

Now, for our purpose, not only the interaction with the radiation field but also the non-coherent relaxation of the occupation of the individual quantum mechanical levels by phonon scattering has to be considered. Statistical processes in the evolution of a quantum system are commonly included via the density matrix formalism. The relaxation of the occupation ρ_{ii} of the quantum state i towards its equilibrium value ρ_{ii}^0 with the relaxation time τ_{life} can be written as in Eqn. 7.7.

$$\frac{d}{dt}\rho_{ii}\left|_{relax}\right. = -\frac{\rho_{ii} - \rho_{ii}^0}{\tau_{life}} \tag{7.7}$$

As no coherence effects were observed in the course of the pump-pump experiments, the quantity of interest for the comparison between simulation and experimental data is the occupation of the levels of the studied system. The interaction of a quantum mechanical system with a reservoir of phonons leads to an inelastic scattering of the charge carriers between the involved quantum levels, and affects the occupation of the levels in an incoherent way. However, even if the coherence times of the investigated system are too short to be experimentally resolved, the excitation process by electromagnetic radiation exhibits coherence over a finite time span, which influences the time evolution of the level occupation. A formalism allowing the combined treatment of both the interaction of a quantum mechanical system with photons and its relaxation by coupling to a phonon bath is formed by the density matrix approach ([126], pp. 29 and [127]).

The density matrix formalism deals with the quantity accessed by the pump-pump experiments, namely the occupation of the individual levels, and its dynamics under the influence of both phonons and photons. For this purpose, a so-called density matrix operator ρ is defined according to Eqn. 7.8.

$$\rho = \sum_{i,j} \rho_{ij} |\phi_i><\phi_j| \tag{7.8}$$

The diagonal elements of the density matrix operator ρ_{ii} represent the population of level i (or, equivalently, its occupation probability). The off-diagonal components of this matrix ρ_{ij} are equivalent to the coherence of a superposition state built up by the eigenstates i and j of the unperturbed hamiltonian H_0. They are thus proportional to the dipole moment associated with the respective states, as induced by the electromagnetic field. The time evolution of the density matrix operator is according to basic quantum mechanics determined by Eqn. 7.9.

$$\frac{d}{dt}\rho = -\frac{i}{\hbar}[H_0,\rho] - \frac{i}{\hbar}[H_d,\rho] + \frac{\partial}{\partial t}\rho\left|_{relax}\right. \tag{7.9}$$

Now, the matrix element of the commutator $[H_0,\rho]$ between the eigenstates $|m>$ and $|n>$ trivially reads as in Eqn. 7.10, where E_n and E_m are the respective eigenenergies.

$$<n|\,[H_0,\rho]\,|m> = \hbar\,\omega_{nm}\,\rho_{nm}$$
$$\hbar\,\omega_{nm} = E_n - E_m \tag{7.10}$$

By inserting Eqn. 7.6, Eqn. 7.11 is gained for the matrix element of commutator $[H_d, \rho]$.

$$<n| [H_d, \rho] |m> = -\boldsymbol{E}(t) \cdot \sum_l (\boldsymbol{\mu}_{nl}\, \rho_{lm} - \rho_{nl}\, \boldsymbol{\mu}_{lm}) \qquad (7.11)$$

Combining Eqns. 7.11, 7.10 and 7.9 leads to the final analytical expression for the time evolution of the density matrix operator as given in Eqn. 7.12.

$$\frac{d}{dt}\rho_{nm} = -\mathrm{i}\,\omega_{nm}\,\rho_{nm} + \frac{\mathrm{i}}{\hbar}\,\boldsymbol{E}(t) \cdot \sum_l (\boldsymbol{\mu}_{nl}\, \rho_{lm} - \rho_{nl}\, \boldsymbol{\mu}_{lm}) +$$
$$+ \frac{\partial}{\partial t}\rho_{nm}\,|_{relax} \qquad (7.12)$$

7.7.2 Model

The density matrix simulations employed to reproduce the experimentally obtained data for sample H019 are based on a four-level-system, as illustrated in Fig. 7.17. In the model, the HH1 ground state and the excited HH2 state of well 1 of structure H019 are represented by the levels 0 and 1, respectively. The valence band continuum is represented by two levels. Level 2 accounts for the continuum states, which can be resonantly reached from the HH2 state by the absorption of an FEL photon of 160 meV. In addition, level 3 represents non-resonant continuum states, into which resonantly excited holes can relax. Level 3 does not interact with the radiation field.

The arrows in Fig. 7.17 represent the processes, which are included in the simulations. The red arrows indicate the absorption and stimulated emission of FEL photons, while the black arrows sketch non-radiative relaxation processes. In consistency with the pump-pump principle discussed in chapter 7.4, the simulation accounts for the interaction of photons with the HH1-HH2 dipole as well as that between the HH2 state and resonant continuum states. The dipole operators between the respective states are represented by $\boldsymbol{\mu}_{10}$ and $\boldsymbol{\mu}_{21}$. Now, the interaction of the quantum mechanical system with the radiation field induces dipole moments by coupling the respective quantum levels and forming superposition states. The coherence of these superposition states between the levels 0 and 1, as well as 1 and 2, is quantified by the density matrix elements $\rho_{10} = \rho_{01}$ and $\rho_{12} = \rho_{21}$, respectively. This coherence decays with the phase relaxation rate Γ_{ij}, following Eqn. 7.13.

$$\frac{\partial}{\partial t}\rho_{10}\,|_{relax} = -\Gamma_{10}\,\rho_{10}$$
$$\frac{\partial}{\partial t}\rho_{21}\,|_{relax} = -\Gamma_{21}\,\rho_{21} \qquad (7.13)$$

As no coherent effects could be observed during the time-resolved experiments, the associated de-phasing time, which is equivalent to $\frac{1}{\Gamma_{ij}}$, has to be short on the scale of the time resolution of the pump-pump setup.

The HH2-HH1 relaxation time is represented by τ_{10} in the model of Fig. 7.17. The depopulation process of level 1 in this figure is equivalent to the relaxation experimentally monitored in the course of the pump-pump measurements, and follows Eqn. 7.14 within the model framework.

$$\frac{\partial}{\partial t}\rho_{11}\big|_{relax} = -\frac{\rho_{11}}{\tau_{10}} \qquad (7.14)$$

Charge carriers excited by the pump 2 pulse from the excited level 1 into the resonant continuum state 2 during the probing process of the experiment, decay into densely packed non-resonant continuum states in the course of generating a photocurrent. In the density matrix model, this decay is represented by its characteristic relaxation time τ_{32}. As this decay is expected to appear on a much smaller time scale than the re-trapping of charge carriers in the resonant level 2 into the quantum well, the relaxation of the level-2-occupation is exclusively determined by this transport decay, as indicated by Eqn. 7.15.

$$\frac{\partial}{\partial t}\rho_{22}\big|_{relax} = -\frac{\rho_{22}}{\tau_{32}} \qquad (7.15)$$

The re-trapping time of continuum carriers into the quantum wells of structure H019 was calculated from its photoconductive gain (between 0.4 and 0.5) as reported in [16] and [20], and was found to be 5-10 ps. This time constant is more than one order of magnitude above the FEL pulse duration, and consequently the re-trapping process is not expected to influence the carrier distribution on the simulated time scale. Re-trapping of continuum carriers is therefore not included in the rate equations. In the frame of the model, charge carriers relaxing into non-resonant continuum states represented by level 3 exhibit no further interaction with the quantum well system. The rate equation of the ρ_{33} element is thus exclusively determined by the filling process via level 2, as given in Eqn. 7.16.

$$\frac{\partial}{\partial t}\rho_{33}\big|_{relax} = \frac{\rho_{22}}{\tau_{32}} \qquad (7.16)$$

In the model, the ground state of the quantum well system is exclusively filled by the relaxation of level-1-carriers, as stated in Eqn. 7.17.

$$\frac{\partial}{\partial t}\rho_{00}\big|_{relax} = \frac{\rho_{11}}{\tau_{10}} \qquad (7.17)$$

The explicit application of Eqn. 7.12 to the model system illustrated in Fig. 7.17 by inserting Eqns. 7.13 to 7.17 leads to an equational system for the density matrix operator as explicitly given in Eqns. 7.18 to 7.26.

$$\frac{d}{dt}\rho_{00}(t) = \frac{\rho_{11}(t)}{\tau_{10}} + \frac{i}{\hbar}\,\boldsymbol{E}(t)\cdot\boldsymbol{\mu}_{10}\,\rho_{10}(t) - \frac{i}{\hbar}\,\boldsymbol{E}(t)\cdot\boldsymbol{\mu}_{10}\,\rho_{01}(t) + \\ + \frac{i}{\hbar}\,\boldsymbol{E}(t)\cdot\boldsymbol{\mu}_{20}\,\rho_{20}(t) - \frac{i}{\hbar}\,\boldsymbol{E}(t)\cdot\boldsymbol{\mu}_{20}\,\rho_{02}(t)$$
(7.18)

$$\frac{d}{dt}\rho_{11}(t) = -\frac{\rho_{11}(t)}{\tau_{10}} - \frac{i}{\hbar}\,\boldsymbol{E}(t)\cdot\boldsymbol{\mu}_{10}\,\rho_{10}(t) + \frac{i}{\hbar}\,\boldsymbol{E}(t)\cdot\boldsymbol{\mu}_{10}\,\rho_{01}(t) + \\ + \frac{i}{\hbar}\,\boldsymbol{E}(t)\cdot\boldsymbol{\mu}_{21}\,\rho_{21}(t) - \frac{i}{\hbar}\,\boldsymbol{E}(t)\cdot\boldsymbol{\mu}_{21}\,\rho_{12}(t)$$
(7.19)

$$\frac{d}{dt}\rho_{22}(t) = -\frac{\rho_{22}(t)}{\tau_{32}} - \frac{i}{\hbar}\,\boldsymbol{E}(t)\cdot\boldsymbol{\mu}_{20}\,\rho_{20}(t) + \frac{i}{\hbar}\,\boldsymbol{E}(t)\cdot\boldsymbol{\mu}_{20}\,\rho_{02}(t) - \\ - \frac{i}{\hbar}\,\boldsymbol{E}(t)\cdot\boldsymbol{\mu}_{21}\,\rho_{21}(t) + \frac{i}{\hbar}\,\boldsymbol{E}(t)\cdot\boldsymbol{\mu}_{21}\,\rho_{12}(t)$$
(7.20)

$$\frac{d}{dt}\rho_{33}(t) = \frac{\rho_{22}(t)}{\tau_{32}}$$
(7.21)

$$\frac{d}{dt}\rho_{01}(t) = -(\Gamma_{10} + i\omega_{01})\,\rho_{01}(t) - \frac{i}{\hbar}\,\boldsymbol{E}(t)\cdot\boldsymbol{\mu}_{21}\,\rho_{02}(t) + \frac{i}{\hbar}\,\boldsymbol{E}(t)\cdot\boldsymbol{\mu}_{20}\,\rho_{21}(t) - \\ - \frac{i}{\hbar}\,\boldsymbol{E}(t)\cdot\boldsymbol{\mu}_{10}\,\rho_{00}(t) + \frac{i}{\hbar}\,\boldsymbol{E}(t)\cdot\boldsymbol{\mu}_{10}\,\rho_{11}(t)$$
(7.22)

$$\frac{d}{dt}\rho_{02}(t) = -(\Gamma_{20} + i\omega_{02})\,\rho_{02}(t) - \frac{i}{\hbar}\,\boldsymbol{E}(t)\cdot\boldsymbol{\mu}_{21}\,\rho_{01}(t) + \frac{i}{\hbar}\,\boldsymbol{E}(t)\cdot\boldsymbol{\mu}_{10}\,\rho_{12}(t) - \\ - \frac{i}{\hbar}\,\boldsymbol{E}(t)\cdot\boldsymbol{\mu}_{20}\,\rho_{00}(t) + \frac{i}{\hbar}\,\boldsymbol{E}(t)\cdot\boldsymbol{\mu}_{20}\,\rho_{22}(t)$$
(7.23)

$$\frac{d}{dt}\rho_{12}(t) = -(\Gamma_{21} + i\omega_{12})\,\rho_{12}(t) + \frac{i}{\hbar}\,\boldsymbol{E}(t)\cdot\boldsymbol{\mu}_{10}\,\rho_{02}(t) - \frac{i}{\hbar}\,\boldsymbol{E}(t)\cdot\boldsymbol{\mu}_{20}\,\rho_{10}(t) - \\ - \frac{i}{\hbar}\,\boldsymbol{E}(t)\cdot\boldsymbol{\mu}_{12}\,\rho_{11}(t) + \frac{i}{\hbar}\,\boldsymbol{E}(t)\cdot\boldsymbol{\mu}_{12}\,\rho_{22}(t)$$
(7.24)

$$\rho_{03}(t) = \rho_{13}(t) = \rho_{23}(t) = 0$$
(7.25)

$$\rho_{ji}(t) = \rho_{ij}^{*}(t)$$
(7.26)

As seen from Eqns. 7.21 and 7.25, the non-resonant continuum state 3 does not interact with the radiation field at all within the model. This is consistent with the fact that the spatially distributed continuum states, into which the charge carriers resonantly excited from the HH2 level decay, exhibit an extremely weak overlap with the bound quantum well states and a consequentially negligible coupling to the FEL photons. As the density matrix operator is Hermitian ($\rho = \rho^\dagger$), Eqn. 7.26 has to be fulfilled.

In order to gain detailed insight into the time dependence of the carrier distribution in the quantum well system of H019, the time evolution of the density matrix elements of the model system in Fig. 7.17 shall be calculated. While the diagonal density matrix elements of the levels 0 and 1 are equivalent to the actual occupation of the bound HH1 and HH2 states in well 1 of structure H019, the interpretation of the elements associated with the levels 2 and 3 is less straight forward. In the photocurrent experiments, charge carriers occupying the bound level 1 are excited by an incoming photon into the continuum, where they are transported to the adjacent quantum well period in the structure. In the model, level-1-carriers can be excited into the resonant continuum level 2, from where they relax into the non-resonant level 3, where they are stuck due to the lack of interaction with the radiation field and the absence of relaxation mechanisms in the model. Level 3 therefore collects any carriers *emitted from* the quantum well system, in analogy to continuum carriers transported away from the well by current flow and collected at the sample contacts. Thus, the number of charge carriers occupying level 3 and 2 at the end of the modelled excitation process can be interpreted as the total number of holes emitted from the quantum well system into the continuum, and is therefore in analogy to the experimentally determined integral photocurrent through the sample. As a consequence, the sum of ρ_{22} and ρ_{33} at the end of an excitation process has to be compared to the value of the measured integral photocurrent in order to simulate the photocurrent experiments.

The electric field component of the FEL radiation pulse shall be given by Eqn. 7.27, where Ω is the radiation frequency, and $\boldsymbol{E}_0(t)$ represents the temporal pulse shape, which modulates the field oscillations by a slowly varying envelope.

$$\boldsymbol{E}(t) = \boldsymbol{E}_0(t) \frac{1}{2}(e^{i\Omega t} + e^{-i\Omega t}) \qquad (7.27)$$

In an unperturbed system in absence of a driving electromagnetic field, the off-diagonal elements of the density matrix perform oscillations with the characteristic frequency $\omega_{nm} = \frac{E_n - E_m}{\hbar}$, as concluded from Eqn. 7.10. Now, in the presence of a driving electromagnetic field oscillating at a frequency $\Omega \sim \omega_{nm}$, the time evolution of the off-diagonal elements of the density matrix with free-evolution-frequencies near resonance can be separated into an oscillation with the driving frequency and a slowly varying amplitude $\sigma_{nm}(t)$. These

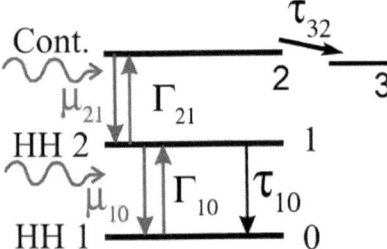

Figure 7.17: Four-level model employed for the density matrix simulations. Level 0 represents the HH1 ground state of well 1 in structure H019, and level 1 its excited HH2 state. Level 2 models the continuum states, to which the transition from level 1 is in resonance with the FEL radiation, while level 3 represents non-resonant continuum states. The interaction with the radiation field is indicated by red arrows, non-radiative relaxation processes are shown by black arrows.

considerations lead to Eqn. 7.28.

$$\begin{aligned}
\rho_{01}(t) &= \sigma_{01}(t)\, e^{i\Omega t} \\
\rho_{10}(t) &= \sigma_{01}^*(t)\, e^{-i\Omega t} \\
\rho_{12}(t) &= \sigma_{12}(t)\, e^{i\Omega t} \\
\rho_{21}(t) &= \sigma_{12}^*(t)\, e^{-i\Omega t}
\end{aligned} \quad (7.28)$$

Now, the unperturbed evolution frequency of ρ_{02} is far off resonance from the driving electromagnetic field. A coherent superposition between the states 0 and 2 can only be induced via ρ_{01} and ρ_{12}. As coherence times are short, this high-order coherence between the states 0 and 2 and thus ρ_{02} can be neglected, leading to Eqn. 7.29.

$$\sigma_{02} = \sigma_{20} = 0 \quad (7.29)$$

The expressions in Eqns. 7.28 and 7.29 are inserted into Eqns. 7.21 to 7.23, which results in a system of differential equations for the correlation amplitudes $\sigma_{nm}(t)$ ($n \neq m$). In order to illustrate the simplifications, which are further applied to this equational system, the resulting expression for $\sigma_{01}(t)$ is shown in Eqn. 7.30.

$$\begin{aligned}
\frac{d}{dt}\sigma_{01}(t) =\ &-\Gamma_{10}\, \sigma_{01}(t) + \frac{i}{\hbar}\, \boldsymbol{E}(t) \cdot \boldsymbol{\mu}_{20}\, \sigma_{21}(t) e^{-i2\Omega t} \\
&- \frac{i}{\hbar}\, \boldsymbol{E}(t) \cdot \boldsymbol{\mu}_{10}\, \rho_{00}(t)\, e^{-i\Omega t} + \frac{i}{\hbar}\, \boldsymbol{E}(t) \cdot \boldsymbol{\mu}_{10}\, \rho_{11}(t)\, e^{-i\Omega t}
\end{aligned} \quad (7.30)$$

After inserting the expression for the electromagnetic field as given in Eqn. 7.27, the equational system for the density matrix components is further simplified by applying the so-called *rotating wave approximation* ([115], pp. 265). In this approximation, the highly oscillatory terms ($\sim e^{\pm i\Omega t}$, $e^{\pm 2i\Omega t}$, $e^{\pm 3i\Omega t}$,...) in the equations of motion for the ρ_{ii} and σ_{ij} are neglected. This simplification is valid, as these terms average out of the time evolution of the monitored occupation probabilities on the time scale of the experiment.

Additionally, the off-diagonal elements of the density matrix operator are approximated adiabatically, which accounts for the fact that no coherence effects could be observed during the pump-pump experiments. This *adiabatic appproximation* consists of setting the time derivative of the correlation amplitude $\sigma_{nm}(t)$ zero. This means that the coherent superposition between quantum states in the system, which is characterized by the non-diagonal elements of the density matrix operator, reaches its quasi-equilibrium instantly on the scale of the experimental time resolution, as expressed in Eqn. 7.31.

$$\frac{d}{dt}\sigma_{nm}(t) = 0 \qquad \text{for } n \neq m \tag{7.31}$$

Applying both the rotating wave and adiabatic approximations to Eqn. 7.30 results in Eqn. 7.32.

$$\begin{aligned}\frac{d}{dt}\sigma_{01}(t) &= -\Gamma_{10}\,\sigma_{01}(t) \\ &\quad - \frac{i}{2\hbar}\,\boldsymbol{E}_0(t)\cdot\boldsymbol{\mu}_{10}\,\rho_{00}(t) + \frac{i}{2\hbar}\,\boldsymbol{E}_0(t)\cdot\boldsymbol{\mu}_{10}\,\rho_{11}(t) \\ &= 0\end{aligned} \tag{7.32}$$

Thus, Eqn. 7.33 results.

$$\sigma_{01}(t) = \frac{i}{2\hbar\Gamma_{10}}\,\boldsymbol{E}_0(t)\cdot\boldsymbol{\mu}_{10}\,[\,\rho_{11}(t) - \rho_{00}(t)\,] \tag{7.33}$$

The rotating wave and adiabatic approximations are applied to the total equational system given by Eqns. 7.18 to 7.26 after inserting the expressions in Eqns. 7.27 and 7.28. The off-diagonal elements can be eliminated from the equational system, leading to the final rate equations for the diagonal elements of the density matrix as presented in Eqns. 7.34 to 7.37.

$$\frac{d}{dt}\rho_{00} = \frac{\rho_{11}}{\tau_{10}} + \frac{2}{\hbar^2} \cdot \frac{1}{\hbar^2 \Gamma_{10}\Gamma_{12}\Gamma_{20} + \Gamma_{10}(\boldsymbol{\mu}_{10} \cdot \boldsymbol{E}_0)^2 + \Gamma_{12}(\boldsymbol{\mu}_{21} \cdot \boldsymbol{E}_0)^2} \cdot$$

$$\cdot [\hbar^2 \Gamma_{12}\Gamma_{20} (\boldsymbol{\mu}_{10} \cdot \boldsymbol{E}_0)^2 (\rho_{11} - \rho_{00}) + (\boldsymbol{\mu}_{10} \cdot \boldsymbol{E}_0)^4 (\rho_{11} - \rho_{00}) +$$

$$+ (\boldsymbol{\mu}_{10} \cdot \boldsymbol{E}_0)^2 (\boldsymbol{\mu}_{21} \cdot \boldsymbol{E}_0)^2 (\rho_{22} - \rho_{11})]$$

(7.34)

$$\frac{d}{dt}\rho_{11} = -\frac{\rho_{11}}{\tau_{10}} + \frac{2}{\hbar^2} \cdot \frac{1}{\hbar^2 \Gamma_{10}\Gamma_{12}\Gamma_{20} + \Gamma_{10}(\boldsymbol{\mu}_{10} \cdot \boldsymbol{E}_0)^2 + \Gamma_{12}(\boldsymbol{\mu}_{21} \cdot \boldsymbol{E}_0)^2} \cdot$$

$$\cdot [\hbar^2 \Gamma_{12}\Gamma_{20} (\boldsymbol{\mu}_{10} \cdot \boldsymbol{E}_0)^2 (\rho_{00} - \rho_{11}) + \hbar^2 \Gamma_{10}\Gamma_{20} (\boldsymbol{\mu}_{21} \cdot \boldsymbol{E}_0)^2 (\rho_{22} - \rho_{11}) +$$

$$+ (\boldsymbol{\mu}_{10} \cdot \boldsymbol{E}_0)^4 (\rho_{00} - \rho_{11}) + (\boldsymbol{\mu}_{21} \cdot \boldsymbol{E}_0)^4 (\rho_{22} - \rho_{11}) +$$

$$+ (\boldsymbol{\mu}_{10} \cdot \boldsymbol{E}_0)^2 (\boldsymbol{\mu}_{21} \cdot \boldsymbol{E}_0)^2 (2\rho_{11} - \rho_{22} - \rho_{00})]$$

(7.35)

$$\frac{d}{dt}\rho_{22} = -\frac{\rho_{22}}{\tau_{32}} + \frac{2}{\hbar^2} \cdot \frac{1}{\hbar^2 \Gamma_{10}\Gamma_{12}\Gamma_{20} + \Gamma_{10}(\boldsymbol{\mu}_{10} \cdot \boldsymbol{E}_0)^2 + \Gamma_{12}(\boldsymbol{\mu}_{21} \cdot \boldsymbol{E}_0)^2} \cdot$$

$$\cdot [\hbar^2 \Gamma_{10}\Gamma_{20} (\boldsymbol{\mu}_{21} \cdot \boldsymbol{E}_0)^2 (\rho_{11} - \rho_{22}) + (\boldsymbol{\mu}_{21} \cdot \boldsymbol{E}_0)^4 (\rho_{11} - \rho_{22}) +$$

$$+ (\boldsymbol{\mu}_{10} \cdot \boldsymbol{E}_0)^2 (\boldsymbol{\mu}_{21} \cdot \boldsymbol{E}_0)^2 (\rho_{00} - \rho_{11})]$$

(7.36)

$$\frac{d}{dt}\rho_{33} = \frac{\rho_{22}}{\tau_{32}} \tag{7.37}$$

For the sake of readability, the explicit declaration of the time dependence of the density matrix elements was abandoned in these equations. The simulations used for reproducing the saturation characteristics presented in Fig. 7.12 do not include current flow through the sample, as such transport simulations would exceed the scope of this work. As a consequence, the total number of charge carriers in the four-level-system is constant, as expressed by Eqn. 7.38., which directly results from Eqns. 7.34 to 7.37.

$$\rho_{00} + \rho_{11} + \rho_{22} + \rho_{33} = N = \text{const.} \tag{7.38}$$

However, the spatial shift of charge carriers drifting away from their quantum well of origination was accounted for by the optical decoupling of the non-resonant continuum level 3 in the model. Charge carriers relaxing into this level are 'transported away' in a sense that they do not interact optically with their ground state.

ρ_{11} can be eliminated from the rate equations in Eqns. 7.34 to 7.37, giving a system of three differential equations for ρ_{00}, ρ_{22} and ρ_{33}, which can be solved numerically.

7.7.3 Simulation results for the FEL power dependence of the photocurrent

In order to simulate the dependence of the integral photocurrent through sample H019 on the energy of the incoming FEL pulse, which was experimentally determined as discussed in chapter 7.6.1, the system of differential equations as given in Eqns. 7.34 to 7.37 was solved for a gauss shaped envelope of the electric field component associated with an FEL pulse centered around $t = 0$. For modelling the experiments presented in chapter 7.6.1 as accurately as possible, the polarization of the electric field was chosen as given in Eqn. 7.39, owing to the polarization distribution of the TM pulse, which is geometrically induced by the sample facette (see chapter 7.5).

$$\boldsymbol{E}_0(t) \;=\; (\frac{1}{2}\boldsymbol{e}_{xy} + \frac{\sqrt{3}}{2}\boldsymbol{e}_z) \cdot \sqrt{I_0}\; e^{-\frac{1}{4}(\frac{t}{\sigma_t})^2} \tag{7.39}$$

$\boldsymbol{E}_0(t)$ as given in Eqn. 7.39 thus represents the shape of the gaussian TM pulse employed for the experimental determination of the integral photocurrent's saturation characteristics as presented in Fig. 7.12. I_0 is the peak intensity of the micropulse, and σ_t characterizes its temporal width. $\boldsymbol{e}_z$ and $\boldsymbol{e}_{xy}$ are the polarization unity vectors normal and parallel to the sample surface, respectively. As already mentioned, due to the sample geometry the pulse labelled 'TM' throughout this work actually exhibits both a strong TM and a weaker TE component. Under actual experimental conditions, however, the TE component is further suppressed by the top metallization of the investigated mesa. This is not considered in the simulations.

Calculating the time evolution of the occupation by solving the differential equations 7.34 to 7.37 requires a set of parameters. Among them are the dipole matrix elements $\boldsymbol{\mu}_{ij}$, which can be deduced from the simulated bandstructure of H019 (see chapter 7.2). A more detailed overview over the bandstructure calculations is given in [20]. For σ_t, a value of 120 fs was chosen in consistency with the FEL configuration as well as the temporal width of the pulse limited excess photocurrent characteristics for positive delays in Fig. 7.15. A σ_t of 120 fs corresponds to a pulse FWHM of 280 fs. The decay times $\frac{1}{\Gamma_{ij}}$ of the coherences were chosen lower than the experimental time resolution according to the non-observance of coherence effects during the experiments. In addition, all of them were set to the same value $\frac{1}{\Gamma}$ as a consequence of the lack of more detailed experimental information. Further, due to the assumption, that the decay of a coherently and resonantly excited continuum state is directly associated with its de-phasing, the decay time τ_{32} was chosen equally to $\frac{1}{\Gamma}$. Finally, for the HH2-HH1 relaxation time τ_{10} the value gained from the pump-pump photocurrent experiments in chapter 7.6.3 was used. A summary of the parameters for the density matrix simulations is given in table 7.5.

The expression in Eqn. 7.39 is inserted into Eqns. 7.34 to 7.37, and the time evolution of the density matrix elements ρ_{ii} is calculated numerically for a series of intensities I_0. For solving the differential equations, a relaxed initial condition of the system is assumed, that means $\rho_{00} = 1$ and $\rho_{11} = \rho_{22} = \rho_{33} = 0$. This is in consistency with the experiment. The system of differential equations was solved using the algorithm *ode45* implemented in *matlab* for a time interval between -500 fs and 1500 fs, where the field-envelope of the interacting FEL pulse is centered around $t = 0$.

The numerical results of the density matrix simulation for a single TM polarized FEL pulse are presented in Figs. 7.18 and 7.19 for a series of micropulse energies, which span the range of the experiment presented in chapter 7.6.1. The plot in Fig. 7.18 (a) presents the time evolution of ρ_{00}, which is equivalent to the occupation probability of level 0, while Fig. 7.18 (b) shows the result for the occupation of level 1. Figure 7.19 (c) presents the summed simulated occupation probability of both continuum levels $\rho_{22} + \rho_{33}$. The red arrows along the time-axis indicate the FWHM of the modelled FEL pulse. In the following paragraphs, the simulation results are discussed and compared to the experiment.

As forced by the initial conditions, all charge carriers of the system occupy the ground level 0 (HH1) before the FEL pulse exerts its impact. As the intensity of the FEL pulse increases, charge carriers are excited from the ground state into level 1 (HH2), and further into the

Table 7.5: Fit parameters employed for the density matrix simulations:

Parameter	Value	Source
σ_t	120 fs	FEL parameter
τ_{10}	560 fs	pump-pump experiments
τ_{32}	180 fs	estimation
$\frac{1}{\Gamma} = \frac{1}{\Gamma_{10}} = \frac{1}{\Gamma_{12}} = \frac{1}{\Gamma_{20}}$	180 fs	estimation
$\frac{\mu_{10}}{e}$	2.6 nm (TM) 0 (TE)	bandstructure calculations
$\frac{\mu_{21}}{e}$	1 nm (TM) 1 nm (TE)	bandstructure calculations

continuum level 2, as seen from the drop of ρ_{00} in Fig. 7.18 (a) and the rise in both ρ_{11} in Fig. 7.18 (b) and $\rho_{22} + \rho_{33}$ in Fig. 7.19 (c). Note that the minimum occupation of level 0 is reached about 150 ps after the FEL pulse intensity peaks, which is due to the fact that charge carriers are still excited from level 0 into 1 by the pulse flank, while the re-occupation of level 0 by the HH2-HH1 relaxation process is still insufficient to compensate this excitation of carriers.

While the temporal position of the minimum in ρ_{00} is independent of the micropulse intensity, the simulated maximum occupation of the experimentally monitored HH2 state depends strongly on the FEL intensity. For low micropulse energies, it is reached simultaneously with the ρ_{00}-minimum. As the FEL intensity increases, the maximum position of ρ_{11} shifts to negative times in Fig. 7.18 (b). This is due to bleaching of the HH1-HH2 transition for higher intensities, as indicated by the equal values of ρ_{00} and ρ_{11} in the plots of Fig. 7.18 (a) and (b). Bleaching is reached as soon as the FEL intensity is high enough to induce a HH1-HH2 excitation rate exceeding its non-radiative relaxation rate characterized by τ_{10}. The bleaching process prevents a further increase in the level-1-occupation, while the FEL intensity has not yet reached its peak value. The maximum (minimum) occupation of $\rho_{11}(\rho_{00}) = 0.4$ shows, that 20% of the holes are excited into the continuum during the time span needed to establish equal populations of levels 0 and 1. From the temporal position of the maximum occupation of the HH2 level on, the value of ρ_{11} even drops before the FEL intensity peaks, which is due to the depopulation of level 1 into the continuum level 2. Note that the HH1-HH2 transition still remains bleached ($\rho_{11} = \rho_{00}$), as the ground state occupation ρ_{00} drops at the same rate as ρ_{11} in this time window.

After the FEL pulse has passed, the value of ρ_{11} decays exponentially with τ_{10}, and ρ_{00} rises at the same rate. This relaxation process in the model represents the HH2-HH1 relaxation, whose characteristic time is studied in this work. During the time span attributed to the FEL pulse, charge carriers occupying level 1 are excited into the resonant continuum level 2, from where they relax into the non-resonant continuum level 3 at a high rate $\frac{1}{\tau_{32}}$. As discussed in section 7.7.2, this process models the emission of charge carriers from the quantum well region of structure H019 and the consequential generation of photocurrent. Thus the sum of all charge carriers emitted from the bound well states into the continuum is equivalent to the *total number* of charge carriers ending up in the levels 2 and 3, and can be compared to the experimentally determined *integral* photocurrent through the sample.

Due to the absence of any depopulation process for charge carriers in level 3, $\rho_{22} + \rho_{33}$ in Fig. 7.19 (c) increases monotonically with time for any given FEL micropulse energy, and finally stays constant. The value of $\rho_{22} + \rho_{33}$ at $t = 1.5$ ps, which is marked for the whole

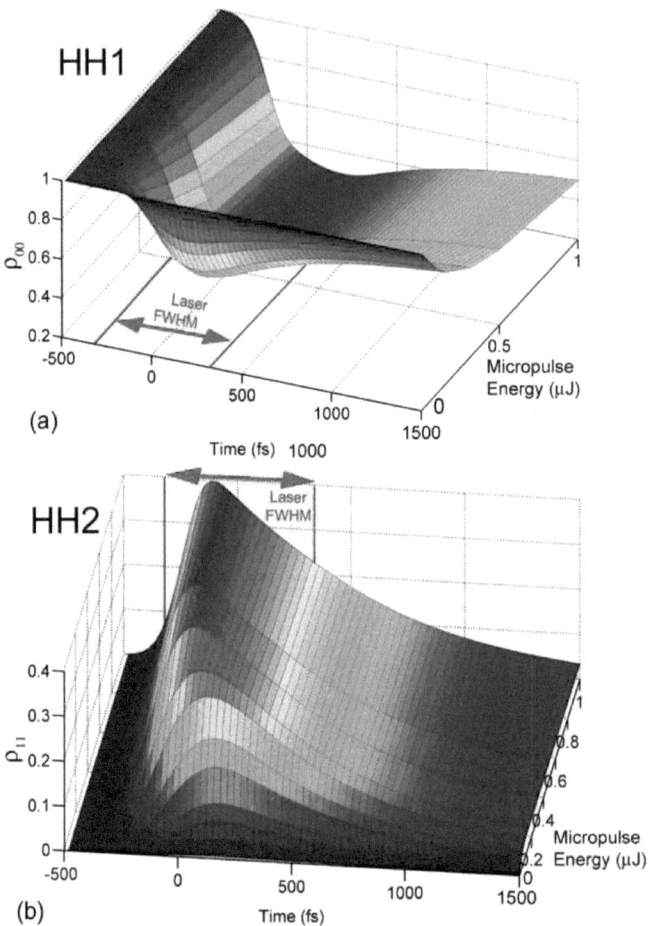

Figure 7.18: Simulated time evolution of the diagonal density matrix elements for the levels 0 and 1 in Fig. 7.17, representing the HH1 and HH2 states of H019. The calculations were performed for a series of micropulse energies, or peak intensities I_0. The red arrows indicate the FWHM of the intenstiy distribution of the gaussian FEL pulse. [124]

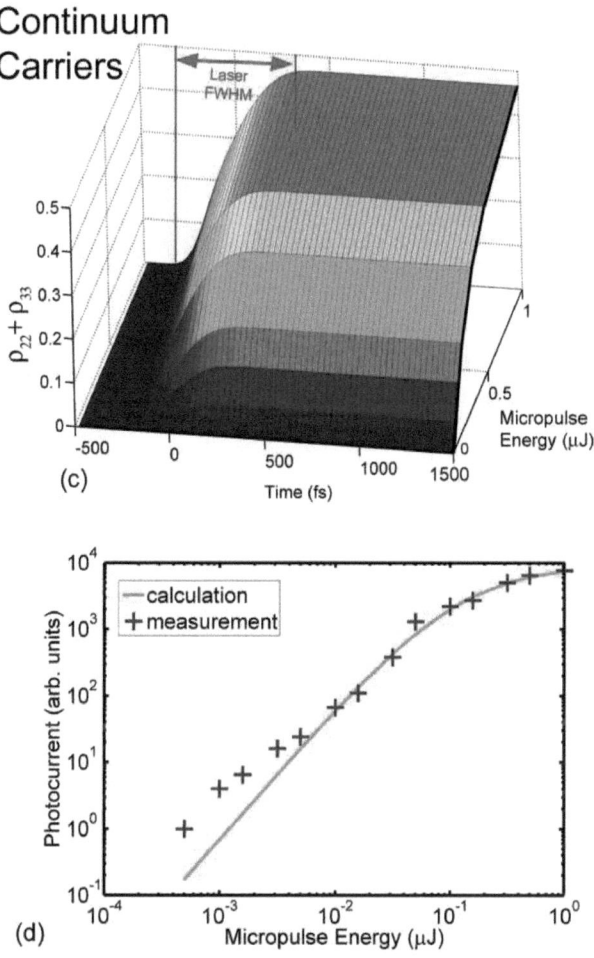

Figure 7.19: Simulated time evolution of the summed diagonal density matrix elements of the levels 2 and 3 in Fig. 7.17, representing the occupation of continuum states in H019 (c). In (d), the simulation results for the final values of $\rho_{22} + \rho_{33}$ at $t = 1.5$ ps are compared to the experimentally determined intensity dependence of the photocurrent through sample H019, as presented in chapter 7.6.1. [124]

sweep of micropulse intensities by the black line in Fig. 7.19 (c), is equal to the integral photocurrent induced by the FEL pulse. The simulated final occupation of levels 2 and 3 can thus be directly compared with the energy dependence of the integral photocurrent in Fig. 7.12, as shown in Fig. 7.19 (d). The green line represents the calculated final continuum occupation at a time of 1.5 ps, normalized to the experimentally determined integral photocurrent.

As seen from the plot in Fig. 7.19 (d), the density matrix simulations based on the model presented in section 7.7.2 reproduce the intensity dependence of the photocurrent accurately over more than four orders of magnitude in both photocurrent and micropulse energy. As concluded from the simulation results, the saturation of the integral photocurrent for high micropulse energies is correlated with the bleaching of the HH1-HH2 transition. The value for $\rho_{22} + \rho_{33}$ at $t = 1.5$ ps begins to saturate with the FEL intensity as soon as ρ_{00} and ρ_{11} become equal around $t = 0$. Moreover, the saturation behavior of the non-linear photocurrent can be simulated with high accuracy using the HH2-HH1 relaxation time extracted from the pump-pump data presented in chapter 7.6 as a key parameter. This reproduction of experimental data independent from the pump-pump measurements by calculations based on the extracted HH2-HH1 relaxation time implies a detailed understanding of the involved processes. Moreover, it strongly indicates the high reliability of the extracted HH2-HH1 relaxation time, which allows a consistent interpretation of two uncorrelated experiments.

7.7.4 Simulation results for the pump-pump characteristics

Apart from reproducing the intensity dependence of the non-linear photocurrent, the density matrix calculations are capable of consistently simulating the pump-pump characteristics presented in chapter 7.6. As the photocurrent pump-pump measurements are based on hitting the sample by differently polarized pulses temporally separated by a delay time τ, the envelope of the electric field component of these two pulses has to be modelled accordingly. Equation 7.40 gives the envelope function of the electric field component used for the density matrix simulation of the pump-pump experiments.

$$\boldsymbol{E}_0(t,\tau) = (\frac{1}{2}\boldsymbol{e}_{xy} + \frac{\sqrt{3}}{2}\boldsymbol{e}_z) \cdot \sqrt{I_0} \, e^{-\frac{1}{4}(\frac{t}{\sigma_t})^2} + \\ + \boldsymbol{e}_{xy} \cdot \sqrt{I_0} \, e^{-\frac{1}{4}(\frac{t+\tau}{\sigma_t})^2} \quad (7.40)$$

In this equation, the first term, which is equal to the envelope function used for the simulation of the intensity dependence in the previous section, accounts for the TM polarized pulse of the pump-pump setup. The second term represents the TE pulse, which arrives a time τ before the TM pulse, and is exclusively built up by an $\boldsymbol{e}_{xy}$ component. The combined field envelope is thus dependent on both t and τ. In consistency with the experiment, the purely

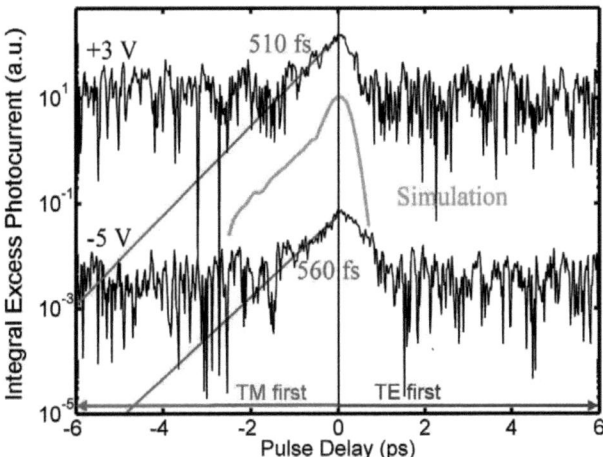

Figure 7.20: Simulated pump-pump characteristics and comparison to the experiment. The green line shows the summed final values for the continuum occupation $\rho_{22}+\rho_{33}$, gained from density matrix calculations for a series of delays τ. The baseline has been subtracted in consistency with the presented data for the measured integral photocurrent, which are shown as black lines. The experimental results are well reproduced by the density matrix simulations. The curves have been offset vertically for clarity.

TE polarized pulse reaches the sample first for positive delays τ.

By inserting Eqn. 7.40 into the differential equational system in Eqns. 7.34 to 7.37, and solving the system within a time interval between -30 ps and +30 ps, the final continuum occupation $\rho_{22} + \rho_{33}$ was calculated at $t \gg \sigma_t+ \mid \tau \mid$. The large time range was chosen in order to ensure a complete relaxation of the system into its final configuration at the end of the simulated interval even for large delays between the pulses. As discussed in the previous section 7.7.3, the value of $\rho_{22} + \rho_{33}$ for times $t \gg \sigma_t+ \mid \tau \mid$, at which both pulses have passed and the system has relaxed into its final configuration, is equivalent to the total number of charge carriers excited from the quantum well system by the pair of pump pulses. Since ρ_{22} decays rapidly into ρ_{33}, at $t \gg \sigma_t+ \mid \tau \mid$ this number is given by $\rho_{33}^{\infty}(\tau) \equiv \rho_{33}(t \gg \sigma_t+ \mid \tau \mid,\tau)$. Thus, $\rho_{33}^{\infty}(\tau)$ can be directly compared to the integral photocurrent measured in the experiment.

Now, for reproducing the result of pump-pump experiments rather than the intensity dependence of the photocurrent, the simulation was carried out for a series of delays τ and a fixed intensity I_0, which was equivalent to a micropulse energy of 1 nJ (see Fig. 7.19(d)). The employed parameters were the same as for the simulations presented in section 7.7.3 and are given in table 7.5. The simulation results for $\rho_{33}^\infty(\tau) - \rho_{33}^\infty(\tau = \infty)$ are shown as a function of the delay by the green line in Fig. 7.20, where the baseline (equivalent to the continuum occupation for large delays) has been subtracted in order to emphasize the exponential shape for negative delays and to compare simulation and experiment. The experimental results are presented as black curves in analogy to Fig. 7.16.

As seen in Fig. 7.20, the shape of the experimentally obtained pump-pump characteristics is accurately reproduced by the density matrix simulations. The small wiggles in the calculated characteristics are attributed to numerical instabilities. The simulated excess photocurrent characteristics feature the strong asymmetry of the experimental data. For negative delays, the curve shape of the simulation results is dominated by the parameter τ_{10}, a behavior which is consistent with the experiment. In case of positive delays, the shape of the simulated characteristics is determined by the temporal overlap of the two pulses in Eqn. 7.40, where once more the measurement results are well reproduced.

Simulated center peak

While the simulated curve in Fig. 7.20 decays exponentially with the HH2-HH1 relaxation time τ_{10} for $\tau < -0.7$ ps, the characteristics exhibit a steeper decay around zero-delay. In this region of small negative delay ($-0.7 < \tau < 0$), the simulated characteristics are pulse-limited, as seen from the symmetry compared to small positive delays. Now, the strength of this simulated feature is found to be strongly dependent on the employed value for the phase relaxation rate Γ, as illustrated by the plots in Fig. 7.21. This figure presents the simulated characteristics for $\rho_{33}^\infty(\tau) - \rho_{33}^\infty(\tau = \infty)$ in analogy to that shown in Fig. 7.20, where the individual curves have been normalized to their maximum value. For the calculations leading to the plots in Fig. 7.21, the same parameters as for that resulting in the green curve in Fig. 7.20 were used, the only exception being the phase coherence time $\frac{1}{\Gamma}$, which was varied. As seen from Fig. 7.21, the narrow center peak feature is more strongly pronounced in case of long phase coherence times, while it becomes less significant with increasing phase relaxation rate Γ.

The reason for this dependence can be deduced from the rate equations, on which the simulations are based. Equation 7.41 presents the expression for the time evolution of the total continuum state occupation $\rho_{22} + \rho_{33}$ as derived from Eqns. 7.36 and 7.37, where C is a

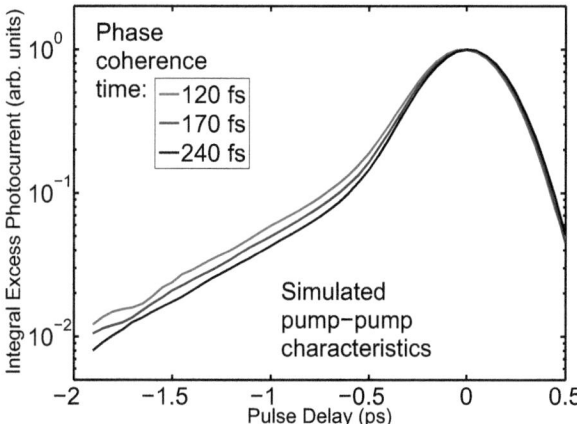

Figure 7.21: Simulated pump-pump characteristics for a series of phase coherence times $\frac{1}{\Gamma}$. The influence of the narrow coherent center peak around zero-delay depends on the phase coherence time, where the feature becomes less pronounced for short coherence times. Apart from the phase relaxation rate Γ, the calculations were based on the same parameters as those leading to the green curve in Fig. 7.20.

time-dependent factor.

$$\frac{d}{dt}(\rho_{22} + \rho_{33}) = C(t) \cdot (\boldsymbol{\mu}_{21} \cdot \boldsymbol{E}_0)^2 (\rho_{11} - \rho_{22}) +$$

$$+ C(t) \cdot \frac{1}{\hbar^2 \Gamma^2} \cdot [(\boldsymbol{\mu}_{21} \cdot \boldsymbol{E}_0)^4 (\rho_{11} - \rho_{22}) + (\boldsymbol{\mu}_{10} \cdot \boldsymbol{E}_0)^2 (\boldsymbol{\mu}_{21} \cdot \boldsymbol{E}_0)^2 (\rho_{00} - \rho_{11})] \quad (7.41)$$

The first term in this equation can be easily identified with the excitation of charge carriers *occupying* level 1 into the continuum. It is linearly depending on the radiation intensity. The terms in the second line of Eqn. 7.41 exhibit a quadratic intensity dependence and are thus associated with two-photon processes. Due to the factor $\frac{1}{\hbar^2 \Gamma^2}$, the contribution of these multi-photon terms strongly depends on the de-phasing rate of the involved states. In order to illustrate the influence of one- and two-photon terms on the pump-pump characteristics, two special cases leading to the absolute domination by one of the two processes shall be discussed in the following.

Dominance of the coherent term

In case of absence of any significant level-1-occupation ($\rho_{11} \sim 0$) and a comparably small de-phasing rate ($\Gamma \ll \frac{\boldsymbol{\mu}_{21} \cdot \boldsymbol{E}_0(0,0)}{\hbar}$ and $\Gamma \ll \frac{\boldsymbol{\mu}_{10} \cdot \boldsymbol{E}_0(0,0)}{\hbar}$), the first term does not contribute significantly to the rate equation, as ρ_{22} is small at any time due to a fast relaxation into non-resonant continuum states. The situation of an insignificant level-1-occupation is e.g. reached, if $\tau_{10} \ll \frac{1}{\Gamma}$. In such a case, the equation of evolution for the continuum occupation is dominated by the term given in Eqn. 7.42, where $\boldsymbol{E}_0$ is given by Eqn. 7.40.

$$\frac{d}{dt}(\rho_{22} + \rho_{33}) \simeq C \cdot \frac{1}{\hbar^2 \Gamma^2} \cdot (\boldsymbol{\mu}_{10} \cdot \boldsymbol{E}_0)^2 (\boldsymbol{\mu}_{21} \cdot \boldsymbol{E}_0)^2 \rho_{00} \qquad \text{for } \rho_{11} \sim 0 \tag{7.42}$$

This term can be associated with a coherent two-photon absorption from the ground level 0 into the continuum level 2. As the coherent continuum level 2 is empty in comparison to level 0 ($\rho_{22} \ll \rho_{00}$), Eqn. 7.42 exclusively depends on ρ_{00}. Due to the absence of any refilling processes of ρ_{00} on the time scale of the time-resolved experiments in case of zero-ρ_{11}-occupation, the level-0-occupation remains constant after the first pulse of the experiment has passed. Therefore, no relaxation processes can be observed in the pump-pump characteristics, and $\rho_{33}^{\infty}(\tau)$ differs from $\rho_{33}^{\infty}(\tau = \infty)$ only in the case of an overlap between the two pulses due to the non-linear dependence of Eqn. 7.42 on $\boldsymbol{E}_0(t)$. Thus, the shape of the pump-pump characteristics $\rho_{33}^{\infty}(\tau) - \rho_{33}^{\infty}(\infty)$ is pulsewidth-limited.

Dominance of the occupation-based term

Now, while the situation described by Eqn. 7.42 requires long coherence times, this subsection discusses the opposite case of strong de-phasing and a significant level-1-occupation. A very short phase coherence time results in a negligible contribution of the terms in Eqn. 7.41 depending quadratically on the intensity of the electromagnetic radiation. As a consequence, the temporal evolution of the continuum occupation $\rho_{22} + \rho_{33}$ is determined by the term given in Eqn. 7.43.

$$\frac{d}{dt}(\rho_{22} + \rho_{33}) \simeq C(t) \cdot (\boldsymbol{\mu}_{21} \cdot \boldsymbol{E}_0)^2 (\rho_{11} - \rho_{22}) \qquad \begin{array}{l} \text{for } \Gamma \gg \dfrac{\boldsymbol{\mu}_{21} \cdot \boldsymbol{E}_0(0,0)}{\hbar} \\ \text{and } \Gamma \gg \dfrac{\boldsymbol{\mu}_{10} \cdot \boldsymbol{E}_0(0,0)}{\hbar} \end{array} \tag{7.43}$$

In case of a negligible phase coherence of the involved quantum states, the generation of photocurrent is thus exclusively based on establishing a non-equilibrium level-1-occupation. Charge carriers in the ground state are only able to contribute to a photocurrent by being excited into level 1 first, and then absorbing a second photon into the continuum. A direct, coherent two-photon absorption is not feasible. As a consequence of the domination of

Eqn. 7.43 by the term depending on $(\rho_{11} - \rho_{22})$ and the strong temporal variation of ρ_{11}, the shape of the pump-pump characteristics $\rho_{33}^{\infty}(\tau) - \rho_{33}^{\infty}(\infty)$ is determined by the decay of ρ_{11}, a situation which is optimal for extracting the studied intersubband relaxation time.

Contribution by both processes

Now, in cases somewhere in between the two extremes described in the previous subsections, any of the terms in Eqn. 7.41 contributes to the differential equation for the continuum occupation. As seen from the simulation result presented in Figs. 7.20 and 7.21, under such conditions the pump-pump characteristics are dominated by the relaxation-based slow decay for large delay values, while a sharper, pulse limited peak originating from coherent two-photon processes rises above this occupation-based contribution in the vicinity of zero-delay. As seen from Fig. 7.21, the strength of the contribution by the coherent center peak to the pump-pump characteristics $\rho_{33}^{\infty}(\tau) - \rho_{33}^{\infty}(\infty)$ is predicted by the simulations to be strongly dependent on the phase relaxation rate Γ. For small phase coherence times (large values of Γ), the coherent two-photon process is too weak to produce an observable excess current.

As no pulse-limited center peak was observed for negative delays during the experiments, it can be assumed that the estimated phase relaxation rate in table 7.5 was chosen too low. However, the deviations between simulation and experiment occur only in close vicinity to zero-delay, and the presence of a pulse-limited center peak originating from coherent two-photon-absorption does not influence the shape and interpretation of the exponential decay with τ_{10} for larger negative delays. It shall be mentioned, that during the photocurrent pump-pump experiments on different samples, a symmetric pulse limited center peak could be observed in addition to the lifetime-induced asymmetry, as will be discussed in chapter 8.5.2.

The fact that the density matrix simulations give results accurately reproducing the experimentally obtained data shows that the model includes all processes relevant for the interpretation of the experiment. The consistency between simulation and experiment strongly supports the model used for extracting the HH2-HH1 intersubband relaxation time and proves the reliability of the obtained values. As demonstrated by simulating both the intensity dependence of the photocurrent in the previous section 7.7.3 and the pump-pump characteristics with *one consistent set of parameters*, the processes leading to the observed experimental results are understood in detail and the interpretation of the measured data is thoroughly founded.

7.8 Summary and conclusion

The photocurrent pump-pump experiments presented in this chapter enabled the first *direct* experimental determination of intersubband relaxation times for transition energies above the optical phonon energy in a SiGe heterostructure. The experiments were carried out on a *p*-type SiGe heterostructure based on the quantum cascade design, for which electroluminescence was reported in [12]. This enabled the time-resolved study of intersubband processes in a structure, which closely resembles an actual emitter device.

The measurements were carried out at the free-electron-laser FELIX at the FOM Rijnhuizen, which is capable of emitting infrared micropulses with a duration of about 200 fs. This temporal pulse width proved sufficient for resolving the studied intersubband relaxation times. Prior to the time-resolved experiments, the HH1-HH2 transition of the deepest quantum well of the investigated structure H019 was characterized by both non-linear photocurrent spectroscopy and voltage-pulsed transmission spectroscopy. The experimentally obtained value of 160 meV was consistent with that gained from bandstructure simulations.

For the pump-pump experiments carried out in the optical configuration of a Michelson interferometer, the photocurrent through the sample was used as the measured correlation quantity, in contrast to transmission pump-probe experiments, where the intensity of the probe beam transmitted through the sample is studied. For this purpose, a bias was applied to the top contact of the investigated mesa in respect to its bottom contact, and the current flowing vertically through the structure was measured. The FEL wavelength was tuned into resonance with the HH1-HH2 transition, where both involved states are of bound nature. A ground state charge carrier can thus exclusively contribute to the photocurrent by absorbing two photons. This non-linearity of the photocurrent response is the basis for the pump-pump experiments. The advantage of measuring the current instead of the transmission characteristics, lies among others in the high sensitivity of this method, resulting in a sufficiently high signal-to-noise ratio of the experimental data.

In order to confirm the non-linear dependence of the photocurrent through the sample on the micropulse energy of the FEL, and to determine the optimal micropulse energy for the time-resolved experiments, the photocurrent was measured in dependence of the FEL intensity. The results exhibit clear superlinear characteristics for low micropulse energies and a saturation behavior for higher intensities.

During the time-resolved experiments, the integral photocurrent through the sample was measured as a function of the delay between two FEL pulses, where one pulse was of mixed

TM and TE polarization, and one exhibited a TE component only. The first (TM) pulse excited a non-equilibrium HH2 occupation, while the second (TE) pulse probed this occupation into the continuum and generated an excess photocurrent. As the HH2 occupation relaxed into the HH1 ground state with the HH2-HH1 intersubband decay time, the amount of excess current generated by the second pulse decayed with the delay between the two FEL pulses. Thus, by fitting a single exponential decay to the obtained excess photocurrent characteristics, an intersubband relaxation time of about 550 fs could be directly extracted from the photocurrent pump-pump data. The measurement was carried out for two different bias values, where no change in the relaxation time could be observed.

By exploiting the selection rules of the HH1-HH2 transition along with the cross-polarization of the two pulses, the time resolution of the experimental setup could be determined. This was done by measuring the pulse-limited excess photocurrent for a configuration, in which the TE pulse hit the sample first, unable to establish a non-equilibrium HH2 occupation. The asymmetry of the photocurrent pump-pump characteristics in respect to the impact order of the differently polarized pulses clearly proves, that the monitored intersubband relaxation time is above the time resolution of the setup.

The simple interpretation of the experimental data, which allowed a direct extraction of the intersubband relaxation time by a single-exponential fit, was confirmed by more sophisticated density matrix simulations. These simulations gave a detailed insight into the temporal evolution of the occupation of the individual quantum states in the system. By numerically solving rate equations derived from the density matrix equations for a four-level-system, *both* the experimentally determined intensity dependence of the photocurrent and the pump-pump characteristics could be reproduced at high accuracy with one consistent set of parameters, one of which was the extracted HH2-HH1 intersubband relaxation time. This not only confirms a detailed understanding of the processes influencing the experiments, but further supports the reliability of the extracted intersubband relaxation time.

In conclusion, the value of 550 fs obtained by the photocurrent pump-pump experiments in this work poses the by far most reliable experimentally determined figure for the intersubband relaxation time in the SiGe system for transition energies above the LO phonon energy. It was *directly* extracted from the experimental data with a good signal-to-noise ratio by a single exponential fit. No hole heating contributions were observed during the experiments. Our experimental results did not require any de-convolution of heating effect, the decay of the measured quantity was exclusively determined by the HH2-HH1 intersubband relaxation process. Furthermore, the experiments were carried out on a biased and current-carrying structure. Therefore, the experimental conditions of this work resembled that of an operating quantum cascade device closer than in any other reported time-resolved experiment on SiGe heterostructures up to date. This fact further increases the reliability of the obtained relaxation time value in respect to the design of biased emitter devices.

The availability of accurate experimental values for intersubband relaxation times in the SiGe system is of crucial importance for the concept development, detailed design and simulation of SiGe quantum cascade emitter devices on the road to a SiGe QCL, especially in respect to the non-resonant behavior of the lifetime-limiting optical phonon scattering in this material system. The value obtained by the photocurrent pump-pump experiments of this work is by a factor two higher than the value reported by Kaindl et al. in [114], which is in favor of reaching population inversion in a SiGe quantum cascade structure.

Chapter 8

Bias-tuning of intersubband relaxation times

QCLs featuring a diagonal active region exhibit non-radiative intersubband relaxation times associated with the lasing transition, which are orders of magnitude higher than those with spatially direct active regions. Diagonal transitions can therefore be employed for relaxing the requirements for population inversion. As discussed in chapter 6.1.4, in the SiGe material system the concept of diagonal transitions forms a potential way towards reaching population inversion at all.

Motivated by this potential, time-resolved experiments were carried out on a series of SiGe quantum well structures by Pidgeon et al. [117]. These experiments aimed at correlating the extracted intersubband relaxation times with the structural difference between the samples, which is expected to result in a varying spatial overlap between two states. As discussed in chapter 6.3.2, the generalization of the results reported in [117] to diagonal transitions as such is non-trivial. In [116], Bormann et al. reported on the design-induced increase of intersubband relaxation times between states energetically separated by more than the LO phonon energy. The reported decay time increase with decreasing overlap between the involved states, which was induced by slightly varying the design of the individual quantum cascade emitter samples, was concluded from comparing the samples' integral electroluminescence intensities. In the course of Bormann's work no time-resolved experiments were carried out. As discussed in chapter 6.1.4, this indirect way of extracting intersubband relaxation times involves a cumbersome amount of assumptions and uncertainties and further reduces the reliability of conclusions drawn for diagonal transitions in general.

For the experiments presented in both [117] and [116], the amount of overlap between the two studied quantum states was controlled via modifying the growth parameters of the respective samples. Even though this control of the spatial overlap via structural design is attractive, it suffers from the intrinsic disadvantage of requiring the comparison of *different* samples, increasing the amount of unknown parameters possibly influencing the experimental results and therefore degrading their reliability. Now, it is well known that for quantum cascade devices based on diagonal transitions, a change in the applied bias has a significant effect on both the emitted wavelength and the spatial overlap between the excited and ground states of the lasing transition [65, 92, 116]. It is thus expected that the voltage applied to such a structure influences the intersubband relaxation time associated with the diagonal transition significantly. Controlling the overlap between two quantum states and the associated relaxation times by the applied bias offers a way to perform novel dynamics experiments and has potential benefits for understanding incoherent laser dynamics. As varying the voltage applied to a SiGe heterostructure offers a potential means of transforming a spatially direct intersubband transition into a diagonal one, bias-dependent time-resolved experiments allow a comparative study of spatially direct and diagonal transitions *in one and the same* structure. The consequential minimization of unknown parameters influencing the experimental results enables a reliable quantitative study of the influence of the overlap between two heterostructure states on the associated intersubband relaxation time. Despite the potential of a bias-tuning of intersubband relaxation times for the study of intersubband dynamics and its possible application for intersubband emitters, up to now no *direct* observation of a bias-tuning of intersubband decay times has been reported.

During the time-resolved measurements carried out on the biased structure H019, as presented in the previous chapter 7, no dependence of the intersubband relaxation time on the applied voltage could be observed. In chapter 7.6.4, this was explained by the insensitivity of the overlap between the probed excited HH2 state and the HH1 ground state to the applied bias. This insensitivity is, in turn, a direct consequence of the stable quantum cascade emitter design, on which the sample structure was based on. Therefore, the introduction of a bias-tunability of intersubband relaxation times into a SiGe heterostructure requires a design approach different from the quantum cascade injector structure of H019. The following section presents the design concepts for two p-type SiGe heterostructures aiming at a band structure of highly bias-sensitive nature, whose voltage-dependent intersubband relaxation can be monitored by photocurrent pump-pump experiments in analogy to the experiments discussed in chapter 7.6.3.

The results presented and discussed in this chapter were published in reference [133].

Table 8.1: MBE growth sequence and doping concentration of T037 and T038:

	T037			T038	
Thickness	Ge concentration	Doping concentration	Thickness	Ge concentration	Doping concentration
10 nm	20 %	0	17.5 nm	18 %	0
14 nm	24 %	0	14 nm	24 %	0
0.8 nm	0	$1 \cdot 10^{18} \text{cm}^{-3}$	0.8 nm	0	$1 \cdot 10^{18} \text{cm}^{-3}$
6.1 nm	31 %	$1 \cdot 10^{18} \text{cm}^{-3}$	6.1 nm	31 %	$1 \cdot 10^{18} \text{cm}^{-3}$
0.8 nm	0	$1 \cdot 10^{18} \text{cm}^{-3}$	0.8 nm	0	$1 \cdot 10^{18} \text{cm}^{-3}$
14 nm	24 %	0	14 nm	24 %	0
10 nm	20 %	0	17.5 nm	18 %	0

8.1 Design and fabrication

8.1.1 Structure design and growth

The fundamental requirement for the design of the two samples T037 and T038, on which the experiments presented in this chapter were performed, was to provide a quantum well system as simple as possible for studying diagonal transitions under biasing. In order to increase the time scale of the studied intersubband transition, an energetic spacing between its excited and ground states below the LO phonon energy was chosen. For transitions of such low energies the associated relaxation times are expected to lie in the range between 10 and 30 ps, and are well accessible by time-resolved experiments [80, 118, 120], as discussed in chapter 6.3.2. In order to simplify the investigated exemplary quantum well system and thus reduce the number of involved levels, the transition between the HH1 and LH1 states is chosen for the intersubband relaxation studies. In contrast to the HH2-HH1 relaxation mechanism of sample H019 in chapter 7, where the most efficient relaxation path is that via the intermediary LH1 state, the LH1-HH1 decay does not exhibit any multi-step relaxation pathes due to the absence of intermediate states.

The key feature of the design for both T037 and T038 is a central quantum well with a LH1-HH1 spacing below the LO phonon energy. The LH1-HH1 relaxation time is to be determined by non-linear photocurrent experiments. Therefore, the LH1 state must not couple to the valence band continuum of the structure, as in such a case charge carriers excited into the LH1 state could directly contribute to the photocurrent without the absorption of an additional photon. Now, in order to spatially shift the excited LH state from the HH ground state, a side quantum well is required. In order to ensure an efficient coupling between the LH state of the central well and that of the side well, the possibility to energetically align

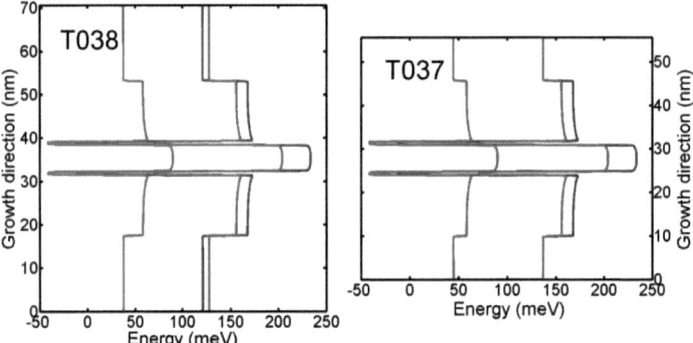

Figure 8.1: Valence band edges of the heavy- (blue), light- (green), and split-off- (red) bands for one period of T037 and T038 at zero applied bias. The band bending is due to charge carrier redistribution. The main differences between the two samples' respective band edges is induced by the lower germanium concentration of the separation layer of T038, resulting in a larger band offset to the side well region, and by its larger width for T038 as compared to T037.

these two states by applying a field to the sample is demanded. Furthermore, the flow of dark current by tunnelling has to be suppressed by de-coupling the quantum wells of subsequent periods by introducing a barrier of sufficient height.

Table 8.1 shows the structural parameters for one period of the two samples T037 and T038, where the design of these heterostructures followed the considerations discussed above. In order to allow strain-symmetrized growth, a $Si_{0.8}Ge_{0.2}$ (100) pseudo-substrate was employed. 15 periods of the sequences given in table 8.1 were grown pseudomorphically on top of a highly boron doped ($4 \cdot 10^{18}$ cm^{-3}) $Si_{0.8}Ge_{0.2}$ bottom contact layer by G. Mussler and D. Grützmacher at the Forschungszentrum Jülich, Germany, using low-temperature MBE. On top of the 15 periods of active region another highly p-type doped contact layer was grown.

Fig. 8.1 presents the band edges of the HH, LH and SO bands induced by the layer sequence of the heterostructure, as calculated by bandstructure simulations (for details on the calculations see [20, 21]). As suggested by the design considerations given above, each structural period forms a deep center quantum well, which is separated from two symmetric shallow quantum wells by a thin barrier formed by a layer of pure silicon. As further demanded by the design concept discussed above, the quantum well regions of neighboring periods are separated by thick barriers of a Ge content roughly matching the pseudosubstrate composition. As seen from the growth parameters in table 8.1 and the band edge profiles in Fig. 8.1, T037 and T038 exclusively differ in the width and energetic height of this separation between subsequent well regions. Sample T037 features a 20 nm wide separation layer of a Ge concentration of 20 %. In contrast, the barrier layer of T038 exhibits a width of 35 nm and a Ge concentration of 18 %. Thus, the barrier of T038 is both spatially wider and energetically higher than that of T037. The effect of this design difference between the two samples on their electro-optical properties is discussed in detail in chapter 8.2.1.

8.1.2 Sample processing

Once more, the preparation of the active structure into contacted, separated mesas of a well-defined surface is required for biasing the samples and to allow a voltage-tuning of the spatial charge carrier distribution, as well as for simultaneously measuring the photocurrent. Both samples subject to the experiments presented in this chapter were processed in analogy to sample H019 in chapter 7.1.2. Apart from using a novel mask aligner model for T037 and T038, the processing differed from that of H019 in the use of an AlSi target for these samples, replacing the evaporation of alternating layers of Si and Al for the top contact metallization. Analogous to the previously used AlSi sandwich, the mixed target containing 1% of Si was employed to prevent the formation of short-cuts between the contact layers by spiking through the active region. For the sake of completeness, the processing work-flow for both samples

Table 8.2: Sample processing steps for T037 and T038

Step No.	Process	Equipment	Parameters
1	cleaning	acetone, methanol	
2	resist deposition	resist 1818, spinner	40 s, 4000/s
3	softbake	oven	90°C, 15 min
4	photolithography mesa	mask aligner EV620	4.5 s exposition
5	developing	developer	1 min
6	mesa etching	reactive-ion-etcher	100% SF_6, 50% O_2, 40 mT, 15% RF, T037B: 4.2 min, 1015 nm T038A: 6.5 min, 1300 nm
7	cleaning	acetone, methanol	
8	resist deposition	resist 1818, spinner	40 s, 4000/s
9	softbake	oven	90°C, 15 min
10	photolithography top contact	mask aligner EV620	4.5 s exposition

Step No.	Process	Equipment	Parameters
11	developing	developer	1 min
12	native oxide removal	hydrofluoric acid	
13	vapor deposition	evaporation chamber, AlSi target	200 nm AlSi
14	lift-off	acetone	
15	cleaning	acetone, methanol	
16	resist deposition	resist 1818, spinner	40 s, 4000/s
17	softbake	oven	90°C, 15 min
18	photolithography bottom contact	mask aligner EV620	4.5 s exposition
19	developing	developer	1 min
20	native oxide removal	hydrofluoric acid	
21	vapor deposition	evaporation chamber, Al target	200 nm Al
22	lift-off	acetone	
23	contact alloying	rapid thermal annealing oven	380°C, 20 sec $N_2 + H_2$
24	cleaning	acetone, methanol	
25	resist deposition	resist 1818, spinner	40 s, 4000/s
26	softbake	oven	90°C, 15 min
27	photolithography gold contact	mask aligner EV620	4.5 s exposition
28	developing	developer	1 min
29	vapor deposition	evaporation chamber, Ti, Au targets	10 nm Ti 100 nm Au
30	lift-off	acetone	
31	backside polishing	polishing wheel paper: 4000, 1000 diamond spray: 1 μm, 0.25 μm	
32	facette polishing	polishing wheel paper: 4000, 1000 diamond spray: 1 μm, 0.25 μm	30° to (100) surface
33	bonding	bonder, Al wire	70°C

T037 and T038 is given in table 8.2.

For sample T037, the experiments presented in this chapter were carried out on a mesa of 400 × 400 μm^2, which featured a frame-shaped top contact leaving the sample surface mostly free of metallization. In case of sample T038, a 900 × 900 μm^2 mesa completely covered by the aluminum top contact was employed for the lifetime studies. The difference in the geometry of the top metallization between the two mesas investigated for the respective samples results in a differing TM:TE ratio of the pump beam labelled 'TM' in Fig. 7.10 at the position of the optically active region, as the metal surface suppresses the TE component of the incoming radiation. However, the differing metallization did not effect the pump-pump characteristics, as seen in chapter 8.5.

8.2 Band structure simulations for T037 and T038

8.2.1 Band structure results

For designing the structures, on which the bias-dependent pump-pump experiments of this chapter were carried out, and whose fabrication is described in chapter 8.1, simulations based on a six band $\bm{k.p}$ envelope function model were performed in analogy to those described several times throughout this work and in more detail in references [20, 21]. The plots in Fig. 8.2 (a) and (b) present the simulation results for T037 and T038, respectively, for similar applied electric fields of 12 kV/cm (T037) and 14 kV/cm (T038). As seen in this figure, the HH1 state of the deep central well forms the ground state of each period. The transition energy from the HH1 ground state of the central well to its excited LH1 state is predicted to be around 30 meV for both structures, well below the lowest LO phonon energy in the SiGe system (38 meV for the Ge-Ge mode). Therefore, the expected range for the LH1-HH1 relaxation times in both studied structures lies between 10 and 30 ps according to references [80, 118, 120]. The LH1-HH1 decay time of the deep well in T037 and T038 should thus be experimentally well accessible (see chapter 6.3.2).

The plots in Fig. 8.2 show that the HH1 ground state of the central well is strongly confined. Now, the structures are designed in a way which enables the coupling of the LH1 state of the central well to that of the wider shallow side well for high applied fields. As a consequence, the confinement of the excited LH1 state is predicted to be strongly dependent on the applied voltage. This is due to the fact that at sufficiently high applied field, charge carriers excited into the LH1 state are expected to tunnel into the continuum via the shallow side well, analogous to the tunnelling processes in sample K091 of chapter 2. In case of K091, the extraction of HH2 carriers via a side well enabled the generation of a linear photocurrent

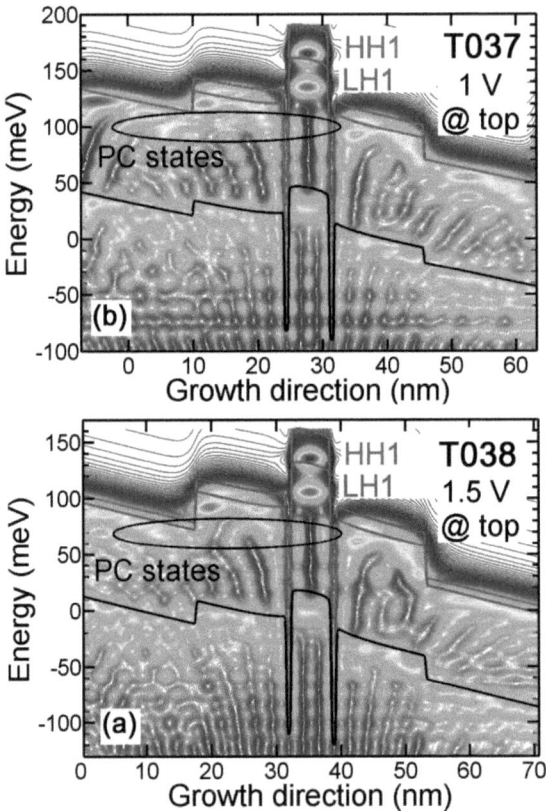

Figure 8.2: Band structure of T037 and T038 at similar applied fields of 12 kV/cm (T037) and 14 kV/cm (T038). The plots show the calculated heavy-, light- and split-off- (red, green, black) hole band edges and a contour plot of the absolute squared wave functions. The spreading of the eigenstates along the energy-axis accounts for their homogenous broadening. The conducting final states, which can be reached from the HH1 ground state by the absorption of two photons of the HH1-LH1 energy, are stronger confined in T038 than in T037 at equal applied fields, which is due to the energetically higher and spatially wider separation barrier of T038. However, the same holds true for the states in the shallow side well. The relevance of the confinement is discussed in the text.

by the optical excitation from HH1 to HH2. The confinement of the LH1 state of the shallow side well is induced by the barrier created by the separation layer. Thus, for the two samples the strength of the LH1 confinement at a given bias depends crucially on the energetic height and width of the separation barrier between the quantum wells of subsequent structural periods. The separation barriers are the only region, in which the samples T037 and T038 differ. The following considerations discuss the motivation for the variation of the barrier dimensions between the two samples and highlight the crucial dependence of the success of photocurrent pump-pump experiments on these parameters.

Role of the difference in the separation barrier between T037 and T038

The basis of any pump-probe experiment independent of the detailed nature of the measured sample property is the non-linear dependence of this observable quantity on the FEL intensity. As discussed in detail in chapters 7.4 and 7.6.4, this non-linear behavior results in an excess photocurrent induced by the correlation between the two pulses, from which the investigated relaxation time can be extracted. In case of photocurrent pump-pump experiments, the basic requirement for non-linear characteristics is a strongly confined excited state of the studied transition. As pointed out above, for samples T037 and T038 the crucial parameters for the fulfillment of this condition are the composition and width of the separation layer.

For considering the influence of the separation layer parameters on the outcome of photocurrent pump-pump experiments in detail, the system presented in Fig. 8.3 shall serve as a model. The relaxation time between the LH1 and the HH1 states is to be determined by photocurrent pump-pump experiments. For this purpose, a non-equilibrium LH1 occupation is excited by the first FEL pump beam 1, which is then probed into the continuum by the pump beam 2 at an average rate Γ_e for the duration of the pulse. Γ_e is strongly dependent on the pulse intensity.

Now, in contrast to the HH2-HH1 transition investigated on sample H019, where the excited state is well confined in the quantum well 1, holes excited to the LH1 state of the model in Fig. 8.3 are able to tunnel into the continuum at a rate Γ_t. In such a case, the photocurrent through the structure exhibits a component linearly dependent on the intensity of an illuminating laser source in resonance with the HH1-LH1 transition. In the following, the influence of such a one-photon component on the result of pump-pump experiments is discussed case by case for different relations between the tunnel rate Γ_t, the investigated intersubband relaxation rate Γ_{10} and the optical emission rate Γ_e into the continuum.

$\boldsymbol{\Gamma_t \gg \Gamma_e, \Gamma_{10}}$: In case of a highly efficient tunnelling of LH1 carriers into the valence band

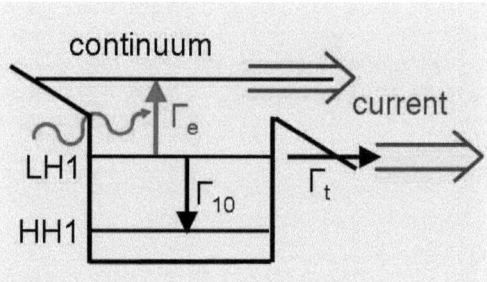

Figure 8.3: Schematics for the illustration of the influence of tunnelling processes on pump-pump photocurrent experiments. In this model, HH1 carriers can contribute to the photocurrent symbolized by the blue arrows by absorbing a FEL photon into the LH1 state and directly tunnelling into the continuum (one-photon-current channel) or absorbing a second photon into the continuum (two-photon-channel). Details are given in the text.

continuum due to a low barrier or a strong applied field, charge carriers are emitted from the quantum well into the continuum by absorbing a single FEL photon in resonance with the HH1-LH1 energy and result in a strong linear photocurrent. According to our assumptions, the emission of LH1 carriers into the continuum is much more efficient than the excitation of these carriers into the continuum by absorbing a second FEL photon. Thus, the two-photon absorption process symbolized by the upper blue arrow in Fig. 8.3 does not contribute significantly to the photocurrent through the sample, and the pump-pump characteristics are *free of any excess photocurrent*, as due to the linearity not even the pulse overlap generates additional current. The only possible non-linear feature in the pump-pump characteristics under such conditions could be a dip in the integral photocurrent near zero-delay, originating from bleaching the HH1-LH1 transitions. These are the conditions, under which a QWIP like that presented in chapter 2 operates.

$\Gamma_e \gg \Gamma_t$ and $\Gamma_t \sim \Gamma_{10}$: In case of a tunnelling rate similar to the LH1-HH1 relaxation rate, charge carriers excited by the pump beam 1 into the LH1 state escape the quantum well by tunnelling into the continuum and contribute to the integral current baseline. However, as the excitation of LH1 carriers into the continuum by the pump beam 2 is according to our assumptions much more efficient than the tunnelling process, an excess current proportionally dependent on the non-equilibrium LH1 population is generated by the second FEL pulse, where the proportionality constant depends on the actual ratio between Γ_t and Γ_e. Thus, the decay of the LH1 occupation *can* be monitored by the generation of an excess current signal by the pump beam 2. However, the total depopulation rate of the LH1 level is in this case given by $\Gamma_t + \Gamma_{10}$. If the influence of the tunnelling process is ignored during data evaluation, the LH1 decay time extracted from experimental data gained under such conditions thus *overestimates* the LH1-HH1 relaxation rate Γ_{10}. As the tunnelling efficiency increases with the field in the structure, the total LH1 depopulation time drops with the applied bias. If furthermore the voltage is increased beyond a point, where the total LH1 depopulation time drops below the time resolution of the experiment, the only remaining excess photocurrent characteristics stem from the overlap of the two pump pulses, as discussed in chapter 7.6.4.

$\Gamma_t \ll \Gamma_e, \Gamma_{10}$: In this case the tunnelling process from the LH1 state into the continuum is too inefficient to influence the LH1 depopulation and therefore the outcome of the photocurrent pump-pump experiments. The photocurrent through the sample does not exhibit any linear component, and the photocurrent baseline is low, ensuring a high signal-to-noise ratio of the excess current characteristics. Thus, inefficient tunnelling through a strong confinement of the excited state of the investigated transition is the ideal configuration for photocurrent pump-pump experiments.

Now, in addition to a strongly *confined* excited state, the successful implementation of photocurrent pump-pump experiments requires *unconfined* continuum states resonantly reached from the LH1 state by the absorption of pump beam 2 radiation. While the FEL photon energy employed for the measurements on H019 as presented in chapter 7 was high enough (160 meV) to fulfill both prerequisites without conflicting requirements, the experiment carried out on T037 and T038 employed a low photon energy of 30 meV. As seen from Fig. 8.2, this low LH1-HH1 resonance energy implies that confining the excited LH1 state well by increasing the energetic height of the separation layer blocks the resonant continuum states. In other words, the need for unconfined continuum states and a confined LH1 state are competing requirements, and an optimal balance has to be sought in order to guarantee reliable conditions for the pump-pump experiments.

As the choice for the spatial and energetic dimensions of the separation layer is expected to decide over the success of the pump-pump experiments, these parameters were varied for the design of T037 and T038. The separation layer of T038 is both thicker and energetically higher than that of T037, resulting in a stronger confinement of both the continuum states and the LH1 state for T038 at a fixed applied field. This fact is reflected in the results of the band structure calculations in Fig. 8.2, where the continuum states resonantly reached from the LH1 state are energetically above the separation barrier for T037, while the barrier approaches this resonant level for T038. Nevertheless, for both samples the pump-pump experiments turned out successfully, even though a separation-induced difference could be observed in their spectral photocurrent characteristics, as discussed in chapter 8.3.

8.2.2 Selection rules

In the course of studying the ultrafast relaxation processes of sample H019 in chapter 7, the selection rules of the involved transitions turned out to be a key issue for unambiguously distinguishing the photocurrent decay induced by the investigated relaxation process from the time resolution of the experimental setup (see chapter 7.6.4). The selection rules exhibited by the structures T037 and T038 shall be exploited in the same way.

In order to predict the dependence of transitions in structure T037 and T038 on the polarization of the involved radiation, once more band structure calculations were performed. The possible final states for an optical transition from the HH1 ground state were mapped in respect to their relevance according to these simulations, as described in chapter 2.2.2. The results of these theoretical considerations are presented in Fig. 8.4. For comparison, Fig. 8.2 presents the total, un-weighted band structure of these samples. The data in Fig. 8.4 implies that the transition between the HH1 and LH1 states is exclusively strong for TE polarized radiation, where it is nearly forbidden in TM polarization. In other words, the transition, whose non-radiative relaxation time is to be determined in this work, can only be excited by TE radiation, which is in contrast to the HH1-HH2 transition investigated on sample H019 in chapter 7.

The simulation results were further used to predict the strength of an optical transition between the excited LH1 state of the central well and the conducting states reached for a transition energy of 30 meV. As seen from Fig. 8.4, weakly confined continuum states can be reached from the HH1 ground state by the absorption of both TM and TE polarized radiation. In case of TM polarization, the transition to weakly bound states of HH character is strong, where the associated transition energy is about 60 meV. As calculations show,

Table 8.3: Allowed transitions in the central quantum well of structure T037 and T038:

Transition	Energy	TE polarization	TM polarization
HH1-LH1	30 meV	strong	negligible
HH1-HH2 continuum	60 meV	negligible	strong
HH1-LH2 continuum	60 meV	moderate	negligible
LH1-HH2 continuum	30 meV	strong	negligible
LH1-LH2 continuum	30 meV	negligible	moderate

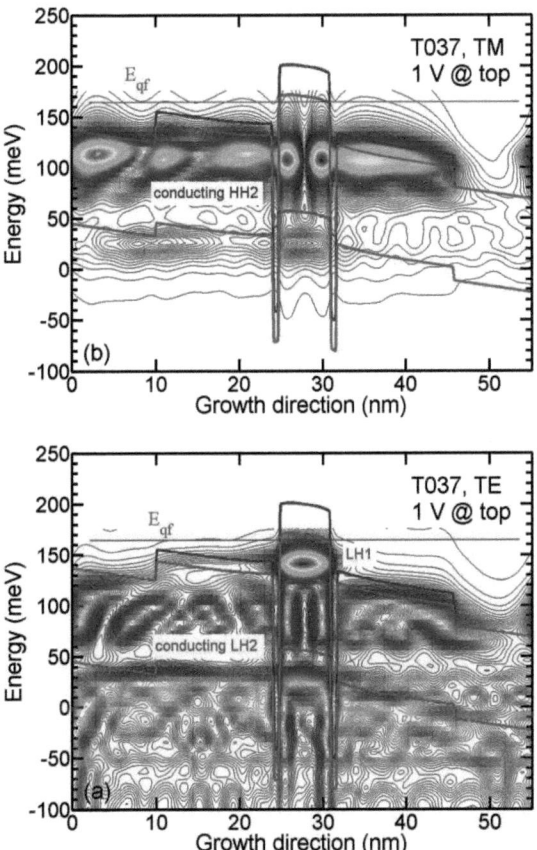

Figure 8.4: Final states relevant for absorption of TE and TM polarized radiation from the HH1 ground state of structure T037 at 1 V bias. The plot is generated by weighting the wavefunctions according to the states' relevance for the absorption of radiation as described in the text. Note that in TE polarization the transition from the HH1 ground state of the central well to its LH1 state is strong. This transition is negligible for TM polarized radiation. Weakly confined conducting states can be reached from the HH1 ground state by the absorption of high energetic photons in both polarizations, though the unconfined LH states reached for TE polarized radiation are located energetically higher.

these HH2 continuum states can also be reached from the LH1 state by the absorption of TE polarized photons of the HH1-LH1 resonance energy. For TE polarized radiation, transitions between the HH1 ground state and LH continuum states are strong, where the energetic spreading of these states is relatively large and involves transition energies larger than 60 meV. As predicted by the simulations, these LH continuum states can be reached from the LH1 state by the absorption of TM polarized photons carrying an energy of 30 meV or more. Thus, calculations show that the transition from the excited LH1 state into weakly confined conducting states is allowed for both TE and TM polarized photons of 30 meV. Table 7.3 gives a summary of the theoretically predicted polarization dependence of the most important transitions in the central well of T037 and T038.

8.3 Spectral characterization of the HH1-LH1 transition

In order to be able to tune the FEL wavelength into resonance with the investigated HH1-LH1 transition of the studied structures T037 and T038, a spectral characterization of this transition was required. As discussed above in section 8.2.1, under ideal conditions for photocurrent pump-pump experiments the excited LH1 state of the investigated transition is strongly bound, and the generation of a photocurrent by exciting a HH1 carrier into LH1 is not feasible. Due to this absence of a linearly generated photocurrent under resonant conditions, the characterization of the HH1-LH1 transition by linear Fourier spectroscopy employing weak radiation sources is not possible. Furthermore, the spectral region of far-infrared, in which the HH1-LH1 transition energy is predicted to lie, is inaccessible by the spectrometer employed for the voltage-modulated transmission experiments on sample H019 (see chapter 7.3.1). Thus, the spectral precharacterization of the transitions in sample T037 and T038 was restricted to nonlinear photocurrent experiments using FELIX.

The FEL photocurrent spectra were recorded in an analogue way as those presented for H019 in chapter 7.3.2 in basically the same setup, in which the time-resolved measurements were performed. This setup is described in chapter 8.4. The spectra were measured for TM (TE pulse blocked) and TE (TM pulse blocked) polarized FEL radiation, and for applied fields (voltages) of 12 kV/cm (1 V) and 14 kV/cm (1.5 V) in case of sample T037 and T038, respectively. In order to compensate for eventual temporal fluctuations in the FEL intensity, the intensity of the beam leaving the beam splitter B2 in Fig. 7.10 normally to the sample direction was measured using a Ge:Ga detector. The photocurrent spectra measured for the sample where then normalized to the simultaneously recorded beam intensity. However, this normalization did not change the basic shape of the determined photocurrent spectra. Additionally, the individual curves have been scaled as discussed later on. The experimental results are presented in Fig. 8.5 (a) and (b).

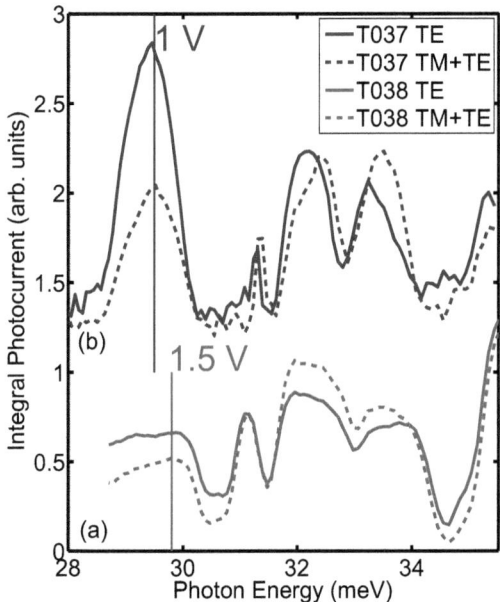

Figure 8.5: Characterization of the HH1-LH1 transition energy of the central well in T037 (b) and T038 (a) by two-photon photocurrent experiments. The photocurrent spectra exhibit a clear resonant enhancement around 30 meV, which is consistent with the HH1-LH1 transition energy predicted by the bandstructure simulations. The resonance is more pronounced in case of purely TE polarized radiation (solid lines) as compared to the data obtained for mixed TM+TE polarization (broken lines). Note that the resonant enhancement is significantly stronger for sample T037 than for T038. The graphs were normalized to the features at higher energies, and were offset for clarity. The vertical lines mark the photon energy, at which the time-resolved experiments were performed.

Each of the spectral characteristics in Fig. 8.5 exhibits a peak around 30 meV, which is associated with the resonantly enhanced two-photon absorption of HH1 carriers into the continuum. As the 30 meV are according to the simulations presented in chapter 8.2 equivalent to the energetic spacing between HH1 states and LH1 states of zero in-plane wavevector, the associated resonant peak is predicted to exhibit the strongest polarization dependence of the features shown in the spectrum. The features in the integral photocurrent spectra at higher photon energies are attributed to slightly off-resonant transitions into higher continuum states, thus involving holes of finite in-plane wave vector as well as coherent two-photon processes without any LH1 occupation. The spectral features at higher energies are thus expected to be less dependent on the polarization of the involved radiation.

A quantitative statement on the polarization dependence of the strength of the respective transitions by the comparison of the obtained photocurrent values is difficult to make due to the influence of boundary conditions on the electric field components near the sample surface. As mentioned before, the top contact metallization of sample T038 significantly suppresses the electric field component parallel to the sample surface at the position of the heterostructure. However, a comparison of the spectral shape of the respective curves shall serve to illustrate the polarization dependence of the respective transitions. For this purpose, the curves given in Fig. 8.5 have been normalized to the features at higher photon energies, as they are expected to be weakly polarization dependent. In case of the two spectra measured on T038 with a FEL micropulse energy of about 160 nJ attenuated by 3 dB, the curve obtained for TM polarized radiation had to be divided by a factor 2.8, which is consistent with the expected suppression of TE polarized radiation. In contrast, for the spectrally resolved data on T037, which was obtained using the same micropulse energy but a different attenuation of 0 dB, the curve measured for TE polarized radiation had to be divided by 1.6. As the top contact of sample T037 was open due to the use of a frame contact, the stronger photocurrent response under the influence of purely TE polarized radiation is consistent with the associated surface boundary conditions.

Fig. 8.5 shows that the resonant enhancement is much stronger for purely TE polarized radiation than in case of mixed polarization, as expected from the selection rules discussed in section 8.2.2. According to the simulated dipole matrix element between the two involved states, the associated transition is strong in TE polarization and negligible in case of TM polarized radiation. However, the resonance can be observed for both orthogonally polarized pump beams, which is due to the waveguide-coupling-induced significant TE component of the pump beam labelled 'TM'.

The comparison of Fig. 8.5 (a) and (b) shows that the integral photocurrent peak associated with the HH1-LH1 resonance is significantly stronger pronounced for sample T037 than for T038. The two investigated samples differ in two features, namely in the structural parameters of the separation layer, as discussed in chapter 8.2.1, and in the top metallization of the respective investigated mesa. Consequently, there are two possible origins for the difference in the spectral photocurrent characteristics between the two samples.

One of the possible explanations for the less pronounced photocurrent peak in Fig. 8.5 (a) as compared to (b) is given by the fact, that the closed top contact metallization of sample T038 suppresses the TE component of the electromagnetic radiation in the vicinity of the sample surface and therefore at the spatial position of the active heterostructure layers. This suppression of the TE field component is expected to reduce the resonantly enhanced photocurrent around 30 meV significantly in comparison to an open mesa like that of sample T037. However, this explanation can be discarded by considering the photocurrent peaks around 32 meV, which are expected to originate from slightly non-resonant two-photon excitation processes into higher continuum states. As in case of purely TE polarized radiation (solid lines) both photocurrent peaks around 30 and 32 meV are expected to depend quadratically on the intensity of the TE field component of the radiation, the change in the ratio between their peak intensities from Fig. 8.5 (a) to (b) cannot be explained by the differing top metallization and the TE field suppression.

The second possible reason for the difference in the photocurrent spectra between T037 and T038 lies in their structural difference. As seen in Fig. 8.2, the continuum state reached by a resonant two-photon absorption from the HH1 level is significantly stronger confined for T038 as compared to T037. Charge carriers resonantly excited into the continuum of structure T037 are energetically well above the barrier induced by the separation layer, and un-hindered transport is possible. In contrast, resonant continuum carriers in T038 face a significant barrier and thus a higher probability of being re-trapped before escaping the well region and contributing to a photocurrent. The fact that states higher up in the continuum of T038 exhibit a much higher escape probability in comparison to states close or below the barrier edge of the separation layer accounts for the damped shape of the photocurrent peak around 30 meV in Fig. 8.5 (a). Thus, the lower resonantly enhanced peak in the photocurrent spectra of T038 is a direct consequence of the stronger continuum confinement by the separation layer of this structure in comparison to T037. The comparison of the non-linear photocurrent spectra of both samples therefore experimentally confirms the design considerations given in chapter 8.2.1 and illustrates the barrier-dependent trade-off between the unfavored continuum confinement and the required LH1 confinement. As a significant resonant photocurrent peak around 30 meV is observed for T037 and T038, both samples fulfill one of the basic requirements for photocurrent pump-pump experiments, namely the possibility of reaching conductive states by the absorption of two photons of the LH1-HH1 resonance energy.

In addition to the photocurrent spectrum obtained under an attenuation of the laser pulse by 3 dB, the spectral characteristics for T038 were recorded for attenuations of 0 dB and 5

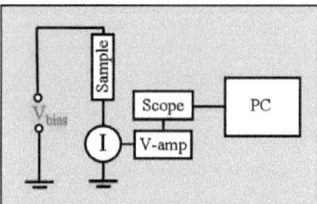

Figure 8.6: Electronic setup for the pump-pump experiments on T037 and T038. The photocurrent induced by the FEL radiation is measured by a current amplifier. During the experiments on T038, the output signal of the current amplifier was further enhanced by a voltage amplifier, as shown in the figure. However, the voltage amplifier was not used for the measurements on sample T037. Details are given in the text.

dB. No change in the spectral shape was observed. However, the integral photocurrent values around 30 meV measured for the different attenuation settings exhibited a clear super-linear dependence on the micropulse energy, or the intensity of the employed FEL beam. In comparison to the value of 0.66 gained from the characteristics measured for T038 under illumination by TE polarized radiation attenuated by 3 dB (shown in Fig. 8.5(a)), the integral photocurrent of 1.84 measured at 0 dB was nearly three times as high. Thus, increasing the FEL intensity by a factor of two led to a current increase by a factor of three, clearly indicating the domination of the photocurrent generation by two-photon processes.

8.4 Pump-pump setup

The time resolved experiments on T037 and T038 were once again performed at the FOM Rijnhuizen, employing the free-electron-laser FELIX. Apart from replacing the MCT reference detector by a Ge:Ga detector, the optical setup for the FEL experiments on these two structures is completely equivalent to that presented in chapter 7.5, including the waveguide geometry, in which the FEL radiation was coupled into the samples. The only significant difference in the setup between the experiments on H019 and those on the samples discussed in the current chapter lies in the operation of the FEL (see chapter 7.5.1). The FEL was tuned into resonance with the HH1-LH1 transition in the central well of the respective structure, using the transition energies determined by the spectral photocurrent measurements presented in the previous section 8.3. For sample T037, a FEL photon energy of 29.5 meV was chosen, while the FEL photon energy was tuned to 29.8 meV for T038. The energies, at which the photocurrent pump-pump experiments on the respective sample were performed, are indicated by the vertical lines in Fig. 8.5. As the HH1-LH1 transition energy lies well below the LO phonon energies, a significantly lower time resolution was required for the pump-pump experiments as compared to those presented in chapter 7. It was therefore possible to employ far longer FEL pulses, where a large temporal FWHM of slightly below 10 ps increased the stability of operation of the FEL 2 line.

Apart from the change in the FEL parameters, the experiments presented in this chapter differed from the measurements on H019 in the electronic setup. In the course of the experiments on H019 in chapter 7, the current through the sample was measured as a voltage drop over a resistor, while the investigated mesa was biased via a battery-driven high-precision voltage source. In order to improve the electronic setup and reduce the system noise, a Femto DLPCA-200 low-noise current amplifier was employed for measuring the current through T037 and T038. For applying an electric field to the respective structure, the biasing capability of the current amplifier was used. In the course of the experiments on sample T037, the signal at the output of the DLPCA-200 was directly fed into a scope. For studying sample T038, in order to scale the signal finally fed into the scope without changing the amplification level of the current amplifier, the voltage signal at its output was further amplified by a LeCroy DA 1820 voltage amplifier. The amplified photocurrent response of the studied sample to an FEL macropulse was recorded by the scope and further processed by a PC, which digitally calculated the integral excess photocurrent using box-car functions replacing the analogue box-car integrator employed for the experiments on sample H019.

A detailed discussion of the principle of photocurrent pump-pump experiments and the employed setup was given in chapters 7.4 and 7.5. One important detail shall be mentioned to conclude this section: In the following experiments on samples T037 and T038, a **negative delay** between the two FEL pulses is equivalent to a situation in which the **TE pulse** hits the sample **first**. This is inconsistent with the definition of the delay sign in chapter 7, but results in a more intuitive interpretation of the plots: For all pump-pump characteristics presented in this work, the data relevant for the extraction of intersubband relaxation times is located on the negative delay side of the plot.

8.5 Time-resolved experiments

The time-resolved experiments aiming at determining the bias-dependence of the LH1-HH1 relaxation time were performed on structures T037 and T038 in the setup described in the previous chapter 8.4. The measurements were carried out for a series of different biases applied to the sample contacts in complete analogy to the experiments presented in chapter 7.

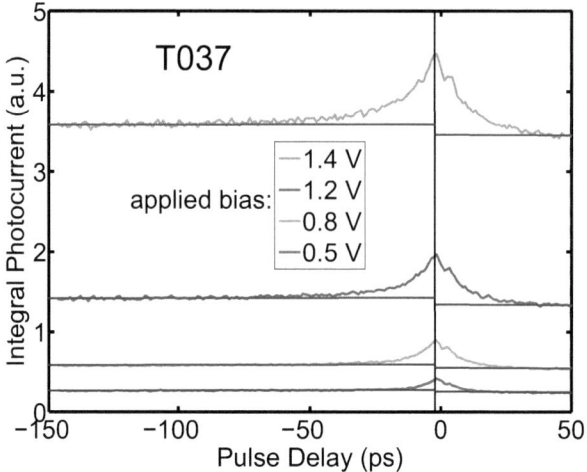

Figure 8.7: Photocurrent pump-pump results for T037 at applied voltages of 0.5, 0.8, 1.2 and 1.4 V. For negative delays, the TE pulse hits the sample first, for positive delays the TM pulse is first. The integral photocurrent through the sample features a clear peak around zero-delay between the two FEL micropulses. Note the offset in the photocurrent baseline between negative and positive delays. Details are given in the text.

8.5.1 Photocurrent pump-pump experiments: Results

T037

As suggested by the considerations in chapter 7.6.2, for the time resolved experiments on sample T037 the FEL micropulse energy was chosen as low as possible while maintaining a sufficient signal-to-noise ratio of the integral excess photocurrent, from which the investigated LH1-HH1 relaxation time is to be extracted. Therefore, the FEL beams featuring a micropulse energy of about 260 nJ each were attenuated by 5 dB.

The photocurrent pump-pump experiments were carried out by sweeping the position of the movable retro-reflector relative to its zero-delay point over a total length of about 30 mm, covering delays between the TM and TE micropulses from -150 ps to +50 ps. For each mirror position, the integral photocurrent through the sample was recorded as described in chapters 7.5 and 8.4, where the average over 16 box-car values was taken. The mirror step width was 150 μm. Simultaneously with the integral photocurrent through the sample, the output of a helium-cooled Ge:Ga detector was recorded. Once again, in order to compensate for eventual temporal fluctuations in the FEL intensity, the obtained photocurrent pump-pump data was divided by the Ge:Ga data.

For investigating the influence of the externally applied field on the relaxation behavior of sample T037, the pump-pump characteristics were recorded for a series of eleven different voltages, which are listed in table 8.4 together with the associated fields. In order to exclude any influence of eventual monotonic temporal drifts in the FEL intensity on the outcome of the experiment, the measurements for the individual applied voltages were performed in a randomly chosen order as given in this table. The amplification setting of the DLPCA amplifier was chosen as high as possible for each measurement, where the used configuration is also given in table 8.4.

Figure 8.7 presents a selection of photocurrent characteristics as measured, where the individual curves are scaled according to the amplifier setting. The curves obtained at an amplification of 10^6 V/A are plotted as resulting from the box-car integration, those gained for a setting of 10^5 V/A are multiplied by a factor of ten. For the sake of clarity, only four of the ten obtained characteristics are shown in this plot. Trivially, the photocurrent through the sample generally rises with increasing bias. All of the curves exhibit a clear excess photocurrent above the integral current baseline, where the maximum photocurrent value exceeds the baseline by up to 20 %. The maximum of the pump-pump characteristics is reached slightly off zero delay, which is attributed to a minor misalignment of the position

Table 8.4: Series of pump-pump experiments on T037 for different voltages performed in the following order:

Voltage	Field	Order	Amplification
0.2 V	2.4 KV/cm	7	10^6 V/A
0.5 V	6 KV/cm	2	10^6 V/A
0.6 V	7.2 KV/cm	8	10^6 V/A
0.7 V	8.4 KV/cm	11	10^6 V/A
0.8 V	9.6 KV/cm	9	10^6 V/A
0.9 V	10.8 KV/cm	10	10^6 V/A
1 V	12 KV/cm	1	10^5 V/A
1.1 V	13.2 KV/cm	5	10^5 V/A
1.2 V	14.4 KV/cm	4	10^5 V/A
1.4 V	16.8 KV/cm	3	10^5 V/A

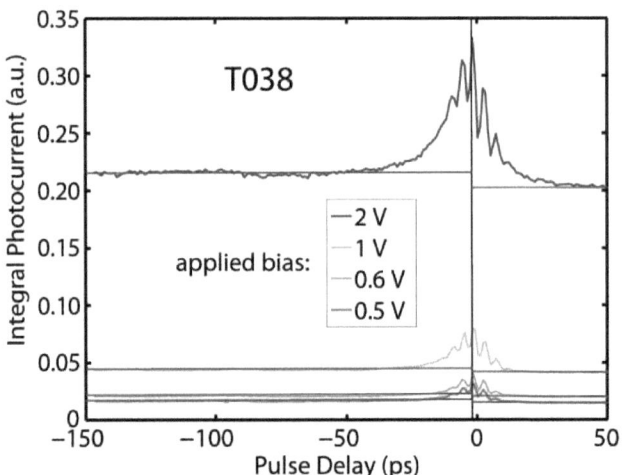

Figure 8.8: Photocurrent pump-pump results for T038 at applied voltages of 0.5, 0.6, 1 and 2 V. For negative delays, the TE pulse hits the sample first, for positive delays the TM pulse is first. The integral photocurrent through the sample features a clear peak around zero-delay. Note the offset in the photocurrent baseline between negative and positive delays. Details are given in the text.

of zero path difference. In addition, a clear offset between the individual current baselines for positive and negative delays is observed (indicated by horizontal lines in Fig. 8.7). The origin of this offset is discussed in chapter 8.5.2.

T038

For the time resolved experiments on sample T038, a micropulse energy of 160 nJ was employed and further attenuated by 3 dB. Apart from the differing FEL intensity and the use of an additional voltage amplifier as described in chapter 8.4, the experiments on T038 were carried out in complete analogy to those on T037. Once again, the measurements were performed for a series of voltages in a random order, which is given in table 8.5 together with the settings for both employed amplifiers. A selection of the resulting pump-pump characteristics is presented in Fig. 8.8, where the curves once again have been scaled according to their amplifier settings. The obtained curves are scaled to a net amplification of 10^6 V/A in consistency with Fig. 8.7.

As those obtained for sample T037, the time-resolved characteristics of T038 exhibit an offset in the current baseline between the regions of positive and negative delay as well as a shift between the excess current maximum and the zero-delay position. The maximum excess current of T038 amounts more than 30 % of the current baseline.

The comparison between Figs. 8.7 and 8.8 shows that the photocurrent through sample T037 is more than one order of magnitude higher than that through T038. This difference in the current baseline cannot be attributed to the higher FEL micropulse energy of 260 nJ used for the experiments on T037 as compared to the 160 nJ employed for T038, as this larger beam intensity is more than counter-balanced by the four times larger mesa area of T038. Thus, the difference in photocurrent is expected to originate from the difference in the top metallization, which suppresses the TE mode in case of T038, and from the more efficient blocking of charge transport by the separation layer of T038, as discussed in chapter 8.2.1.

The characteristics in Fig. 8.8 exhibit oscillations on a time scale of a few picoseconds, which might originate from coherent effects. However, the experimental data collected during the experiments on sample T038 do not allow a definite attribution of the observed photocurrent modulations to coherent effects. In order to clarify the matter, pump-pump experiments of a higher time resolution would be necessary, which could not be carried out in the course of this work due to a limited amount of available beam time. The interpretation of the oscillations in the pump-pump characteristics as coherent effects would consistently explain their absence during the experiments on T037. Continuum states in this sample are expected to de-phase quickly due to a lack of confinement (see chapter 8.2.1), whereas the energetically higher separation layer of T038 might confine the final states of a two-photon process sufficiently well to allow a slower de-phasing and the observation of coherent effects.

Table 8.5: Series of pump-pump experiments on T038 for different voltages performed in the following order:

Voltage	Field	Order	DLPCA amplification	LeCroy amplification
0.3 V	2.8 KV/cm	4	10^6 V/A	×100
0.5 V	4.7 KV/cm	3	10^6 V/A	×100
0.6 V	5.7 KV/cm	6	10^6 V/A	×100
0.7 V	6.6 KV/cm	5	10^6 V/A	×100
1 V	9.4 KV/cm	1	10^6 V/A	×10
1.5 V	14.1 KV/cm	7	10^6 V/A	×10
2 V	18.9 KV/cm	2	10^6 V/A	×10

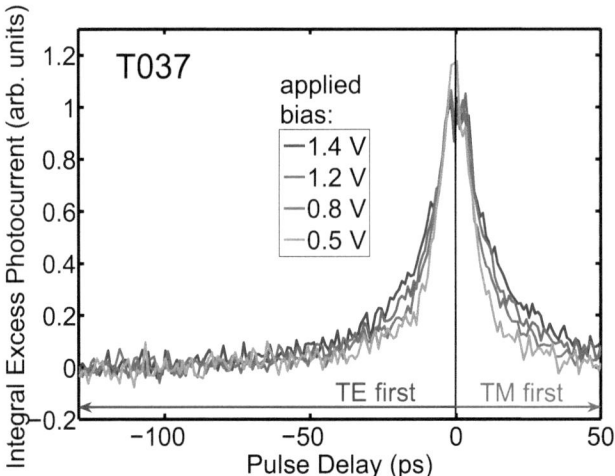

Figure 8.9: Normalized excess photocurrent characteristics for T037 at applied voltages of 0.5, 0.8, 1.2 and 1.4 V. The plots where gained from the data presented in Fig. 8.7 by subtracting different baseline current values for positive and negative delays and normalizing the excess current characteristics to their maximum value. The integral photocurrent through the sample features a peak at zero-delay between the two FEL micropulses. As clearly seen from this plot, the excess photocurrent signal for negative delays decays slower and slower with increasing applied bias.

8.5.2 Photocurrent pump-pump experiments: Interpretation

The interpretation of the photocurrent pump-pump characteristics measured on T037 and T038 is analogous to that presented in chapter 7.6.4 for sample H019. Once again, the integral current baselines in Figs. 8.7 and 8.8 correspond to the current generated by two uncorrelated pump pulses, that means in a long-delay-situation. The fraction of the measured integral photocurrent relevant for the extraction of relaxation time values is the excess photocurrent observed for smaller delays. Once more, a clear and intuitively accessible plot of the excess photocurrent requires the subtraction of the current baseline values.

Current baseline offset

As already mentioned in chapter 8.5.1 and seen in Figs. 8.7 and 8.8, the baseline in the photocurrent characteristics exhibits an offset between the values for large positive and negative

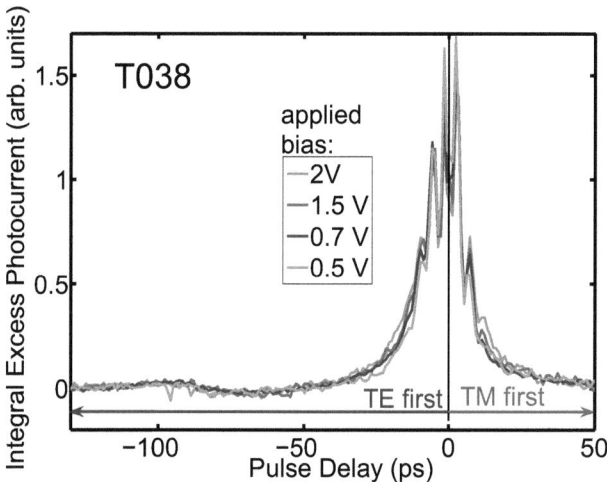

Figure 8.10: Normalized excess photocurrent characteristics for T038 at applied voltages of 0.5, 0.7, 1.5 and 2 V. The plots where gained from the data presented in Fig. 8.8 by subtracting different baseline current values for positive and negative delays and normalizing the excess current characteristics to their maximum value. The integral photocurrent through the sample features a peak at zero-delay between the two FEL micropulses. As clearly seen from this plot, the excess photocurrent signal for negative delays decays slower and slower with increasing applied bias.

delays. This offset is attributed to heating effects with long characteristic relaxation times. The absorption of FEL radiation by the charge carriers in the investigated samples results in their redistribution in k-space, where the consequential occupation of higher k-states results in a change in the photoconducting properties of the structures. As concluded from the difference in the integral photocurrent induced by the two uncorrelated FEL pulses between the cases of long negative and positive delays, the change in the photoconductive properties is dependent on the order of the two pulses. The detailed cause of the current offset is, however, a complex matter, and is probably based on an interplay between the polarization dependence and the power dependence of the involved transitions. One possible reason for the baseline offset is discussed in the following.

Let us first consider the case of positive delays, in which the pulse labelled TM in Fig. 7.10 hits the sample first. At long delays within the experimental range any occupation of higher subband states is expected to have relaxed into the HH1 ground state. In case of a significant persistent heating, the only change in the sample properties prevailing at the impact of the second, purely TE polarized pulse is based on the occupation of HH1 states of high in-plane wave vectors. As according to the bandstructure simulations presented in chapter 8.2 the HH1-LH1 transition for TE polarized radiation is weaker for states of finite in-plane wave vectors than for those with $k_\parallel = 0$, the pump beam 2 generates less photocurrent than it would in absence of carrier heating.

For negative delays, that means in the case of the purely TE polarized beam hitting the sample first, the heating of charge carriers induced by the pump beam 1 is predicted to have the opposite effect on the integral photocurrent. Bandstructure calculations indicate, that the HH1-LH1 transition by the absorption of TM polarized radiation is forbidden for $k_\parallel = 0$ and *increases* in strength with the in-plane wavevector of the involved states. As the polarization of pump beam 2 is dominated by TM components, its contribution to the total photocurrent is stronger in case of hot HH1 holes than for a completely relaxed, cool system.

As concluded from the considerations above, hole heating is expected to increase the integral photocurrent through the investigated structures in case of negative delays as compared to the cool carrier system, and to decrease the pump-pump signal for large positive delays. This is consistent with the behavior observed during the time-resolved experiments on samples T037 and T038. Whether this simple explanation is the dominating reason of the observed baseline offset, or whether more complex effects exert a stronger influence cannot be determined with certainty from the measurement data.

However, the change in the system configuration caused by carrier heating induced by the respective pump beam 1 does not exhibit any relaxation behavior on the time scale of the experiment. Furthermore, as discussed in chapter 7.4, the heating of HH1 carriers does not effect the probing process in photocurrent pump-pump experiments at all, as the occupation-induced excess current exclusively stems from transitions between the LH1 state and the continuum. Thus heating effects exclusively induce a change in the integral current baseline without any observable influence on the decay characteristics of the excess current. Put differently, as long as the data regions of positive and negative delays are treated separately, carrier heating does not effect the extraction of the LH1-HH1 relaxation time. The excess current for positive and negative delays is consequentially defined in respect to different baseline values. Figures 8.9 and 8.10 present the data given in Figs. 8.7 and 8.8 after subtraction of these different values for positive and negative delays, where the origin of the delay-axis was additionally shifted to the position of the excess current maximum. Further, for the plots in Figs. 8.9 and 8.10 the excess photocurrent characteristics have been normalized to their zero-delay value.

Curve shape

For any of the biases applied in the course of the time-resolved experiments, the excess current through the studied samples reached a maximum for small delays, originating from the correlation of the two FEL pulses as discussed in chapters 7.4 and 7.6.4. In contrast to the experiments carried out on sample H019, the pump-pump characteristics of T037 and T038 clearly exhibit a two-stage decay. In close vicinity to zero-delay, the integral excess photocurrent signal decays rapidly, while the relaxation behavior flattens for longer delay time values. This two-stage decay of the excess photocurrent with the delay between the two FEL pulses is observed for positive delays as well as for negative ones.

The sharp center peak of the characteristics is associated with the coherent two-photon-excitation of HH1 carriers into conducting continuum states, as discussed in chapter 7.7.4. According to the pump-pump simulations presented there, this non-linear process results in an excess current peak of a FWHM determined by the FEL pulse width, which is *not* based on establishing a non-equilibrium occupation of an excited state by the first absorbed photon. While the measurement data obtained for T037 and T038 feature a sharp center peak for both positive and negative delays, the results for H019 in Fig. 7.20 did not exhibit a pulse-limited center peak for negative delays, where the pump-pump characteristics were exclusively determined by the intersubband relaxation process. However, as discussed in chapter 7.7.4 the observability of the coherent center peak in addition to the slow decay induced by the intersubband relaxation process depends on the phase coherence time $\frac{1}{\Gamma}$ of the involved transitions (see chapter 7.7.2). For small phase coherence times (large values of Γ), the coherent two-photon-absorption process is too weak to produce an observable excess current, which is the case for sample H019. In case of T037 and T038, the phase coherence time of the respective states is sufficiently long to enable the generation of an excess current by coherent excitation, which contributes significantly to the pump-pump characteristics. Thus, the influence of coherent two-photon excitation on the pump-pump characteristics has to be taken into account when extracting the intersubband relaxation times from the experimentally obtained data on T037 and T038.

In analogy to the experimental results on H019 presented in Fig. 7.15, the pump-pump characteristics obtained for T037 and T038 exhibit an asymmetry in respect to the sign of the delay between the two FEL pulses, even though the asymmetry for these two samples is less pronounced than that observed for H019. In case of negative delays, which correspond to the TE-first-situation, the slowly decaying excess photocurrent signal attributed to the occupation of the LH1 state is much stronger than for positive delays, where the characteristics are to a large extent dominated by the pulse-limited center peak. Once more, the observed asymmetry is associated with the selection rules of the involved optical transitions, and its interpretation is in complete analogy with that of the data obtained for H019 (see chapter 7.6.4 and Fig. 7.6).

Negative delay: TE first

In case of negative delays in Figs. 8.9 and 8.10, the purely TE polarized pulse hits the sample first. Note that the sign of the delay was chosen differently from the definition during the experiments on H019. Prior to the impact of pump pulse 1, charge carriers in the system exclusively occupy the HH1 ground state. According to the theoretical and experimental data presented in table 8.3 and Fig. 8.5, the HH1-LH1 transition is strong in TE polarization, which is in contrast to the HH1-HH2 transition in H019. Thus, the case of negative delays for the samples presented in this chapter is equal to that discussed in chapter 7.6.4 for H019. The TE polarized pump beam 1 is allowed to excite a non-equilibrium LH1 occupation, which is subsequently probed into the continuum by the pump beam 2. The processes enabling the monitoring of the LH1-HH1 relaxation process in T037 and T038 can in principle be illustrated by the left branch of Fig. 7.6 in analogy to H019, where the HH1 state of H019 has to be replaced by the LH1 level for T037 and T038. In addition, the TE and TM labels have to be exchanged. However, there is one difference between the detailed processes during the pump-pump data collection for negative delays during the experiments on H019 and those on T037 and T038. While in the course of the experiments on H019 the pump beam 2 was of pure TE polarization and therefore could not excite any ground state carriers into the excited HH2 state, the pump beam 2 is in case of the measurements presented in this chapter of mixed TE and TM polarization. It is therefore allowed to induce HH1-LH1 transitions and generate an additional photocurrent proportional to the number of ground state carriers, a process which is not illustrated in Fig. 7.6 due to its absence for sample H019.

Now, the contribution by this additional channel to the photocurrent generated by the second FEL pulse is much weaker than that originating from the non-equilibrium LH1 occupation. This is due to the fact that the excitation of ground state carriers into the continuum by the second pulse requires the absorption of two photons of this pulse, while the probing of LH1 carriers into the continuum is a linear process. Any time-dependent contribution to the pump-pump characteristics by the probing of the ground state occupation into the continuum by the second FEL pulse is thus based on a three-photon process, and can be neglected.

Therefore, the interpretation of the slow decay for large negative delays in Figs. 8.9 and 8.10 as a directly monitored relaxation of the LH1 occupation is in complete analogy to the interpretation of the data on sample H019. In other words, for large negative delays the excess photocurrent signal measured for both samples and a series of applied voltages decays exponentially with the investigated LH1-HH1 relaxation time.

Positive delay: TM first

During the experiments on sample H019, pump beam 1 was purely TE polarized in case of positive delays and thus forbidden to excite any HH2 occupation. This lead to the absence of any relaxation-induced characteristics in the pump-pump curves for negative delays, reducing the curve shape to a pulse-limited excess current peak. As discussed in chapter 7.6.4, this enabled an unambiguous distinction between the data relevant for the extraction of the intersubband relaxation time and the pulse limited characteristics. As seen in Fig. 8.9 and 8.10, the experimentally obtained pump-pump data on samples T037 and T038 exhibit a slowly relaxing excess current for both negative and positive delays. The reason for the presence of relaxation-induced characteristics for both signs of the delay can be easily understood by considering the experimental setup.

In case of positive delays in Figs. 8.9 and 8.10, the pulse exhibiting both TE and TM components hits the sample first, where the TE component is able to induce optical HH1-LH1 transitions allowed by the selection rules. In contrast to the experiments on H019, the pump beam 1 is thus able to establish a non-equilibrium LH1 occupation for both signs of the pulse delay. As a consequence, for long positive delays the excess photocurrent decays with the LH1-HH1 relaxation time, as it does for negative delays. However, due to the TE:TM intensity ratio of 1:3 induced by the coupling geometry, the photocurrent signal originating from the occupation of the LH1 state is much weaker than that for negative delays, for which all of the pump beam 1 intensity is carried by TE modes. This results in a relatively weak signal-to-noise ratio of the excess photocurrent signal for long positive delays. Therefore, for the extraction of the intersubband relaxation time only the pump-pump characteristics obtained for negative delays are considered during the data evaluation process for T037 and T038.

A pronounced asymmetry was essential for the interpretation of the data on sample H019 in chapter 7.6.4 due to the ultrashort relaxation time of the investigated transition, which was close to the time resolution of the measurement system. However, as the LH1-HH1 relaxation time studied during the experiments on T037 and T038 lies in the range of ten picoseconds, the data interpretation is not dependent on a strongly pronounced asymmetry. Moreover, the clearly observable coherent center peak of the pump-pump characteristics shown in Figs. 8.9 and 8.10 delivers a clear measure for the time resolution of the setup. While during the experiments on H019 the relaxation based slow decay and the pulse-limited narrow center peak were each observed for one particular sign of the delay, a two-stage decay featuring both of the characteristic temporal widths was observed in case of T037 and T038 for both signs of the delay.

To conclude, the observation of a two-stage decay in the pump-pump characteristics of T037 and T038 delivers a definite indication for the relevance of the relaxation-based excess photocurrent characteristics for the extraction of intersubband relaxation times, which are clearly distinguishable from the experimental time resolution defined by the width of the coherent center peak. Moreover, the consistency between the observed slight asymmetry in the experimentally obtained excess current characteristics and the theoretically predicted selection rules strongly supports the suggested data interpretation and indicates the consideration of all processes relevant for a detailed understanding of the performed experiments.

Voltage dependence

The ultimate motivation for the time-resolved experiments presented in this chapter was the investigation of the voltage dependence of intersubband relaxation efficiencies in the quantum well samples T037 and T038. For this purpose, the photocurrent pump-pump experiments were carried out for a series of applied biases. Already from the linearly plotted curves in Figs. 8.9 and 8.10 a clear systematic dependence of the current decay slope on the strength of the applied field can be observed. In case of negative delays, which according to the discussion in the previous subsections features a stronger contribution by the non-equilibrium LH occupation to the excess current, the integral photocurrent decays slower and slower with increasing applied voltage for both studied samples. According to the discussion given in this section, the decay of the excess photocurrent for longer negative delays can be directly associated with the LH1-HH1 relaxation process in the sample's quantum well structure. Thus, the experimentally observed trend directly indicates a monotonic increase of the LH-HH intersubband relaxation time with the externally applied bias. In order to quantify this qualitative observation, values for the LH1-HH1 relaxation time are to be extracted from the experimentally obtained pump-pump data by a fit procedure, which is presented in the following section.

8.5.3 Relaxation time extraction

Fitting model

As discussed in the previous section, the pump-pump characteristics of T037 and T038 are determined by two distinguishable influences. For small delays, the characteristics are dominated by the excess photocurrent based on the non-linearity of the coherent excitation of HH1 carriers into the continuum without any intermediate occupation of the LH1 state. For large delays, the characteristics are dominated by the decay process of the non-equilibrium LH1 occupation, allowing the extraction of the LH1-HH1 intersubband relaxation times under investigation. Due to the more complex shape of the pump-pump characteristics of T037 and T038 as compared to those of H019 in chapter 7, which is caused by the influence of the coherent center peak, a single exponential fit as that used for extracting the relaxation times of H019 (see Eqn. 7.5) is insufficient for an accurate determination of the decay times of T037 and T038. In principle, the pump-pump characteristics gained for T037 and T038 are fully described by the rate equations 7.34 to 7.37 gained from applying the density matrix formalism to a four-level model, which is discussed in chapter 7.7.2. However, the exact numerical evaluation of these rate equations for simulating the pump-pump experiments as done for H019 in chapter 7.7.4 is too elaborate and time-consuming to allow basing a fit procedure on such simulations. Therefore, in order to gain a fit function describing the pump-pump characteristics of T037 and T038 with sufficient accuracy while still allowing the implementation of a stable and reasonably fast fitting procedure, several assumptions and approximations have to be applied to the rate equations in Eqns. 7.34 to 7.37. The result of these approximations can also be reached by a more intuitive approach, which allows a modelling of the pump-pump characteristics in Figs. 8.7 and 8.8 based on the insight gained on the general shape of such characteristics by the simulations in chapter 7.7.4. The section at hand discusses this intuitive approach, as it leads to a transparent understanding of the physical processes, by which the shape of the pump-pump characteristics is determined. However, as demonstrated in the next section, the fit function used for the relaxation time extraction applied to the data on T037 and T038 can as well be derived from Eqns. 7.34 to 7.37.

In consistency with the assumption of a gaussian temporal pulse shape of the FEL as expressed in Eqn. 7.39, the time evolution of the intensity of one FEL pulse centered at $t = 0$ shall be given by Eqn. 8.1.

$$I_p(t) = I_0 \, e^{-\frac{1}{2}(\frac{t}{\sigma_t})^2} \tag{8.1}$$

Now, with a well-defined delay τ between two non-interfering FEL pulses hitting the samples during the pump-pump experiment, the total temporal intensity profile $I_t(t)$ reads as in Eqn. 8.2.

$$I_t(t) = I_p(t) + I_p(t-\tau) \tag{8.2}$$

As the coherent excitation of HH1 carriers into the valence band continuum is based on the absorption of two photons, the current generated by the occupation of continuum states, $j_c(t)$

is quadratically dependent on the total FEL intensity $I_t(t)$, as seen in Eqn. 8.3.

$$j_c(t) \sim I_t(t)^2$$

$$\sim I_p^2(t) + I_p^2(t-\tau) + 2 \cdot I_p(t) \, I_p(t-\tau) \tag{8.3}$$

In the course of the pump-pump experiments, the *integral* photocurrent through the studied sample is measured as a function of the delay τ. The contribution of the coherently excited holes to the measured quantity is given by Eqn. 8.4, where Δt_{MAP} is the macropulse duration, Δt_{MP} represents the time window attributed to each micropulse within the macropulse and N_{MAP} is the number of micropulses in each macropulse. Note that $\Delta t_{MP} \gg \tau$ holds for each experimental value of τ.

$$J_c(\tau) = \int_{\Delta t_{MAP}} j_c(\tau, t) \, dt = N_{MAP} \cdot \int_{\Delta t_{MP}} j_c(\tau, t) \, dt$$

$$= J_0^c + c_1 \cdot \int_{\Delta t_{MP}} I_p(t) \, I_p(t-\tau) \tag{8.4}$$

As the contribution of the first two terms in Eqn. 8.3 to J_c is independent of the delay time τ, these terms do not add to the integral excess current, but exclusively deliver a contribution J_0^c to the current baseline. The last term, however, adds a delay-dependent contribution to the integral current and is thus equal to the excess photocurrent generated in case of overlapping pulses due to the non-linearity of the coherent two-photon absorption. As seen from the rate equation in Eqn. 7.42, which is associated with the process of coherent photocurrent generation, the proportionality constant c_1 between the integral photocurrent and the convolution of the two intensity functions $I_p(t)$ depends on the dipole matrix elements $\boldsymbol{\mu}_{10}$ and $\boldsymbol{\mu}_{21}$ as well as the polarization of the involved FEL radiation.

In case of larger delays between the pulses, at which the overlap between them is too low to produce a significant excess current via coherent excitation, the shape of the pump-pump characteristics is dominated by photocurrent generated by the second FEL pulse by exciting LH1 carriers into the continuum. This region of the pump-pump curves essentially allows the extraction of the LH1-HH1 intersubband relaxation time, as the excess photocurrent decays with increasing delay along with the relaxation of the excited state occupation into the ground state. The photocurrent generated by the excitation of LH1 carriers via the absorption of a pump-2-photon is proportional to the amount of charge carriers in this state.

The non-equilibrium LH1 occupation n_{LH} is induced by the pump beam 1 at a rate proportional to the pulse intensity. In a time interval dt' around t' the pump beam 1 centered at $t=0$ thus excites a number of dn_{LH}^e holes into the LH1 state, following Eqn. 8.5.

$$dn_{LH}^e = c_2 \cdot I_p(t') \, dt' \tag{8.5}$$

The proportionality constant between the excitation rate and the pulse intensity depends on the dipole matrix element μ_{10} as well as the polarization of the involved FEL radiation, as concluded from Eqn. 7.43. As far as the amount of excess current generated by the pump beam 2 from this LH1 occupation is regarded, the fraction of dn^e_{LH} still remaining at the time of its incidence is relevant. As the LH1 occupation decays exponentially with the LH1-HH1 intersubband relaxation time τ_{decay}, the amount of carriers generated within dt' around t' still remaining at time t is given by Eqn. 8.6.

$$dn^r_{LH}(t',t) = c_2 \cdot I_p(t') \cdot e^{-\frac{t-t'}{\tau_{decay}}} dt' \tag{8.6}$$

In order to obtain the total LH1 occupation at t, the remaining contributions generated at t' have to be summed up over all $t' < t$, resulting in Eqn. 8.7.

$$n_{LH}(t) = c_2 \cdot \int_{-\infty}^{t} I_p(t') \cdot e^{-\frac{t-t'}{\tau_{decay}}} dt' \tag{8.7}$$

Following the interpretation discussed in the previous section, the excited LH occupation is probed into the continuum and produces a photocurrent based on a single-photon-absorption. The second FEL pulse centered around $t = \tau$ exhibits a temporal intensity distribution $I_p(t - \tau)$. Note that in the discussion at hand the FEL pulse characterized by $I_p(t - \tau)$ is the second to hit the sample, and thus $\tau > 0$ holds. However, the considerations can be easily generalized to follow the sign convention of the data presented in Figs. 8.7 and 8.8. The photocurrent generated from the charge carriers remaining in the LH state at t by the intensity of the second FEL pulse present at this time therefore follows Eqn. 8.8.

$$j_{oc}(t) \sim n_{LH}(t) \cdot I_p(t - \tau) \tag{8.8}$$

Again, the quantity relevant for fitting the experimental results is the integral photocurrent through the device, where the component originating from the decaying LH1 occupation is expressed in Eqn. 8.9.

$$J_{oc}(\tau) = c_3 \cdot \int_{\Delta t_{MP}} n_{LH}(t) \cdot I_p(t - \tau) \, dt \tag{8.9}$$

Once more, the proportionality constant c_3 depends on one of the dipole matrix elements, μ_{21}, and the FEL polarization.

Now, the mirror step width between two subsequent individual data points during the pump-pump experiments was 150 μm, which is equal to a pulse delay of 1 ps. Thus, the precise point of zero delay can be experimentally determined with an accuracy of $\pm$ 0.5 ps only. In order to account for this small uncertainty in the point of zero-delay, a variable delay-time offset τ_0 is introduced into the model function and is allowed to vary between -0.5 ps and 0.5 ps as a fit parameter.

Summing up the two integral excess current contributions in Eqns. 8.4 and 8.9 leads to the final expression accounting for the total curve shape of the pump-pump characteristics, as given in Eqns. 8.10 to 8.13.

$$J(\tau) = J_0 + \int_{\Delta t_{MP}} I_p(t - \tau - \tau_0) \cdot [\, c_3 \cdot n_{LH}(t) + c_1 \cdot I_p(t)\,] \, dt \tag{8.10}$$

$$= J_0 + C_1 \cdot \int_{\Delta t_{MP}} \frac{I_p(t - \tau - \tau_0)}{I_0} \cdot [\, \frac{I_p(t)}{I_0} + C_2 \cdot \int_{-\infty}^{t} \frac{I_p(t')}{I_0} \cdot e^{-\frac{t-t'}{\tau_{decay}}} \, dt'\,] \, dt \tag{8.11}$$

$$C_1 = c_1 \, I_0^2 \tag{8.12}$$

$$C_2 = \frac{c_3 \, c_2}{c_1} \tag{8.13}$$

In this equation, J_0 represents the sum of all photocurrent contributions independent of the delay between the two FEL pulses and thus the integral photocurrent baseline.

As shown in Eqn. 8.14, the term representing the integral photocurrent contribution by the non-equilibrium LH1 occupation in Eqn. 8.11 can be brought into a form, which allows an analytical interpretation by inserting Eqn. 8.1 and by two times completing the squares.

$$\int_{\Delta t_{MP}} \frac{I_p(t - \tau - \tau_0)}{I_0} \int_{-\infty}^{t} \frac{I_p(t')}{I_0} \cdot e^{-\frac{t-t'}{\tau_{decay}}} \, dt' \, dt =$$

$$= \int_{\Delta t_{MP}} e^{-\frac{1}{2}(\frac{t-\tau-\tau_0}{\sigma_t})^2} \, e^{\frac{1}{2}(\frac{\sigma_t}{\tau_{decay}})^2 - \frac{t}{\tau_{decay}}} \int_{-\infty}^{t} e^{-\frac{1}{2\sigma_t^2}(t' - \frac{\sigma_t^2}{\tau_{decay}})^2} \, dt' \, dt =$$

$$= e^{\frac{1}{2}(\frac{\sigma_t}{\tau_{decay}})^2} \int_{\Delta t_{MP}} e^{-\frac{1}{2}(\frac{t-\tau-\tau_0}{\sigma_t})^2 - \frac{t}{\tau_{decay}}} \cdot \frac{\sqrt{\pi}}{2} \mathrm{erf}(\frac{t - \frac{\sigma_t^2}{\tau_{decay}}}{\sqrt{2}\,\sigma_t}) \, dt =$$

$$= \sqrt{\frac{\pi}{2}}\, \sigma_t\, e^{-\frac{\tau + \tau_0}{\tau_{decay}}} \cdot e^{(\frac{\sigma_t}{\tau_{decay}})^2} \int_{\Delta t''_{MP}} e^{-(t'' - \frac{\tau+\tau_0}{\sqrt{2}\sigma_t} + \frac{\sqrt{2}\sigma_t}{\tau_{decay}})^2} \cdot \mathrm{erf}(t'') \, dt''$$

$$\text{with } t'' = \frac{t - \frac{\sigma_t^2}{\tau_{decay}}}{\sqrt{2}\,\sigma_t} \tag{8.14}$$

In this equation, $\mathrm{erf}(t'')$ represents the Gaussian error function. As seen from Eqn. 8.14, the delay-dependence of the integral excess photocurrent originating from a non-equilibrium LH1 occupation is determined by the product between two terms, namely between the exponential function $e^{-\frac{\tau + \tau_0}{\tau_{decay}}}$ and a delay-dependent integral. The integrand of the latter is composed of a product between the error function and a Gaussian centered around $t'' = \frac{\tau+\tau_0}{\sqrt{2}\sigma_t} - \frac{\sqrt{2}\sigma_t}{\tau_{decay}}$. This term is, however, delay-dependent in case of small delays only. For $t'' > 2$, the value of the error function erf(t'') is in good approximation equal to one. Thus, in case of large delay times τ, the peak of the Gaussian function in the integrand shifts into the constant region of the error function, and Eqn. 8.14 simplifies into Eqn. 8.15, where $\widetilde{C}$ is constant.

$$\int_{\Delta t_{MP}} \frac{I_p(t-\tau-\tau_0)}{I_0} \int_{-\infty}^{t} \frac{I_p(t')}{I_0} \cdot e^{-\frac{t-t'}{\tau_{decay}}} \, dt' \, dt =$$

$$= \sqrt{\frac{\pi}{2}} \, \sigma_t \, e^{-\frac{\tau+\tau_0}{\tau_{decay}}} \cdot e^{(\frac{\sigma_t}{\tau_{decay}})^2} \int_{\Delta t''_{MP}} e^{-(t'' - \frac{\tau+\tau_0}{\sqrt{2}\sigma_t} + \frac{\sqrt{2}\sigma_t}{\tau_{decay}})^2} \, dt'' =$$

$$= \sqrt{\frac{\pi}{2}} \, \sigma_t \, e^{-\frac{\tau+\tau_0}{\tau_{decay}}} \cdot e^{(\frac{\sigma_t}{\tau_{decay}})^2} \int_{\Delta t''_{MP}} e^{-t''^2} \, dt'' =$$

$$= \widetilde{C} \cdot e^{-\frac{\tau}{\tau_{decay}}}$$

(8.15)

The physical interpretation of this behavior is simple. For delay times up to $\sqrt{2}\sigma_t$, the two FEL pulses overlap, and the first FEL pulse still excites ground state carriers into the LH1 state at the impact time of the second pulse. The delay dependence is in such a case determined by both the excitation of additional carriers and the decay of already excited ones. However, for delays at which both pulses are temporally clearly separated, the delay-dependence of the integral photocurrent is exclusively determined by the exponential relaxation of the total number of charge carriers excited into the LH1 state by first pulse. Thus, for delay times exceeding the FEL pulse duration, the integral photocurrent decays exponentially with the intersubband relaxation time τ_{decay}. Note that the interpretation of the data on sample H019 in chapter 7 and thus the fit function employed for the extraction of the HH2-HH1 relaxation time as given in Eqn. 7.5 is based on this behavior. The exponential decay of the integral photocurrent with long delays exceeding the FEL pulse width is also reflected in the simulated pump-pump characteristics as shown in Fig. 7.20.

Derivation of the fit function from the density-matrix-based rate equations

As already mentioned, the fit function given by Eqn. 8.11 can be directly derived from the rate equations in Eqn. 7.34 to 7.37, which were gained from the density matrix equations in adiabatic and rotating wave approximation. As already discussed on several occasions, the quantity measured during the pump-pump experiments is directly associated with the quantity $\rho_{22} + \rho_{33}$ as featured in the density matrix formalism. A rate equation for this total occupation of the continuum states 2 and 3 is gained by summing Eqn. 7.36 and 7.37, as done in Eqn. 8.16.

$$\frac{d}{dt}(\rho_{22} + \rho_{33}) = \frac{2}{\hbar^2} \cdot \frac{1}{\hbar^2 \, \Gamma_{10}\Gamma_{12}\Gamma_{20} + \Gamma_{10}\,(\boldsymbol{\mu}_{10} \cdot \boldsymbol{E}_0)^2 + \Gamma_{12}(\boldsymbol{\mu}_{21} \cdot \boldsymbol{E}_0)^2} \cdot$$

$$\cdot [\hbar^2 \, \Gamma_{10}\Gamma_{20} \, (\boldsymbol{\mu}_{21} \cdot \boldsymbol{E}_0)^2 \, (\rho_{11} - \rho_{22}) + (\boldsymbol{\mu}_{21} \cdot \boldsymbol{E}_0)^4 \, (\rho_{11} - \rho_{22}) +$$

$$+ (\boldsymbol{\mu}_{10} \cdot \boldsymbol{E}_0)^2 \, (\boldsymbol{\mu}_{21} \cdot \boldsymbol{E}_0)^2 \, (\rho_{00} - \rho_{11})]$$

(8.16)

$\boldsymbol{E}_0(t, \tau)$ is given by Eqn. 8.17, where $\boldsymbol{p}_1$ and $\boldsymbol{p}_2$ are the polarization vectors of the electric field associated with the first and second FEL pulse, respectively. $\boldsymbol{p}_1$ and $\boldsymbol{p}_2$ further include the factor between the absolute of the electric field and the square root of the associated intensity, $\sqrt{\frac{2}{n\epsilon_0 c}}$. τ represents the delay between the two pulses in consistency with the previous section.

$$\boldsymbol{E}_0(t, \tau) = \boldsymbol{p}_1 \cdot \sqrt{I_0} \, e^{-\frac{1}{4}(\frac{t}{\sigma_t})^2} + \boldsymbol{p}_2 \cdot \sqrt{I_0} \, e^{-\frac{1}{4}(\frac{t-\tau}{\sigma_t})^2} \qquad (8.17)$$

In case of high de-phasing rates $\Gamma = \Gamma_{10} = \Gamma_{12} = \Gamma_{20} \gg \frac{\boldsymbol{\mu}_{10} \cdot \boldsymbol{E}_0(0,0)}{\hbar}$ and $\Gamma \gg \frac{\boldsymbol{\mu}_{21} \cdot \boldsymbol{E}_0(0,0)}{\hbar}$ and far from bleaching conditions of the 0-1 transition ($\rho_{00} \gg \rho_{11}$), Eqn. 8.16 simplifies into Eqn. 8.18.

$$\frac{d}{dt}(\rho_{22} + \rho_{33}) \approx \frac{2}{\hbar^4 \, \Gamma^3} \cdot [\hbar^2 \, \Gamma^2 \, (\boldsymbol{\mu}_{21} \cdot \boldsymbol{E}_0)^2 \, (\rho_{11} - \rho_{22}) +$$

$$+ (\boldsymbol{\mu}_{10} \cdot \boldsymbol{E}_0)^2 \, (\boldsymbol{\mu}_{21} \cdot \boldsymbol{E}_0)^2 \, \rho_{00}]$$

(8.18)

It shall be further assumed that the ground state occupation changes only sightly during the pump-pump experiment ($\rho_{00} \sim 1$), which is the case for weak excitation intensities, and that the occupation of the continuum level 2 decays very rapidly into the non-resonant level 3 (low τ_{32} in Eqns. 7.34 to 7.37) and is thus much smaller than the level-1-occupation ($\rho_{22} \ll \rho_{11}$). These approximations lead to Eqn. 8.19.

$$\frac{d}{dt}(\rho_{22} + \rho_{33}) \approx \frac{2}{\hbar^2 \, \Gamma} \cdot (\boldsymbol{\mu}_{21} \cdot \boldsymbol{E}_0)^2 \, \rho_{11} + \frac{2}{\hbar^4 \, \Gamma^3} \cdot (\boldsymbol{\mu}_{10} \cdot \boldsymbol{E}_0)^2 \, (\boldsymbol{\mu}_{21} \cdot \boldsymbol{E}_0)^2$$

(8.19)

Further approximations can be made by considering the polarization of the two FEL pulses. The case of the purely TE polarized beam hitting the sample first shall be discussed in the following, as this case is according to the considerations given in chapter 8.5.2 relevant for the relaxation time extraction. As in this case the second pulse is mostly composed of a TM component, Eqns. 8.20 and 8.21 hold in good approximation of the experimental conditions.

$$\boldsymbol{\mu}_{10} \cdot \boldsymbol{E}_0 \approx (\boldsymbol{\mu}_{10} \cdot \boldsymbol{p}_1) \sqrt{I_0}\, e^{-\frac{1}{4}(\frac{t}{\sigma_t})^2}$$

$$\approx (\boldsymbol{\mu}_{10} \cdot \boldsymbol{p}_1) \sqrt{I_p(t)} \qquad (8.20)$$

$$\boldsymbol{\mu}_{21} \cdot \boldsymbol{E}_0 \approx (\boldsymbol{\mu}_{21} \cdot \boldsymbol{p}_2) \sqrt{I_0}\, e^{-\frac{1}{4}(\frac{t-\tau}{\sigma_t})^2}$$

$$\approx (\boldsymbol{\mu}_{21} \cdot \boldsymbol{p}_2) \sqrt{I_p(t-\tau)} \qquad (8.21)$$

In these equations, $I_p(t)$ is defined in consistency with Eqn. 8.1. Inserting Eqns. 8.20 and 8.21 into Eqn. 8.19 leads to Eqn. 8.22.

$$\frac{d}{dt}(\rho_{22} + \rho_{33}) \approx \frac{2}{\hbar^2\,\Gamma} \cdot (\boldsymbol{\mu}_{21} \cdot \boldsymbol{p}_2)^2\, I_p(t-\tau)\, \rho_{11} +$$

$$+ \frac{2}{\hbar^4\,\Gamma^3} \cdot (\boldsymbol{\mu}_{10} \cdot \boldsymbol{p}_1)^2\, (\boldsymbol{\mu}_{21} \cdot \boldsymbol{p}_2)^2\, I_p(t-\tau)\, I_p(t)$$

$$(8.22)$$

$$\approx c_3\, I_p(t-\tau)\, \rho_{11} + c_1\, I_p(t-\tau)\, I_p(t)$$

$$(8.23)$$

With the definition of the factors c_1 and c_3 as given by Eqns. 8.31 and 8.33, the integration of Eqn. 8.23 carried out between a time before the impact of the first FEL pulse ($t \ll -\sigma_t$) and a time, at which the system has completely relaxed after the second FEL pulse ($t \gg \sigma_t + |\tau|$), leads to the expression for the delay-dependent integral photocurrent as given by Eqn. 8.24.

$$J(\tau) = \rho_{33}^\infty(\tau) = \int_{t \ll -\sigma_t}^{t \gg \sigma_t + |\tau|} \frac{d}{dt}(\rho_{22} + \rho_{33})\, dt$$

$$\approx J_0 + \int_{\Delta t_{MP}} I_p(t-\tau) \cdot [\, c_3\, \rho_{11}(t) + c_1\, I_p(t)\,]\, dt \qquad (8.24)$$

In this equation, $\rho_{33}^\infty(\tau)$ as defined in chapter 7.7.4 is equal to the final value of $\rho_{22} + \rho_{33}$ at a time when both FEL pulses have exerted their influence on the quantum well system ($t \gg \sigma_t + |\tau|$). This expression, which has been derived in the density matrix formalism, is equal to the model function for $J(\tau)$ based on an intuitive approach, as given in Eqn. 8.10. The integration window Δt_{MP} is chosen in a way to ensure that the upper integration limit fulfills $t \gg \sigma_t + |\tau|$ for any delay time value in the range of the experiment. Therefore, the integration is carried out over the total time span between two subsequent FEL micropulses in the FEL macropulse (40 ns). In order to gain a final analytic expression for the integral photocurrent through the sample, the rate equation for ρ_{11} as given by Eqn. 7.35 has to be solved. By once more applying the approximations discussed above ($\Gamma = \Gamma_{10} = \Gamma_{12} = \Gamma_{20} \gg \frac{\boldsymbol{\mu}_{10} \cdot \boldsymbol{E}_0(0,0)}{\hbar}$, $\Gamma \gg \frac{\boldsymbol{\mu}_{21} \cdot \boldsymbol{E}_0(0,0)}{\hbar}$, $\rho_{00} \gg \rho_{11}$ and $\rho_{22} \ll \rho_{11}$), Eqn. 7.35 simplifies into Eqn. 8.25.

$$\frac{d}{dt}\rho_{11} \approx -\frac{\rho_{11}}{\tau_{10}} + \frac{2}{\hbar^2 \Gamma} \cdot [(\boldsymbol{\mu}_{10} \cdot \boldsymbol{E}_0)^2 \rho_{00} - (\boldsymbol{\mu}_{21} \cdot \boldsymbol{E}_0)^2 \rho_{11}] \tag{8.25}$$

The approximations given by Eqn. 8.21 and $\Gamma \gg \frac{\boldsymbol{\mu}_{21} \cdot \boldsymbol{E}_0(0,0)}{\hbar}$ together with the assumption of an intersubband relaxation rate $\frac{1}{\tau_{10}}$, by far exceeding the excitation rate of state-1-carriers into the continuum ($\frac{1}{\tau_{10}} \gg \frac{\boldsymbol{\mu}_{21} \cdot \boldsymbol{E}_0(0,0)}{\hbar}$) leads to Eqn. 8.26. The latter approximation is valid for a carrier distribution far from bleaching conditions of the 0-1-transition, as in such a case $\frac{1}{\tau_{10}} \gg \frac{\boldsymbol{\mu}_{10} \cdot \boldsymbol{E}_0(0,0)}{\hbar}$ is fulfilled. As further the dipole matrix element between the weakly confined continuum state 2 and the excited state 1 is much smaller than that between the strongly confined states 1 and 0, $\frac{1}{\tau_{10}} \gg \frac{\boldsymbol{\mu}_{10} \cdot \boldsymbol{E}_0(0,0)}{\hbar} \gg \frac{\boldsymbol{\mu}_{21} \cdot \boldsymbol{E}_0(0,0)}{\hbar}$ holds.

$$\frac{d}{dt}\rho_{11} \approx -\frac{\rho_{11}}{\tau_{10}} + \frac{2}{\hbar^2 \Gamma} \cdot (\boldsymbol{\mu}_{10} \cdot \boldsymbol{p}_1)^2 I_p(t) \tag{8.26}$$

Eqn. 8.26 can be solved by using an integrative factor $e^{\frac{t}{\tau_{10}}}$, as shown by Eqns. 8.27 to 8.29.

$$e^{\frac{t}{\tau_{10}}} \cdot \frac{d}{dt}\rho_{11} + \frac{\rho_{11}}{\tau_{10}} \cdot e^{\frac{t}{\tau_{10}}} \approx \frac{2}{\hbar^2 \Gamma} \cdot (\boldsymbol{\mu}_{10} \cdot \boldsymbol{p}_1)^2 I_p(t) e^{\frac{t}{\tau_{10}}} \tag{8.27}$$

$$\frac{d}{dt}[\rho_{11} \cdot e^{\frac{t}{\tau_{10}}}] \approx \frac{2}{\hbar^2 \Gamma} \cdot (\boldsymbol{\mu}_{10} \cdot \boldsymbol{p}_1)^2 I_p(t) e^{\frac{t}{\tau_{10}}} \tag{8.28}$$

$$\rho_{11} \cdot e^{\frac{t}{\tau_{10}}} \approx \frac{2}{\hbar^2 \Gamma} \cdot (\boldsymbol{\mu}_{10} \cdot \boldsymbol{p}_1)^2 \int_{-\infty}^{t} I_p(t') e^{\frac{t'}{\tau_{10}}} dt' \tag{8.29}$$

The final result for the time evolution of the state-1-occupation ρ_{11}, as derived from the density-matrix-based rate equations under a series of approximations, is given by Eqn. 8.30.

$$\begin{aligned}\rho_{11} &\approx \frac{2}{\hbar^2 \Gamma} \cdot (\boldsymbol{\mu}_{10} \cdot \boldsymbol{p}_1)^2 \int_{-\infty}^{t} I_p(t') e^{\frac{t'-t}{\tau_{10}}} dt' \\ &\approx c_2 \cdot \int_{-\infty}^{t} I_p(t') e^{\frac{t'-t}{\tau_{10}}} dt'\end{aligned} \tag{8.30}$$

The three constants entering the expression for $J(\tau)$ in Eqn. 8.24 are thus given by the following equations.

$$c_1 = \frac{2}{\hbar^4 \Gamma^3} \cdot (\boldsymbol{\mu}_{10} \cdot \boldsymbol{p}_1)^2 (\boldsymbol{\mu}_{21} \cdot \boldsymbol{p}_2)^2 \tag{8.31}$$

$$c_2 = \frac{2}{\hbar^2 \Gamma} \cdot (\boldsymbol{\mu}_{10} \cdot \boldsymbol{p}_1)^2 \tag{8.32}$$

$$c_3 = \frac{2}{\hbar^2 \Gamma} \cdot (\boldsymbol{\mu}_{21} \cdot \boldsymbol{p}_2)^2 \tag{8.33}$$

Note that with the definition of the factor c_2 as given by Eqn. 8.32, Eqn. 8.30 is equivalent to the model function given by Eqn. 8.7. Thus it has been shown, that the fit function given by Eqn. 8.11 can be directly derived from the density matrix equations in Eqns. 7.18 to 7.26. This derivation further allows the physical interpretation of the fitting parameters C_1 and C_2 of the fit function in Eqn. 8.11, as shown in Eqn. 8.34 and 8.35.

$$C_1 \;=\; \frac{2}{\hbar^4 \, \Gamma^3} \cdot (\boldsymbol{\mu}_{10} \cdot \boldsymbol{p}_1)^2 \, (\boldsymbol{\mu}_{21} \cdot \boldsymbol{p}_2)^2 \, I_0^2 \qquad (8.34)$$

$$C_2 \;=\; 2\,\Gamma \qquad (8.35)$$

While the detailed interpretation of C_1 is cumbersome due to a number of physical quantities entering Eqn. 8.34, C_2 can be directly associated with the doubled phase relaxation rate Γ. This is consistent with the numerical observation, that the strength of the coherent center peak in the pump-pump characteristics crucially depends on the de-phasing rate Γ, as discussed in chapter 7.7.4 and shown by Fig. 7.21. As indicated by the derivation above, the ratio between the integral photocurrent contribution caused by coherent two-photon-excitation and that based on a non-equilibrium LH1 occupation is exclusively dependent on the de-phasing rate Γ. In case of a low phase coherence (high Γ), the pump-pump characteristics are dominated by the intersubband relaxation process, as experimentally observed for sample H019 in chapter 7. In such a case, the coherences between the quantum states induced by the FEL radiation decay too quickly to allow a significant coherent excitation of state-0-carriers into the continuum state 2. However, as a significant contribution of the coherent center peak to the pump-pump characteristics was observed for T037 and T038, a finite value for $\frac{1}{\Gamma}$ has to be assumed for fitting the data on these samples.

Fitting procedure

In Figs. 8.9 and 8.9, a monotonic bias dependence of the LH1-HH1 relaxation time can be qualitatively observed. However, in order to gain a quantitative measure for the voltage dependence of the studied intersubband relaxation time, the model function defined in Eqn. 8.11 was used to fit the experimentally obtained data presented in Figs. 8.7 and 8.8. In this equation, the intersubband relaxation time of interest is represented by the variable τ_{decay}.

The model function is dependent on a series of additional parameters, among them the FEL pulse width σ_t (see Eqn. 8.1) and the integral current baseline J_0. The parameter C_1 represents the amplitude scaling factor of the excess photocurrent and gives a value for the strength of the excess current contribution to the total measured signal, thus determining the signal-to-noise ratio of the data relevant for the extraction of the intersubband relaxation time. As seen from Eqn. 8.34, it depends on the dipole matrix elements between the involved states as well as the FEL peak intensity. C_2 is equivalent to the ratio between the coherent component of the excess photocurrent and the contribution based on establishing a finite LH occupation around zero-delay, as discussed in the previous subsection. It is dependent on the phase relaxation rate Γ, as seen from Eqn. 8.35.

The model parameters, which are expected to vary with the applied bias, are J_0, C_1 and, as investigated in this work, τ_{decay}. The current baseline trivially rises with increasing applied bias, as seen in Figs. 8.7 and 8.8. As the dipole matrix elements between the individual quantum states are expected to change with the bias due to a voltage tuning of their wave functions, C_1 may vary with the voltage applied to the sample.

In contrast, the model parameters σ_t and C_2 are expected to be independent of the bias applied to the structure. σ_t is trivially independent of the sample parameters, and its constant nature is directly linked to the stability of the FEL operation for the duration of the experimental series. As already mentioned in section 8.5.1 and documented in detail in tables 8.4 and 8.5, the voltage series of pump-pump characteristics was recorded in random order, thus excluding the possibility of observing a trend in the fit-parameters induced by a monotonic change in the FEL properties. For fitting the series of pump-pump characteristics in Figs. 8.9 and 8.9 consistently, a common FEL pulse width was assumed for each voltage-series of curves. However, as the two series of experiments on T037 and T038 were carried out on different days and thus involved slightly different parameters for stable FEL operation, two different global values for the FEL pulse width had to be employed, one for each series of pump-pump characteristics. The method of choosing a value for the global parameter σ_t is discussed later in this section.

The second parameter, which was chosen globally for each bias series of experiments due to its expected insensitivity to a change in the applied sample bias, is the ratio between the coherent and the occupation-based photocurrent contribution, C_2. As seen from Eqn. 8.35, this ratio is fundamentally dependent on the phase coherence time of the involved states. Due to the fact that the de-phasing rate of the quantum well states is not expected to show a significant bias dependence, C_2 is treated as a global parameter for each of the two pump-pump data sets.

To sum up, for evaluating the experimentally obtained pump-pump data on T037 and T038, C_2 and σ_t in the model function given by Eqn. 8.10 are treated as global parameters. This means, that a common value is assigned to these parameters for fitting each bias series of integral photocurrent characteristics. The remaining parameters C_1, J_0, τ_0 and, of highest interest, τ_{delay}, are used as fit parameters and are allowed to vary freely while performing

Table 8.6: Global parameters employed for fitting the data on T037 and T038:

	T037 TE first	T038 TE first
σ_t	3.2 ps	5 ps
$C_2 = 2\Gamma$	20 THz	10 THz

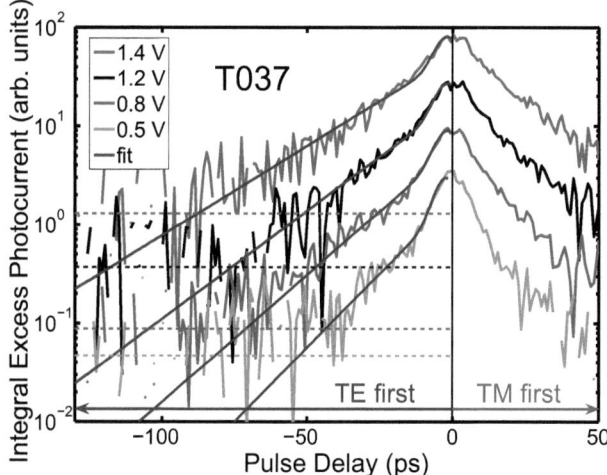

Figure 8.11: Logarithmic plot of the normalized excess photocurrent characteristics as shown in Fig. 8.9 together with the respective fits for T037. The individual curves have been offset vertically for clarity. The broken horizontal lines mark the noise level of the respective measurement.

a least-squares-fit of the model function to each of the characteristics presented in Figs. 8.7 and 8.8. The global parameters are chosen iteratively. Starting with estimated values for C_2 and σ_t, the standard deviations of all experimentally obtained curves from their respective fit are summed up, and this *total deviation* is minimized by manually varying the values for the global variables while repeatedly performing the least-squares-fit.

The fitting procedure was applied to the data acquired on T037 and T038, where the position of zero-delay was shifted to the maximum of the integral excess photocurrent, as done for the plots in Figs. 8.9 and 8.10. The characteristics for the negative delay side were treated separately from the data for positive delays. For fitting the data fraction with negative delays, τ was set to the absolute of the experimental delay values in order to consistently maintain the sign conventions in Eqns. 8.10 to 8.13. As already discussed in section 8.5.2, the amount of charge carriers occupying the excited LH state directly after pump beam 1 differs in case of positive delays from that at negative delays due to selection rules. This leads to an asymmetry in the occupation-based contribution to the photocurrent, where this contribution to the total integral photocurrent is much weaker in case of positive delays as compared to negative ones. Put differently, the signal-to-noise ratio of the data relevant for the relaxation time extraction obtained for negative delays is superior to that for positive delays. Additionally, in order to reduce measurement time, which was highly limited due to the restricted access to the FEL, less data were collected for positive delays than for the negative delay side. This results in a further degradation of the reliability of the data extraction process for positive delays. Thus, exclusively the parameters obtained from the data recorded at negative delays are discussed in the following. The optimal set of global parameters, which was employed for fitting the data on T037 and T038 gained for negative delays, is presented in table 8.6.

Note that the optimal values for C_2 are equivalent to 20 THz and 10 THz for the fit of the data obtained on T037 and T038, respectively. According to Eqn. 8.35, these values are equivalent to phase coherence times of $\frac{1}{\Gamma} = 100$ fs (T037) and 200 fs (T038). These times are far below any of the intersubband relaxation times τ_{decay} extracted by the fitting procedure, as seen in the following section.

As observed during the fitting procedure, the resulting figures for τ_{decay} exhibit a monotonic dependence on the voltage, under which the respective fitted data were obtained, relatively independent of the detailed choice for the global parameters. This reflects the qualitative observation of a monotonically increasing photocurrent relaxation time with the applied field in Figs. 8.9 and 8.10. However, the quantitative values for the intersubband relaxation time obtained by fitting the measurement data *do* depend on the detailed choice of C_2 and σ_t.

Relaxation time results and discussion

The results of the fitting procedure performed on the data for negative delay values in Figs. 8.7 and 8.8 are shown in Figs. 8.11 and 8.12 by the thin blue lines. These figures show a version with a logarithmic y-axis of the plots in Figs. 8.9 and 8.10, where the subtracted current baseline values are gained from the performed least-squares-fits of the model function presented in this section. These values are trivially depending on the voltage, at which the respective data was obtained, and differ for the regions of positive and negative delays, as already discussed in section 8.5.2. For the plots, the individual pump-pump characteris-

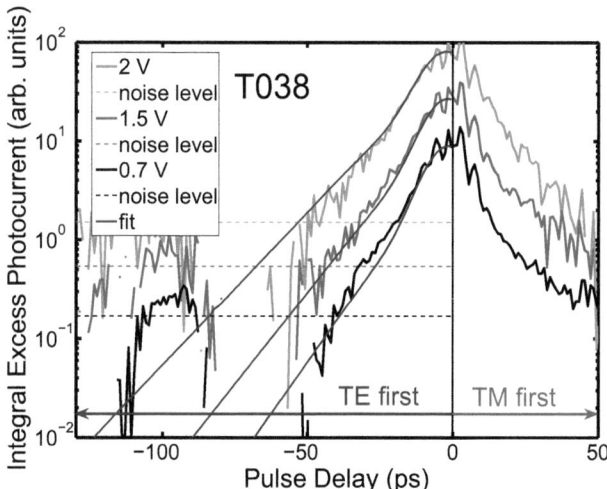

Figure 8.12: Logarithmic plot of the normalized excess photocurrent characteristics as shown in Fig. 8.10 together with the respective fits for T038. The individual curves have been offset vertically for clarity. The broken horizontal lines mark the noise level of the respective measurement.

tics have been offset vertically for clarity. The noise levels of the obtained experimental data are shown relatively to the plotted characteristics by the dashed horizontal lines (and are thus also shifted vertically). It can be clearly seen that the integral excess photocurrent drops below the respective noise level at larger and larger delays as the applied bias increases.

The blue lines in Figs. 8.11 and 8.12 present the results of the model function in Eqn. 8.10 for the parameter values gained by the fitting procedure. As seen from this figures, the pump-pump characteristics are well reproduced over two orders of magnitude, where both the sharp coherent and the slowly decaying LH-based photocurrent contributions are accounted for by the fit function with high accuracy. The individual plots show, that both the experimentally obtained characteristics and the modelled excess photocurrent decay exponentially (indicated by the linear curve shape in the logarithmic plot) once the coherent center peak drops to insignificance for larger delays.

Now, the ultimate purpose of the fitting procedure was the extraction of the decay time of the excess photocurrent originating from a finite occupation of the LH1 state, equivalent to the LH1-HH1 intersubband relaxation time. Figs. 8.13 and 8.14 present the relaxation

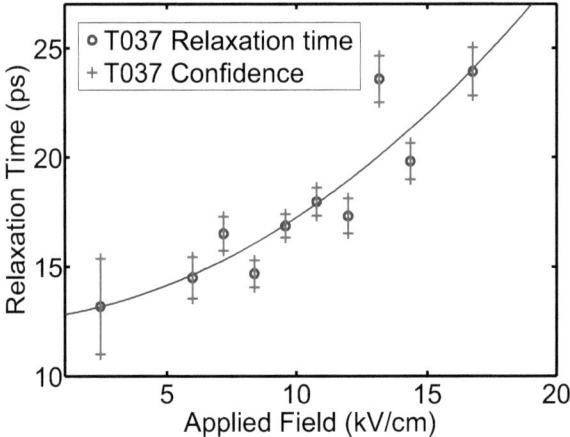

Figure 8.13: Bias dependence of the LH1-HH1 relaxation time for sample T037. As indicated by the blue dots, the intersubband relaxation time extracted from the pump-pump data exhibits a clear, monotonic rise with the voltage applied to the sample. The red errors bars indicate the 68% confidence limit of the fitted decay times.

time values gained from the total series of pump-pump characteristics for T037 and T038, respectively, where the fit was performed on the data of negative pulse delay as discussed above. The extracted values for τ_{decay} are represented as blue and black circles for T037 and T038, respectively, and the solid lines through the data point ensembles serve as a guide for the eye.

The values for the LH-HH intersubband relaxation time presented in Figs. 8.13 and 8.14 exhibit a clear, monotonic dependence on the corresponding field applied to the sample. For both samples, an increase in the field from about 5 kV/cm to 18 kV/cm leads to an increase of the LH-HH relaxation time by a factor of two, where the decay time values change continuously. In case of sample T037, the relaxation time could be tuned between 13 and 24 ps, while the decay time range observed for T038 under equivalent biasing spans 8 to 14 ps. The observed intersubband relaxation times, which are associated with an LH-HH transition energy of 30 meV according to the simulations in section 8.2 and the spectral characterization in section 8.3, are well within the range expected for transition energies below the LO phonon energy, as discussed in chapter 6.3.2 (see table 6.2). According to references [104] and [109],

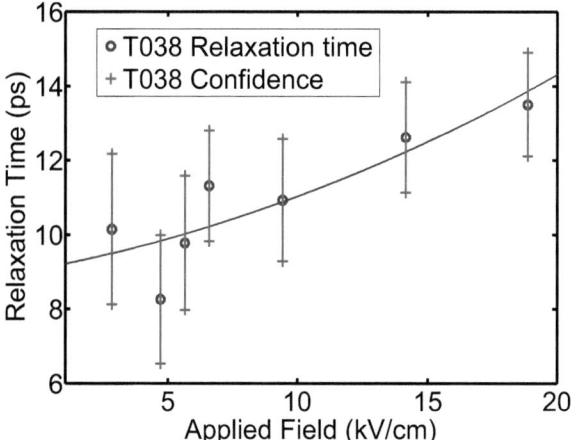

Figure 8.14: Bias dependence of the LH1-HH1 relaxation time for sample T038. As indicated by the blue dots, the intersubband relaxation time extracted from the pump-pump data exhibits a clear, monotonic rise with the voltage applied to the sample. The red errors bars indicate the 68% confidence limit of the fitted decay times.

alloy scattering processes limit the intersubband relaxation times in SiGe heterostructures in case of transition energies below the LO phonon energy (see chapter 6.2.5).

The comparison between Figs. 8.13 and 8.14 shows that the intersubband relaxation times of T037 differ in their actual values from those gained for T038. The detailed reason for this difference in the relaxation times could not be clearly singled out of the several potential explanations. One of the probable reasons for the differing results on T037 and T038 is given by the possible inequality of the two samples in respect to growth quality and alloying and an associated difference in the actual scattering mechanisms. As up to now the scattering processes for transition energies below the LO phonon energy lack the detailed understanding of the mechanisms determining the relaxation above the LO phonon energy, this issue is difficult to address and reaches beyond the scope of this work. A second potential reason for the different lifetime values extracted from the experimental data on T037 and T038 at equal biasing lies in the process of data evaluation itself. As discussed above, the precise values for the extracted fit parameters and therefore τ_{decay} depend on the detailed choice of the global parameters of the fitting model. Thus, an inaccuracy in the determination of these global parameters might result in relaxation time results slightly off their actual values. However, the trend in τ_{decay} of increasing with the sample bias is *independent* of the chosen global parameter values. Therefore, both of the potential explanations given above do not weaken the basic observation illustrated by Figs. 8.13 and 8.14, namely the increase of the intersubband relaxation times in T037 and T038 with the applied electric field. As the fit parameters and thus the extracted relaxation times further exhibit a finite uncertainty in their obtained values, as will be discussed in the next subsection, a detailed interpretation of the difference between the observed lifetime ranges in Figs. 8.13 and 8.14 is non-trivial and would require additional experimental studies beyond the scope of this work.

Now, the significance of the monotonic trend in the relaxation time characteristics clearly seen in Figs. 8.13 and 8.14 shall be quantified by determining the confidence limits of the extracted fit parameter τ_{decay} as discussed in the following section.

Confidence limits

In order to obtain a measure for the reliability of the extracted voltage-dependent LH-HH intersubband relaxation time, the confidence region of the determined optimal fit parameter τ_{decay} was calculated. For this purpose, the principle of the least-squares fit method is discussed briefly. The purpose of a fit is to determine the parameters $\boldsymbol{\beta}$ of a model function $f(x_i, \boldsymbol{\beta})$, for which this function optimally reproduces a number of n discrete data points (x_i, y_i). The number of model parameters and therefore the number of components of $\boldsymbol{\beta}$ be m. A perfect fit between the model function and the experimental data fulfills Eqn. 8.36.

$$y_i = f(x_i, \boldsymbol{\beta}) \qquad i = 1...n \; , \; \dim(\boldsymbol{\beta}) = m \tag{8.36}$$

However, in most of the cases there is a discrepancy between model and experiment even for an optimally chosen set of parameters. The deviation of each experimental data point from the fit function at a given parameter set $\boldsymbol{\beta}$ is given in Eqn. 8.37.

$$r_i = y_i - f(x_i, \boldsymbol{\beta}) \tag{8.37}$$

The fitting method of least squares now targets at approaching the experiment as closely as possible by determining the parameter set, for which the sum of all squared deviations, $S(\boldsymbol{\beta})$ in Eqn. 8.38, is minimal.

$$S(\boldsymbol{\beta}) = \sum_{i=1}^{n} r_i(\boldsymbol{\beta})^2 \tag{8.38}$$

This fitting procedure requires the numerical evaluation of the Jacobi determinant $X_{ij}(\boldsymbol{\beta})$, which is defined by Eqn. 8.39 and is built up by the partial derivatives of the fit function with respect to the individual parameters.

$$X_{ij}(\boldsymbol{\beta}) = -\frac{\partial r_i}{\partial \beta_j} = \frac{\partial f(x_i, \boldsymbol{\beta})}{\partial \beta_j} \qquad (8.39)$$

After obtaining an optimized set of parameters by numerically minimizing S in Eqn. 8.38, the reliability of the individual parameters can be determined by calculating the matrix $\boldsymbol{M^\beta}$ as defined in Eqn. 8.40.

$$\boldsymbol{M^\beta} = \frac{S}{n-m}(\boldsymbol{X}^\mathrm{T}\boldsymbol{X})^{-1} \qquad (8.40)$$

The off-diagonal elements of this matrix are equivalent to the correlation between the individual fit parameters, while the diagonal elements give their autocorrelation, or variance. The standard deviation of the optimized fit parameter $\sigma^2(\beta_i)$ in Eqn. 8.41 gives a measure for the confidence region of the extracted fit parameter β_i, which usually resembles a quantity of physical relevance.

$$\sigma^2(\beta_i) = \boldsymbol{M^\beta_{ii}} \qquad (8.41)$$

Put differently, the actual value of the physical quantity, which is represented by the parameter β_i in the model, lies within $\beta_i^f \pm \sigma(\beta_i^f)$ with a probability of 68%. Here, a normal distribution of the scattering of the experimental data points y_i is assumed. β_i^f is the value attributed to this physical quantity by fitting an ensemble of experimental data points. The 68% confidence region for the fit parameter representing the LH-HH relaxation time τ_{decay} during the evaluation of the data on T037 and T038 is indicated by the vertical lines confined by short horizontal bars in Figs. 8.13 and 8.14. The width of the confidence region was determined by calculating the Jacobi determinant of the fit function in Eqn. 8.10 and interpreting its values for the optimal fit parameter set according to the discussion above.

As indicated by the error bars defining the 68% confidence region for τ_{decay} in Figs. 8.13 and 8.14, the observed monotonic increase of the intersubband relaxation time is of high significance for both studied samples. In particular the values extracted for T037 exhibit a very narrow confidence region, where the standard deviation decreases while the field increases from 2.4 kV/cm to 8.4 kV/cm. This behavior can be associated with the rise of the current signal with the applied bias under fixed amplifier settings (see table 8.4). For low applied voltages, the noise of the measurement system contributes significantly to the total output noise, limits its signal-to-noise ratio and thus the confidence region of the extracted fit parameters. As the bias increases, the influence of the system noise is reduced due to the increase in the current signal and the noise associated with the device itself. As a consequence, the signal-to-noise ratio of the measured integral photocurrent increases and the confidence region of the fit parameters narrows. This increase in the signal-to-noise ratio of the measured integral photocurrent when raising the applied field from 6 kV/cm to 9.6 kV/cm is also indicated by the noise level markers in Fig. 8.11.

The comparison of the error bars in Figs. 8.13 and 8.14 suggests, that the uncertainty in the extracted intersubband relaxation time is significantly higher for sample T038 than for T037. The reason for this observation is reflected in the value of the optimal global fit parameter C_2, which is given in table 8.6. As already discussed above, this parameter gives a measure for the contribution of the photocurrent component originating from a finite LH1 occupation to the total signal, and thus is a measure for the signal-to-noise ratio of the integral excess photocurrent. The significantly higher value of C_2 for T037 indicates that the photocurrent component relevant for the relaxation time extraction contributes stronger to the total measured signal and thus results in a higher accuracy for the determination of τ_{decay} as compared to T038.

Calculating the variance as given by Eqn. 8.41 for J_0 in Eqn. 8.10 shows that the confidence region of this model parameter constitutes less than 1% of its fit value. Put differently, the voltage-dependent current baseline of the pump-pump characteristics in Figs. 8.7 and 8.8 represented by this fit parameter could be determined with very high accuracy.

Summary of the experimental results

By employing photocurrent pump-pump experiments at a free-electron-laser, the LH1-HH1 intersubband relaxation times in two SiGe heterostructure samples were monitored under biasing. As expected from the structures' LH1-HH1 transition energy below the LO phonon energy, the observed relaxation times were in the 10 ps regime. By extracting the relaxation time values from a series of pump-pump characteristics measured for different bias voltages, a continuous voltage tunability of these intersubband relaxation times could be observed. For both samples, the decay times could be increased by a factor of two by raising the externally applied field. The origin of this voltage-tunability of intersubband relaxation times can be directly related to bias-induced changes in the bandstructure of the samples, as shown in the next chapter, which presents a detailed interpretation of the results shown in Figs. 8.13 and 8.14

8.5.4 Interpretation of the relaxation time tuning

As will be shown in this section, the bias tunablility of intersubband relaxation times observed during the experiments presented in chapters 8.5.1 to 8.5.3 can be associated with a bias-induced change from a spatially direct to a diagonal intersubband transition. In [16,20,132] as well as in chapters 2 and 3 of this work, the voltage-tuning of the responsivity onset of quantum well infrared photodetectors based on cascade-like SiGe heterostructures was reported. For these structures, the bias-induced change in the optical properties could be associated with the tuned occupation of different quantum states and was thoroughly understood on the basis of bandstructure calculations. Chapter 2.2.2 discusses, how charge carriers can be transferred between a shallow and a deep quantum well in sample K091 by changing the relative energetic position of the HH ground states of these two wells via altering the externally applied field. The concept of the voltage tunable two-color detectors is thus based on inducing a change of the *ground state alignment* between different quantum wells in a heterostructure by varying the applied bias.

Now, each heterostructure period of T037 and T038 exhibits a well-defined HH ground state strongly localized in the central quantum well of the potential landscape presented in Fig. 8.1, independent of the applied field within the experimentally accessed bias range. In contrast to the two-color QWIPs discussed in chapters 8.5.1 to 8.5.3, the ground state configuration of T037 and T038 is completely stable in respect to the externally applied field. On the other hand, the spatial position of the first *excited* state of each quantum well period is strongly dependent on the applied bias, as will be shown in the following, and thus allows a voltage-tuning of the overlap with the stable ground state. Therefore, while the spectral tuning of the responsivity of the two-color QWIP samples was enabled by a bias-induced change in the ground state configuration, the change in the relaxation properties of T037 and T038 can be associated with a voltage-tuning of the excited state wave function. In analogy to the interpretation of the bias-tunable properties of K090 and K091, band structure calculations can be employed to thoroughly understand the field-dependence of the LH1-HH1 intersubband relaxation time of T037 and T038.

Bias dependence of the bandstructure

Figure 8.15 shows the results of 6-band $\mathbf{k \cdot p}$ calculations such as those discussed on several occasions during this work. Basically, this figure presents the simulated band structures in analogy to those shown in Fig. 8.2. However, unlike Fig. 8.2, Fig. 8.15 presents the results of the bandstructure simulations not as a contour plot, but merely shows the calculated eigenenergies as horizontal lines at their energetic position in the band edge profile. In order to illustrate the shape of the spatial carrier density distribution of the respective eigenstates, the absolute squares of the associated wavefunctions are plotted using their eigenenergy line as x-axis. Note that these squared wavefunction are plotted on an arbitrary y-axis, but are scaled consistently in respect to each other. For the sake of clarity, the plots in Fig. 8.15 concentrate on the lowest HH and LH states of each structural period, since the pump-pump

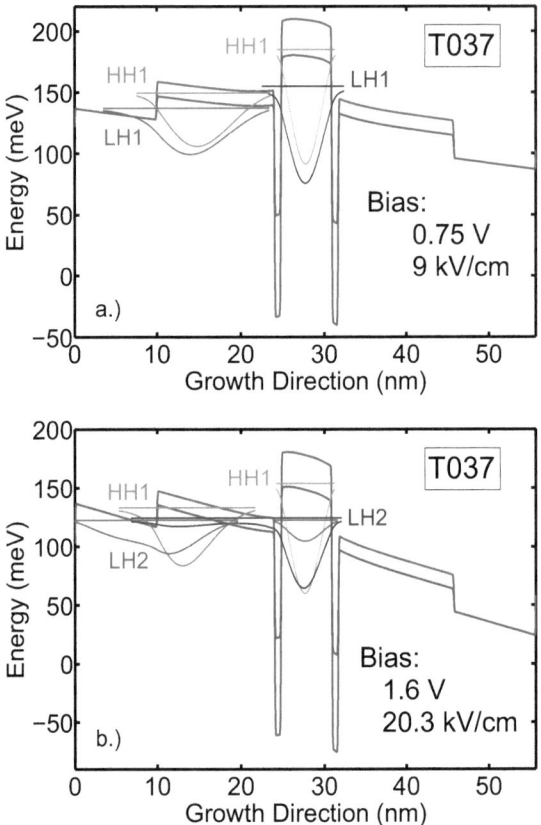

Figure 8.15: Change in the band structure of T037 with the applied field. The plots present the calculated HH and LH band edges as red and green lines, respectively, together with the absolute squared wave functions at their eigenenergy values. In (a), the first excited LH1 state is confined to the central well, while it couples to the LH state of the side well in (b). The resulting change in the overlap between the first excited state and the ground state of the structural period affects the relaxation efficiency, as discussed in the text. Note that y-axis shows the energy scale for electrons. Lowering the energy of a hole state is thus equivalent to an increase in the energy shown on the scale.

experiments reported in this chapter monitored the intersubband relaxation processes between exactly these states. Highly excited hole states were omitted in the plots, as they are not relevant for the interpretation of the voltage-tunability of the intersubband relaxation time.

Figure 8.15 presents the results of the bandstructure simulations for the structure of T037 under applied fields of 9 kV/cm and 20.3 kV/cm. The high value of 20.3 kV/cm in Fig. 8.15 (b) represents the upper limit of the experimentally accessed voltage range of the relaxation time characteristics. It was chosen for the bandstructure simulations, as it allows a clear illustration of the change in the bandstructure induced by a strong increase in the applied electric field. Samples T037 and T038 differ exclusively in the separation layer between subsequent quantum well regions, as discussed in section 8.2.1. Thus the tuning of the intersubband relaxation time is for both samples based on the same changes in the bandstructure, enabling the analogous interpretation of both Figs. 8.13 and 8.14 on the basis of the bandstructure plots in Fig. 8.15.

As seen from the plots, the HH1 ground state of each period is strongly confined to the deep central well, independent of the applied bias within the discussed range. Under low applied biases, represented by 0.75 V in Fig. 8.15 (a), the excited LH1 state is also confined to the central well of the structure, as indicated by the blue curve in the plot. As the voltage applied to the structure is increased, one of the two symmetric shallow side quantum wells energetically lowers in respect to the central well. While lowering the side well more and more by increasing the bias, the LH1 state of the central quantum well energetically approaches the level of the LH1 state of the side well. Along with their alignment, these two states couple. This coupling of the LH states of the two wells results in a common lowest LH state, which exhibits significant components in both wells. Such a common first excited LH1 state is observed in the bandstructure plot of Fig. 8.15 (b), which represents the case of a strong sample biasing. The lowest LH state of the structure represented by a blue line clearly features components in both wells, and exhibits a center of mass shifted into the shallow side well. Now, the higher the applied field, the stronger the side-well-component of the common lowest LH state is. Put differently, by increasing the field externally applied to the structure, the charge carrier distribution associated with the common first excited LH1 state shifts more and more from the central well to the spatial position of the side well.

Influence on the relaxation time

This continuous shift of the first excited state's charge distribution into the side well with increasing voltage directly affects the relaxation time between the excited LH1 state and the

HH1 ground state of the structure. For both T037 and T038, the transition energy associated with the LH1-HH1 transition is well below the LO phonon energy, leading to a domination of the intersubband relaxation process by alloy scattering. As discussed in detail in chapter 6.2.7 and expressed by Eqn. 6.61, the alloy scattering rate between two states crucially depends on the overlap between their charge carrier densities. Thus, the quantity fundamentally influencing the intersubband relaxation time experimentally investigated in this work, is the charge density overlap between the HH1 ground state and the first excited LH1 state of the studied structures. The spatial position of the HH1 ground state in the central well is unaffected by an increase in the bias. As the charge carrier distribution of the LH1 state shifts more and more into the shallow side well while raising the applied voltage, the HH1-LH1 overlap decreases with increasing bias, and so does the efficiency of the alloy scattering process. This, in turn, leads to an increase in the LH1-HH1 intersubband relaxation time with the applied electric field, as observed in Figs. 8.13 and 8.14. Due to the continuous shift of the LH1 charge carrier distribution into the side well with the applied voltage, the charge density overlap as well as the intersubband relaxation time changes *continuously* with the bias.

Thus, the short relaxation times for low biasing in Figs. 8.13 and 8.14 can be directly related to the bandstructure configuration shown in Fig. 8.15 (a). At a bias of 9 kV/cm, both the HH1 and the first excited LH1 state are located in the central well, resulting in a strong carrier density overlap between them. According to Eqn. 6.61, the LH1-HH1 relaxation based on alloy scattering is very efficient for this configuration, and the resulting intersubband relaxation times are as low as 15 ps (T037) and 10 ps (T038). This overlap configuration at low biasing is equivalent to a spatially direct intersubband transition.

On the other hand, the long relaxation times given for high voltages in Figs. 8.13 and 8.14 can be associated with the configuration shown in Fig. 8.15 (b). At strong biasing, the first excited LH state couples to the side quantum well, and the charge carrier density associated with this state shifts away from the central well. This spatial separation of the excited LH state, whose population was monitored experimentally, and the ground state results in a decreased overlap between the respective charge carrier densities. As a consequence of this reduced density overlap between the two states, the associated scattering efficiency is low, and the resulting LH1-HH1 relaxation times are as high as 24 ps (T037) and 14 ps (T038). Put differently, for high applied voltages the quantum well structure features a diagonal transition between the LH1 and HH1 state, along with an increased associated relaxation time.

To conclude, by simulating the bandstructure of the studied structures for different voltages, the experimentally monitored intersubband relaxation properties of both samples could be related to the bias-induced change in the bandstructure. The voltage-tunability of the LH1-HH1 relaxation time experimentally observed in the course of this work for both samples T037 and T038 is based on the bias-induced change from a spatially direct transition into a diagonal one. By transforming the monitored quantum well transition from a direct into an indirect one, the intersubband relaxation time could be increased by a factor of two. This experimental finding strongly supports the concept of increasing intersubband lifetimes in the SiGe system via diagonal transitions. The results presented in this work suggest, that diagonal transitions possess a strong potential as a means of manipulating intersubband decay times in this material system in order to tailor them as required for achieving population inversion. It shall be once more pointed out, that the transition between a spatially direct and a diagonal configuration was exclusively induced by changing the externally applied bias, where the resulting lifetime prolongation was directly monitored via pump-pump experiments.

8.6 Summary and conclusions

Summary

The chapter at hand presented the first experimental data on bias-manipulation of intersubband relaxation times in a SiGe heterostructure, as well as the first *direct* monitoring of the difference in the associated decay times between spatially direct transitions and diagonal ones. The presented experimental results are thus of high relevance for determining the suitability of the concept of diagonal transitions for the realization of quantum cascade lasers in the SiGe system.

The two p-type SiGe quantum well samples designed and fabricated in the course of this work served as a simple model system for investigating the dependence of the intersubband relaxation time between two quantum well states on their spatial density overlap. Each period of the structure featured a central, deep quantum well and two shallow side wells. According to bandstructure calculations, the HH1 ground state of each structural period is strongly confined to the central quantum well, as is the first excited LH1 state in case of zero-biasing. As the field externally applied to the sample is increased, the LH1 state couples to the LH state of the shallow side well, resulting in a shift of the excited state's carrier density away from the central well. This bias-induced spatial separation of ground and excited state is the basis of the tunability of the associated relaxation time. The structures were grown by MBE and processed into QWIP mesas, in order to be able to apply a bias and measure the photocurrent through the sample.

The time-resolved experiments, which were carried out at the free-electron-laser FELIX, aimed at the determination of the LH1-HH1 relaxation time in both heterostructure samples. For this purpose, the photocurrent pump-pump method discussed in detail in chapter 7 was employed. As predicted by bandstructure simulations and confirmed by spectrally resolved photocurrent characteristics based on two-photon absorption, the LH1-HH1 transition energy was equal to 30 meV for both samples, a value well below the LO phonon energies. An FEL pulse duration of the order of a few picoseconds was sufficient for achieving the required time resolution. The pump-pump experiments were carried out at a series of biases for each sample in order to investigate the influence of the applied electric field on the intersubband relaxation time. The relaxation time values were extracted from the experimental data by fitting a model for the excess photocurrent characteristics, where the lifetime results valued between 8 and 24 ps and were thus within the range expected for a domination of the relaxation process by alloy scattering.

For both samples, the decay time results obtained for a series of applied voltages show a clear, monotonic rise of the LH1-HH1 intersubband relaxation time with increasing bias. By raising the externally applied electric field from 5 kV/cm to 20 kV/cm, the relaxation time could be continuously increased by a factor of two. This continuous voltage-tunability of the LH1-HH1 intersubband relaxation time could be directly associated with a bias-induced gradual spatial separation of the two involved states and the consequential decrease of the alloy scattering efficiency. In other words, by increasing the applied bias, the character of the LH1-HH1 transition was continuously shifted from spatially direct to indirect, or diagonal. Therefore, the photocurrent pump-pump experiments on T037 and T038 constitute a *direct* study of the difference in relaxation time between a spatially direct transition and a diagonal one.

Conclusions

One of the main difficulties for achieving population inversion in a SiGe heterostructure lies in the absence of a resonant scattering mechanism. As a consequence, the SiGe system lacks the possibility of manipulating intersubband relaxation times by tailoring the associated transition energies. The clear design rules applicable for QCLs in the III-V system thus cannot be adapted for the SiGe material, and the successful development of a SiGe QCL crucially depends on the implementation of alternative design concepts. As discussed in 6.1.4, diagonal transitions are one of the few remaining potential ways for manipulating the intersubband relaxation times in a SiGe heterostructure with the aim of reaching population inversion. Conventionally, the overlap between the two quantum well states of the lasing transition in III-V QCLs is manipulated via structural design.

Up to now, the influence of a spatial separation of two states in a SiGe heterostructure on their associated relaxation efficiency has been studied by comparing experimental data on *differently grown* samples [116, 117] (see chapters 6.3.2 and 6.3.3). In the course of the photocurrent pump-pump experiments carried out on T037 and T038, the spatial separation between the final and initial state of a transition could be tuned by the applied bias, enabling the change from a spatially direct to a diagonal transition within one and the same structure. The experimental method employed in the course of this work thus tremendously increases the reliability of the conclusions drawn from the obtained data, especially as it gave access to a large number of data points for each sample, whereas the conventional comparison of different structures is usually limited by the huge effort of growing a large number of samples.

In addition to enabling the continuous transformation of a transition's character from spatially direct to diagonal within one and the same sample, the method of photocurrent pump-pump experiments on biased samples allowed the *direct* monitoring of the relaxation time between the involved states, a feature which was already highlighted in the discussion of the experiments on sample H019 in chapter 7.8. Therefore, the experimental work presented in this chapter enabled a *direct* monitoring of the lifetime increasing effect of diagonal transitions in a heterostructure within one and the same sample (appropriate of course also for III-V QCLs).

The increase of the intersubband relaxation time by a factor of two, as observed for both samples studied in the course of this work, strongly supports the concept of diagonal transitions as a means of increasing the lifetimes in SiGe heterostructures. On the way to a SiGe quantum cascade laser, concepts for manipulating the intersubband relaxation efficiencies are inevitably needed in order achieve population inversion and finally lasing. As shown by the results obtained by the time-resolved experiments, diagonal transitions in SiGe heterostructures possess the potential of increasing intersubband relaxation times strongly, be it by structural design or, as pursued within this work, by alternative control mechanisms such as the applied bias. By considering the quantitative relaxation time characteristics presented in this chapter, the detailed tailoring of intersubband relaxation times by bandstructure design aiming at population inversion increases in accuracy.

The time resolved studies on SiGe heterostructures clearly indicate, that the experimental access to intersubband relaxation times relevant for the design of operating, electrically driven quantum cascade structures strictly requires data acquisition on *biased* structures. Any time resolved experiment on unbiased devices, such as transmission pump-probe measurements on unprocessed heterostructures, inevitably delivers decay time figures of low reliability, as a strong dependence of intersubband relaxation times on the present fields has to be expected in complex quantum well systems according to the insights gained in the course of this work. Therefore, the experimental method of determining relaxation times under applied fields in the range of those relevant for operating devices gives accurate access to physical quantities, whose knowledge tremendously increase the reliability of simulations predicting the operation of SiGe emitter devices.

To conclude, the work highlighted in this chapter presented experimental methods, concrete results for intersubband relaxation times as well as alternative concepts for manipulating intersubband relaxation times in SiGe heterostructures, which are of high relevance for the design and characterization of SiGe quantum cascade devices on the road to a SiGe quantum cascade laser.

Chapter 9

Conclusions and outlook

This thesis presented experimental studies on silicon-based optoelectronic devices, ranging from two-color mid-infrared QWIPs to blocked-impurity band detectors operating in the terahertz regime. The main part of this work, however, is composed of a thorough experimental study of intersubband relaxation times in p-type SiGe heterostructures, which aimed at determining the intersubband scattering rates relevant for the development of a SiGe quantum cascade laser and concluding on the feasibility of different transition energy ranges and the concept of diagonal transitions for the implementation of QCL structures in this material system.

In the course of the time-resolved studies on a SiGe heterostructure based on a quantum cascade emitter device, the non-radiative intersubband relaxation time between the HH2 and HH1 states of this p-type structure was directly measured employing photocurrent pump-pump experiments (see chapter 7). With a HH2-HH1 transition energy of 160 meV, which is far above the LO phonon energies in the SiGe system, the associated intersubband relaxation time was found to be around 550 fs [123, 124]. Thus, the HH2-HH1 intersubband relaxation via non-polar optical phonon scattering in the SiGe system was found to be of comparable efficiency as the electron scattering by the emission of polar optical phonons in the III-V material system. The ultrafast nature of the deformation potential scattering by LO phonons in SiGe heterostructures can be theoretically explained by the efficient relaxation of HH2 carriers into the HH1 states via the LH1 intermediate state, as the deformation potential interaction is predicted to allow an extremely efficient scattering between different hole bands [114].

In contrast to the ultrafast HH2-HH1 intersubband relaxation times for transition energies above the LO phonon energy in the sub-picosecond regime, time-resolved experiments on LH1-HH1 transitions with associated energies below the LO phonon energies (see chapter 8) revealed relaxation times of the order of ten picoseconds [124, 133]. For a LH1-HH1 transition energy of 30 meV, associated intersubband relaxation times between 8 and 24 ps were determined by photocurrent pump-pump experiments. In this transition energy range, the ultra-efficient emission of LO phonons is no longer possible, and the time scale of intersubband scattering processes is predicted to be determined by alloy scattering [104]. By performing the time resolved experiments for a series of voltages applied to the investigated p-type SiGe heterostructures, the influence of the change from a spatially direct to a diagonal LH1-HH1 transition on the associated intersubband relaxation time was studied. The experiments carried out in the course of this work demonstrated, that a bias-induced change from a spatially direct to an indirect transition can increase the LH1-HH1 intersubband relaxation time by a factor of two.

The main part of this work leads to two major conclusions, which are of relevance for future attempts on the development of an electrically pumped group-IV laser.

The first conclusion is that the usability of HH2-HH1 intersubband transitions with energies above the LO phonon energy (and thus in the mid-infrared region) as an emitting transition in a QCL structure is crucially degraded by the ultrafast non-radiative relaxation processes present in this energetic regime. The observed ultra-short HH2-HH1 relaxation times in the SiGe system (comparable to those induced by polar optical phonon scattering in III-V structures) in combination with the lack of any resonant intersubband relaxation processes in this system drastically reduces the chance of reaching population inversion in any quantum cascade structure based on HH2-HH1 transitions in the mid-infrared. What is more, as an inherent feature of the band structure of any SiGe quantum well, light hole states are to be found between two subsequent HH subbands. Thus, even if the energies of the HH2-LH1 and LH1-HH1 transitions of a quantum well are pushed below the LO phonon energy by means of bandstructure design, there is *always* an additional relaxation channel via the intermediate LH1 state available for the depopulation of the HH2 state. And even though the absence of LO phonon emission reduces the efficiency of this additional channel in case of low transition energies, the relaxation process given by this additional channel has to be compensated for by an efficient depopulation mechanism emptying the ground state of the lasing transition. This renders the HH2-HH1 transition highly unsuitable as an emitting transition in future designs aiming at the implementation of the QCL concept in p-type SiGe.

The second conclusion drawn from this work is, that the concept of diagonal transitions forms a highly promising means for manipulating the intersubband hole relaxation times in p-type SiGe heterostructure devices, particularly when applied to LH-HH transitions in the terahertz regime, as the associated intersubband relaxation times are two orders of magnitude higher than that for transition energies above the LO phonon energy and are thus in a feasible range for quantum cascade emitters. Considering the absence of resonant scattering mechanisms in the SiGe system, diagonal transitions might prove the *only* way of adjusting intersubband relaxation times according to the requirements for population inversion. What is more, apart from the favorable scattering times intrinsic to the transition energy regime below the LO phonon energy, the terahertz emission region is highly attractive for future attempts on the development of SiGe quantum cascade emitters due to the growing need for sources in this spectral regime.

The potential of diagonal transitions is, however, not restricted to SiGe structures based on valence band transitions. Both of the recently proposed concepts for n-type SiGe QCLs are based on diagonal lasing transitions [99, 128]. The fundamental property of SiGe heterostructures, which up to now prevented the implementation of n-type intersubband devices in this material system, is the small maximal overall conduction band offset in the range of a few 10 meV featured by such structures. And even though the requirements regarding the confinement of charge carriers in optical intersubband devices are more relaxed in the terahertz regime than in the mid-infrared region of operation, novel concepts are needed to overcome the principal limitations of conduction band structures in the SiGe system.

One of the two concepts for n-type SiGe QCLs is based on growing the device on a substrate with (111) orientation [99, 100]. In such a structure, there would be no energetic splitting of the Δ-valleys due to the strain induced by pseudomorphic growth. As a result of this preservation of the Δ-valley degeneracy, the total Δ band offset of about 150 meV could be used for realizing a quantum well structure. In [99], Lever et al. presented the design of an [111] n-type SiGe QCL based on a diagonal lasing transition with an emission energy around 15 meV.

The second concept for a conduction band QCL in the SiGe system is based on growing a Ge/Si$_{1-x}$Ge$_x$ structure on a (001) Si$_{1-y}$Ge$_y$ pseudosubstrate (with $y > x$,) and on using the confinement of L-valley electrons for the design of a laser device [98, 128]. As in the concept involving growth in (111) direction, the L-valley approach is based on a preservation of the degeneracy of the valley employed for the formation of an electronic confinement, where the maximal L-valley band offset in such a structure would be about 130 meV. However, due to the induced strain, the degeneracy of the six electronic Δ-valleys is lifted, where electrons of the so-called $\Delta 2$ valleys are confined oppositely to the L-valley carriers and are thus confined in the *barriers* of the quantum well structure. Therefore, in order to prevent the formation of a $\Delta 2$-valley quasicontinuum outside of the L-valley quantum wells with an onset energetically below the confined L-state in the well, where scattering between the two valleys would drastically reduce the confinement of L-electrons, the Si$_{1-x}$Ge$_x$ barriers in the QCL structure have to be very thin (around 10 Å). Thin barriers push the $\Delta 2$-valley quasicontinuum energetically above the confined L-valley-states, and due to a consequential reduction of intervalley scattering efficiency between L and $\Delta 2$ states, the full L-valley confinement can be exploited for the design of a QCL structure. In reference [98], designs of Ge/$Si_{0.22}Ge_{0.78}$ QCLs on a $Si_{0.12}Ge_{0.88}$ pseudosubstrate with emission wavelengths of 25 μm and 50 μm were presented. According to this publication, the quantization effective mass of L-valley electrons in the presented structures would be as low as 0.19.

Even though the design concepts of n-type QCLs presented in references [99] and [128] are promising, the realization of the proposed devices by means of epitaxial growth has yet to be demonstrated. This is due to the high technological demands intrinsic to both approaches to n-type SiGe QCLs, like the growth of high-quality heterostructures on an (111) substrate, or the fabrication of ultra-thin $Si_{1-x}Ge_x$ barriers of thicknesses around 10 Å.

In addition to the attempted implementation of the QCL concept in SiGe on the road to realizing the first electrically pumped group-IV laser source, a different approach based on optical interband transitions in n-type tensile-strained Ge brought this ultimate goal closer to realization very recently [129–131]. Like Si, Ge is an indirect semiconductor, rendering interband transitions between the conduction band minimum and the valence band maximum highly inefficient. However, Ge is classified as a pseudodirect bandgap semiconductor, as the minimum of the indirect L-valley and the direct Γ-valley are only separated by 136 meV in bulk material. In reference [129] the group of C. Kimerling demonstrated, that in a highly n-doped layer of Ge under tensile strain the electronic occupation of the direct Γ valley is sufficiently high for observing direct band gap photoluminescence at room temperature. The tensile strain induced by the difference in the thermal expansion coefficient between the Si substrate and the Ge layer lowers the energetic difference between the Γ- and L-valleys and relaxes the conditions for the Γ-valley occupation. In reference [130], the group reported on the first observation of optical gain in such a system. And according to very recent information, the group has even achieved lasing in a tensile-strained Ge layer on Si [131]. As the gain window demonstrated in [129] lies around a wavelength of 1.6 μm, the spectral regions of operation of a potential Ge interband device and SiGe quantum cascade emitters do not overlap. But even though both types of devices aim at covering complementary regimes in the infrared spectrum, the successful implementation of a Ge-on-Si interband laser would without doubt prove highly stimulating for future work on SiGe quantum cascade emitters.

As indicated by the lack of any operating SiGe QCL device, both the p-type as well as the n-type approaches require further extensive theoretical and experimental studies of intersubband transitions in the SiGe system for finally achieving population inversion and lasing in such a structure. For both approaches, a number of promising concepts are still to be pursued, all of which involve diagonal lasing transitions. The work on n-type QCLs started only very recently and consequently faces a number of technological difficulties still keeping it from exceeding the conceptual phase. The clear advantage of using p-type heterostructures in a future attempt to realize a SiGe QCL lies in the less demanding structural requirements expected for most of the p-type design concepts as compared to the n-type proposals, and in the large number of successfully implemented quantum cascade emitter concepts reported in a series of publications. Put differently, future attempts in the field of p-type quantum cascade structures can be founded on an significant amount of experimental data gathered in the course of several research projects. And even though a number of different design concepts have been pursued by several research groups, the wide parameter space available for the design of quantum cascade structures still allows a variety of unstudied approaches, especially for employing diagonal LH-HH transitions in the terahertz regime. The potential of SiGe quantum cascade structures for covering the terahertz gap of the electromagnetic spectrum as a laser source intrinsically holds great promise to both science and industry. Thus, independent of the possible future success in implementing alternative group-IV laser concepts covering the mid- to far-infrared spectral regions, like that based on interband transitions in strained Ge, SiGe QCLs continue to hold great potential as a means of achieving electrically pumped group-IV lasing in the terahertz regime.

Bibliography

[1] H. C. Liu et al., *Appl. Phys. Lett.* **79** (25), 4237 (2001).

[2] A. Carbone et al., *Appl. Phys. Lett.* **82** (24), 4292 (2003).

[3] H. Lu et al., *Appl. Phys. Lett.* **89**, 131903 (2006).

[4] S. Ehret et al., *Appl. Phys. Lett.* **69** (7), 931 (1996).

[5] S. Steinkogler et al., *Appl. Phys. Lett.* **81** (18), 3401 (2002).

[6] D. Stehr et al., *Appl. Phys. Lett.* **92**, 51104 (2008).

[7] C. V.-B. Grimm et al., *Appl. Phys. Lett.* **91**, 191121 (2007).

[8] H. Schneider et al., *Appl. Phys. Lett.* **89**, 133508 (2006).

[9] J. Jiang et al., *Appl. Phys. Lett.* **85**, 3614 (2004).

[10] T. Maier et al., *Appl. Phys. Lett.* **84** (25), 5162 (2004).

[11] Y. H. Kuo et al., *Nature* **437** (27), 1334 (2005).

[12] G. Dehlinger et al.,*Science* **290**, 2277 (2000).

[13] S. A. Lynch et al.,*Appl. Phys. Lett.* **81** (9), 1543 (2002).

[14] I. Bormann et al., *Appl. Phys. Lett.* **80** (13), 2260 (2002).

[15] P. Kruck et al., *Appl. Phys. Lett.* **69** (22), 3372 (1996).

[16] P. Rauter et al., *Appl. Phys. Lett.* **83** (19), 3879 (2003).

[17] L. Diehl et al., *Appl. Phys. Lett.* **80** (18), 3274 (2002).

[18] L. Diehl et al., *Appl. Phys. Lett.* **84** (14), 2497 (2004).

[19] T. Fromherz et al., *J. Appl. Phys.* **98**, 044501 (2005).

[20] P. Rauter, *Intraband Absorption and Photospectroscopy of SiGe Quantum Cascades*, Diploma Thesis (2003).

[21] T. Fromherz et al., *Phys. Rev. B* **50**, 15073 (1994).

[22] T. Fromherz, *Infrared Spectroscopy of Electronic and Vibrational Excitations in Semiconductor Quantum Wells and Superlattices*, Thesis (1994).

[23] F. Schwabl, *Quantenmechanik*, 5. Auflage, Springer (1998).

[24] S. L. Chuang, *Phyiscs of optoelectronic devices*, Wiley (1995).

[25] H. C. Liu: *Optoelectronic Properties of Semiconductors and Superlattices: Long Wavelength Infrared Detectors* edited by M. O. Manasreh, Gordon and Breach Science Publishers, 1996.

[26] H. Nyquist, *Phys. Rev.* **32**, 110 (1928).

[27] W. A. Beck et al., *Appl. Phys. Lett.* **63**, 3589 (1993).

[28] M. Ershov, H. C. Liu, *J. Appl. Phys.* **86** (11) 6580 (1999).

[29] C. Schönbein et al., *Appl. Phys. Lett.* **73** (9), 1251 (1998).

[30] R. People et al., *Appl. Phys. Lett.* **47** (3), 322 (1985).

[31] M. Ershov, H. C. Liu, *J. Appl. Phys.* **86** (11) 6580 (1999).

[32] R. Köhler et al., *Nature* **417** 156 (2002).

[33] M. Brucherseifer et al., *Appl. Phys. Lett.* **77**, 4049 (2000).

[34] P. Taday et al., *J. Pharm. Science* **92** 831 (2003).

[35] M. D. Petroff, M. G. Stapelbroek, US Patent No. 568,960 (1986).

[36] M. D. Petroff et al., *Appl. Phys. Lett.* **51** (6), 406 (1987).

[37] J. Leotin, *Infrared Phys. Technol.* **40**, 153 (1999).

[38] D. M. Watson et al., *Appl. Phys. Lett* **52**, 1602 (1988).

[39] J. Bandaru et al., *Appl. Phys. Lett.* **80** (19), 3536 (2002).

[40] B. L. Cardozo et al., *Appl. Phys. Lett.* **83** (19), 3990 (2003).

[41] J. C. Garcia et al., *Appl. Phys. Lett.* **87**, 043502 (2005).

[42] J. W. Beeman et al., *Infrared Phys. Technol.* **51**, 60 (2007).

[43] G. H. Rieke, *Detection of Light: From the Ultraviolet to the Submillimeter*, Cambridge University Press, pp. 92 (1994).

[44] B. A. Aronzon et al., *Semiconductors* **32** (2), 174 (1998).

[45] A. J. van Roosmalen et al., *Dry Etching for VLSI*, Plenum Press, pp. 102 (1991).

[46] S. Winnerl, private correspondence

[47] M. J. Madou, *Fundamentals of Microfabrication*, second edition, CRC Press, pp. 104 (2002).

[48] F. Laermer et al., *U.S. patent 5,501,893*, March 26 (1996).

[49] D. G. Esaev et al., *Semiconductors* **33**, 915 (1999).

[50] F. Szmulowicz et al., *J. Appl. Phys* **62**, 2533 (1987).

[51] F. Szmulowicz et al., *J. Appl. Phys* **63**, 5583 (1988).

[52] N. M. Haegel et al., *Appl. Phys. Lett.* **77**, 4389 (2000).

[53] V. D. Shadrin et al., *Appl. Phys. Lett.* **63**, 75 (1993).

[54] D. G. Esaev et al., *Semiconductors* **35**, 459 (2001).

[55] D. D. Coon et al., *Phys. Rev. B* **33**, 8228 (1986).

[56] M. Bohr, *The new era of scaling in an SoC world*, plenary talk at the *International Solid-State Circuits Conference 2009*.

[57] A. Liu et al., *Nature* **427**, 615 (2004).

[58] O. Boyraz et al., *Opt. Express* **12**, 5269 (2004).

[59] S. Pavlov et al., *Appl. Phys. Lett.* **80**, 4717 (2002).

[60] J. Faist et al., *Science* **264**, 553 (1994).

[61] C. Sirtori et al., *Appl. Phys. Lett.* **73**, 3486 (1998).

[62] J. Devenson et al., *Appl. Phys. Lett.* **89**, 191115 (2006).

[63] M. Beck et al., *Science* **295**, 301 (2002).

[64] R. Köhler et al., *Nature* **417**, 156 (2002).

[65] C. Gmachl et al., *Rep. Prog. Phys* **64**, 1533 (2001).

[66] Special issue on QCL, *IEEE J. Quantum Electron* **38**, 509 (2002).

[67] J. Faist, *Fundamentals of QCL: Active region and wave guides*, plenary talk at the *Poise Summer School 2006*.

[68] C. Sirtori et al., *Appl. Phys. Lett.* **66**, 3242 (1995).

[69] A. Tredicucci et al., *Appl. Phys. Lett.* **77**, 2286 (2000).

[70] C. Sirtori et al., *Opt. Lett.* **23**, 1366 (1998).

[71] J. Faist et al., *Appl. Phys. Lett.* **70**, 2670 (1997).

[72] J. Faist et al., *Appl. Phys. Lett.* **69**, 2456 (1996).

[73] P. Y. Yu and M. Cardona, *Fundamentals of Semiconductors*, Springer Verlag (1996).

[74] P. Ibach and H. Lüth, *Festkörperphysik*, 5th edition, Springer Verlag (1999).

[75] R. A. Soref et al., *Superlattices Microstruct.*, **23**, 427 (1995).

[76] G. Sun et al., *Appl. Phys. Lett.*, **66**, 3425 (1995).

[77] R. J. Collin and H. Y. Fan, *Phys. Rev.*, **93**, 647 (1954).

[78] M. Lax and E. Burstein, *Phys. Rev.*, **97**, 39 (1955).

[79] R. Bates et al.,*Appl. Phys. Lett.* **83** (20), 4092 (2003).

[80] P. Murzyn et al., *Appl. Phys. Lett.* **80**, 1456 (2002).

[81] B. N. Murdin et al., *Phys. Rev. B* **55**, 5171 (1997).

[82] P. K. Basu: *Theory of optical processes in semiconductors: Bulk and microstructures*, Oxford University Press, 1997.

[83] E. Mujagić, et al., *Appl. Phys. Lett.*, **93**, 161101 (2008).

[84] D. J. Paul et al., *Physica E* **16**, 147 (2003).

[85] H. Sigg et al., *Physica E* **11** (20), 240 (2001).

[86] I. Bormann et al., *Appl. Phys. Lett.* **83** (26), 5371 (2003).

[87] L. Diehl et al., *Physica E* **16**, 315 (2003).

[88] G. Dehlinger et al., *Mater. Sci. Eng.* **B89**, 30 (2002).

[89] L. Diehl et al., *Physica E* **13**, 829 (2002).

[90] D. J. Paul et al., *Physica E* **16**, 309 (2003).

[91] R. W. Kelsall et al., *Opt. Mater.* **27**, 851 (2005).

[92] J. Faist et al., *Nature* **387**, 777 (1997).

[93] Z. Ikonic et al., *Physica E* **21**, 907 (2004).

[94] D. Indjin et al., *Semicond. Sci. Technol.* **20**, S237 (2005).

[95] Z. Ikonic et al., *Phys. Rev. B* **69**, 235308-1 (2004).

[96] Z. Ikonic et al., *Semicond. Sci. Technol.* **19**, 76 (2004).

[97] Z. Ikonic et al., *J. Lumin.* **121**, 311 (2006).

[98] K. Driscoll et al., *J. Appl. Phys.* **102**, 093103-1 (2007).

[99] L. Lever et al., *Appl. Phys. Lett.* **92**, 021124 (2008).

[100] A. Valavanis et al., *Phys. Rev. B* **78**, 035420-1 (2008).

[101] G. Dolling, *Proceedings Symposium on Inelastic Scattering Neutrons in Solids and Liquids* **2**, 37 (1963).

[102] R. Tubino, *J. Chem. Phys.* **56**, 1022 (1972).

[103] W. Weber, *Phys. Rev. B* **15**, 4789 (1977).

[104] Z. Ikonic et al., *Phys. Rev. B* **64**, 245311 (2001).

[105] M. Woerner et al, *Physica E* **13**, 485 (2002).

[106] K. Reimann et al., *Phys. Rev. B* **65**, 045302 (2001).

[107] R.W. Kelsall et al., *Phys. Rev. B* **71**, 115326-1 (2005).

[108] G. Sun et al., *Phys. Rev. B* **53**, 3966 (1996).

[109] Z. Ikonic et al., *Mater. Sci. Eng.* **B89**, 84 (2002).

[110] D.C. Look et al., *J. Appl. Phys.* **71**, 260 (1991).

[111] T. Kubis et al., *Phys. Rev. B* **79**, 195323-1 (2009).

[112] A. Valavanis et al., *Phys. Rev. B* **77**, 075312-1 (2008).

[113] S. Tsujino et al., *Appl. Phys. Lett.* **86**, 062113-1 (2005).

[114] R.A. Kaindl et al., *Phys. Rev. Lett.* **86** (6), 1122 (2001).

[115] C. Rulliere (Ed.), *Femtosecond Laser Pulses*, Springer (1998).

[116] I. Bormann et al., *Physica E* **21**, 779 (2004).

[117] C. R. Pidgeon et al., *Semicond. Sci. Technol.* **20**, L50 (2005).

[118] W. Heiss et al., *Appl. Phys. Lett.* **66** (24), 3313 (1995).

[119] J. Shah *Ultrafast spectroscopy of semiconductors and semiconductor nanostructures*, Springer, 1999.

[120] C. R. Pidgeon et al., *Physica E* **13**, 904 (2002).

[121] P. Boucaud et al., *Appl. Phys. Lett.* **69** (20), 3069 (1996).

[122] R. A. Kaindl et al., *Physica B* **314**, 255 (2002).

[123] P. Rauter et al., *Appl. Phys. Lett.* **89**, 211111 (2006).

[124] P. Rauter et al., *New J. Phys.* **9** (5), 128 (2007).

[125] P. G. O'Shea et al., *Science* **292**, 1853 (2001).

[126] J. Shah, *Ultrafast Spectroscopy of Semiconductors and Semiconductor Nanostructures*, 2nd edition, Springer (1999).

[127] A. Yariv, *Quantum Electronics*, Wiley (1989).

[128] K. Driscoll et al., *Appl. Phys. Lett.* **89**, 191110-1 (2006).

[129] X. Sun et al., *Appl. Phys. Lett.* **95**, 011911-1 (2009).

[130] J. Liu et al., *Opt. Lett.* **34**, 1738 (2009).

[131] J. Liu et al., *Opt. Lett.*, accepted (2010).

[132] P. Rauter et al., *Appl. Phys. Lett.* **94**, 081115 (2009).

[133] P. Rauter et al., *Phys. Rev. Lett.* **102**, 147401 (2009).

List of publications

Reviewed Articles

P. Rauter, T. Fromherz, N.Q. Vinh, B.N. Murdin, G. Mussler, D. Grützmacher, G. Bauer:
"Continuous voltage tunability of intersubband relaxation times in coupled SiGe quantum well structures using ultrafast spectroscopy",
Phys. Rev. Lett. **102** (14), 147401 (2009).

P. Rauter, T. Fromherz, C. Falub, D. Grützmacher, G. Bauer:
"SiGe quantum well infrared photodetectors on pseudosubstrate",
Appl. Phys. Lett. **94**, 081115 (2009).

P. Rauter, T. Fromherz, S. Winnerl, M. Zier, A. Kolitsch, M. Helm, G. Bauer:
"Terahertz Si:B blocked-impurity-band detectors defined by nonepitaxial methods",
Appl. Phys. Lett. **93** (26), 261104 (2008).

J. Kasberger, P. Rauter, B. Jakoby:
"Wavelength Selectivity of a Thermal IRAbsorber as Part of a Fully Integrated IRAbsorption Sensor",
Proc. IEEE Sensors Conference 2008, 996 (2008).

P. Rauter, T. Fromherz, G. Bauer, N.Q. Vinh, B. Murdin, J.P. Phillips, C.R. Pidgeon, L. Diehl, G. Dehlinger, D. Grützmacher, Ming Zhao, Wei-Xin Ni:
"Direct determination of ultrafast intersubband hole relaxation times in voltage biased SiGe quantum wells by a density matrix interpretation of femtosecond resolved photocurrent experiments",
New J. Phys. **9** (5), 128 (2007).

P. Rauter, T. Fromherz, G. Bauer, N.Q. Vinh, B. Murdin, J.P. Phillips, C.R. Pidgeon, L. Diehl, G. Dehlinger, D. Grützmacher:
"Direct monitoring of the excited state population in biased SiGe valence band quantum wells by femtosecond resolved photocurrent experiments",
Appl. Phys. Lett. **89** (21), 211111-1-3 (2006).

M. Grydlik, P. Rauter, T. Fromherz, G. Bauer, C. Falub, D. Grützmacher, G. Isella:
"Resonator fabrication for cavity enhanced, tunable Si/Ge quantum cascade detectors",
Physica E **32** (1-2), 313 (2006).

P. Rauter, T. Fromherz, G. Bauer, L. Diehl, G. Dehlinger, H. Sigg, D. Grützmacher, H. Schneider:
"Voltage tuneable, two-band MIR detection based on Si/SiGe quantum cascade injector structures",
Appl. Phys. Lett. **83** (19), 3879 (2003).

Talks

P. Rauter, T. Fromherz, G. Bauer, N.Q. Vinh, B. Murdin, G. Mussler, D. Grützmacher:
"Control of the QW subband relaxation time by an applied electric field",
10th International Conference on Intersubband Transitions in Quantum Wells, Montreal, Canada, September 6-11, 2009.

P. Rauter:
"Voltage-Tuning of Intersubband Relaxation Times in SiGe",
Annual Meeting of the Austrian Physical Society 2009, Innsbruck, Austria, September 2-4, 2009.

P. Rauter, T. Fromherz, G. Bauer, S. Winnerl, M. Zier, A. Kolitsch, M. Helm:
"Terahertz Si:B blocked-impurity-band detectors by ion implantation",
Annual Meeting of the Austrian Physical Society 2009, Innsbruck, Austria, September 2-4, 2009.

P. Rauter, T. Fromherz, G. Bauer, N.Q. Vinh, B. Murdin, G. Mussler, D. Grützmacher:
"Continuous Voltage Tunability of Intersubband Relaxation Times in SiGe Quantum Well Structures",
29th International Conference on the Physics of Semiconductors, Rio de Janeiro, Brazil, July 27 - August 1, 2008.

P. Rauter, T. Fromherz, G. Bauer, N.Q. Vinh, B. Murdin, J.P. Phillips, C.R. Pidgeon, L. Diehl, G. Dehlinger, D. Grützmacher:
"Direct Monitoring of the Excited State Population of Optical Transitions in SiGe Quantum Well Structures by Femtosecond Resolved Pump-Pump Photocurrent Spectroscopy",
7th International Conference on Physics of Light-Matter Coupling in Nanostructures, Havana, Cuba, April 12-17, 2007.

T. Fromherz, M. Brehm, P. Rauter, G. Bauer, N.Q. Vinh, B. Murdin, J.P. Phillips, C.R. Pidgeon, Z. Zhong, G. Chen, J. Novak, J. Stangl, D. Grützmacher:
"Optoelectronic Properties and Bandstructure of SiGe Quantum Dot and Cascade Structures",
2006 MSR Fall Meeting, Boston, USA, November 27 - December 1, 2006 (invited).

P. Rauter, T. Fromherz, G. Bauer, N.Q. Vinh, B. Murdin, J.P. Phillips, C.R. Pidgeon, L. Diehl, G. Dehlinger, D. Grützmacher:
"Direct Measurement of Optical Phonon Limited Hole Intersubband Lifetimes in SiGe Quantum Cascade Structures",
28th International Conference on the Physics of Semiconductors, Vienna, Austria, July 24-28, 2006.

P. Rauter, T. Fromherz, G. Bauer, N.Q. Vinh, B. Murdin, J.P. Phillips, C.R. Pidgeon, L. Diehl, G. Dehlinger, D. Grützmacher:
"Optical Phonon Limited Hole Lifetime in a SiGe Cascade Emitter Structure: A Direct Measurement via Photocurrent Pump-Pump Measurements",
Physics of Intersubband Semiconductor Emitters Summer School, Cortona, Italy, June 25-30, 2006.

P. Rauter T. Fromherz, G. Bauer, N.Q. Vinh, B. Murdin, J.P. Phillips, C.R. Pidgeon, L. Diehl, G. Dehlinger, D. Grützmacher:
"Direct Measurement of HH2-HH1 Intersubband Lifetimes in SiGe Quantum Cascade Structures",
3rd International SiGe Technology and Device Meeting, Princeton, USA, May 15-16, 2006.

P. Rauter, T. Fromherz, M. Grydlik, C. Falub, G. Dehling, H. Sigg, D. Grützmacher, G. Bauer:
"SiGe Quantum Cascade Structures for Voltage-Tunable Mid-Infrared Detection",
Nanoforum 2005, Linz, Austria, 2005.

T. Fromherz, M. Grydlik, P. Rauter, M. Meduna, C. Falub, L. Diehl, G. Dehlinger, H. Sigg, D. Grützmacher, H. Schneider, G. Bauer:
"Si/SiGe QWIPs for voltage-tunable, resonator-enhanced, two-colour detection in the MIR",
3rd International Workshop on Quantum Well Infrared Photodetectors, Kananaskis, Canada, August 7-13, 2004.

T. Fromherz, P. Rauter, G. Bauer, L. Diehl, G. Dehlinger, H. Sigg, D. Grützmacher:
"Voltage-tuneable QWIPs based on Si/SiGe cascade injector structures for two-colour detection",
7th International Conference on Intersubband Transition in Quantum Wells (ITQW2003), Evoléne, Switzerland, 1-5 September, 2003.

P. Rauter, T. Fromherz, G. Bauer, L. Diehl, G. Dehlinger, H. Sigg, D. Grützmacher:
"Intraband absorption and photocurrent spectroscopy of Si/SiGe quantum cascades",
4th International Conference on Low Dimensional Structures and Devices (LDSD), Fortaleza, Brazil, December 8-13, 2002.

Posters

P. Rauter, T. Fromherz, G. Bauer, S. Winnerl, M. Zier, A. Kolitsch, M. Helm:
"Terahertz Si:B blocked-impurity-band detectors by ion-implantation",
14th International Conference on Modulated Semiconductor Structures, Kobe, Japan, July 19-24, 2009.

P. Rauter, T. Fromherz, G. Bauer, N.Q. Vinh, B. Murdin, G. Mussler, D. Grützmacher:
"Bias-induced relaxation-time manipulation in SiGe quantum well structures",
14th International Conference on Modulated Semiconductor Structures, Kobe, Japan, July 19-24, 2009.

P. Rauter, T. Fromherz, G. Bauer, N.Q. Vinh, G. Mussler, D. Grützmacher:
"Voltage Tunability of Intersubband Lifetimes in SiGe Quantum Well Structures",
15th International Winterschool on New Developments in Solid State Physics, Bad Hofgastein, Austria, February 18-22, 2008.

J. Kasberger, P. Rauter, B. Jakoby:
"Wavelength selectivity of a thermal IR-absorber as part of a fully integrated IR-absorption sensor",
IEEE Sensors 2008, Lecce, Italy, October 26-29, 2008.

P. Rauter, T. Fromherz, G. Bauer, N.Q. Vinh, B. Murdin, J.P. Phillips, C.R. Pidgeon, L. Diehl, G. Dehlinger, D. Grützmacher:
"Direct Monitoring of the Excited State Population of Holes in Voltage Biased SiGe Quantum Wells by a Density Matrix Interpretation of Femtosecond Resolved Photocurrent",
13th International Conference on Modulated Semiconductor Structures, Genova, Italy, July 15-20, 2007.

M. Grydlik, P. Rauter, T. Fromherz, C. Falub, G. Isella, D. Grützmacher, G. Bauer L. Diehl, G. Dehlinger, E. Müller, H. Sigg, H. Schneider, D. Grützmacher:
"Membrane Based Resonators for Enhancing Detectivity of SiGe QWIPs",
13th International Conference on Modulated Semiconductor Structures, Genova, Italy, July 15-20, 2007.

M. Grydlik, P. Rauter T. Fromherz, C. Falub, G. Isella, D. Grützmacher, G. Bauer:
"Resonator Fabrication for Switchable Two-Color MIR Detectors Based on p-type SiGe Quantum Cascade Injectors",
28th International Conference on the Physics of Semiconductors, Vienna, Austria, July 24-28, 2006.

P. Rauter T. Fromherz, G. Bauer, N.Q. Vinh, B. Murdin, J.P. Phillips, C.R. Pidgeon, L. Diehl, G. Dehlinger, D. Grützmacher:
"Direct Measurement of Optical Phonon Limited HH2-HH1 Intersubband Lifetimes in SiGe Quantum Cascade Structures",
14th International Winterschool on New Developments in Solid State Physics, Mauterndorf, Austria, 2006.

M. Grydlik, P. Rauter T. Fromherz, C. Falub, G. Isella, D. Grützmacher, G. Bauer:
"Resonator fabrication for switchable two-color MIR detection based on SiGe quantum cascade infrared photodetectors",
12th International Conference on Modulated Semiconductor Structures, Albuquerque, USA, July 10-15, 2005.

T. Fromherz, P. Rauter, G. Bauer, L. Diehl, G. Dehlinger, H. Sigg, D. Grützmacher:
"Si/SiGe cascade injector structures for voltage-tuneable two-colour detection in the 3 - 6 μm spectral range",
11th International Conference on Modulated Semiconductor Structures (MSS11), Nara, Japan, July 14-18, 2003.

Acknowledgements

The final pages of this thesis are devoted to all the people to whom I owe more than my deepest gratitude for accompanying me during my thesis and for making this work possible, more productive or just more cheerful by sharing their knowledge, their time, their technical skills, their enthusiasm or their friendship with me. My devotion is not only restricted to those that are mentioned below, but is owed to each member of this institute and to all of my friends and colleagues. Thanks a lot, in particular to:

- Prof. Günther Bauer for supporting me in any possible and impossible means, for guiding me on the way into the scientific community, for sharing his time and his seemingly infinite source of knowledge, and for enriching all of this with a sense of humor among the finest I have come across.

- Thomas Fromherz for his dedicated supervision, for always finding time to join me in the lab in order to share the treasure of his experimental experience, for teaming up with me during endless shifts on FEL beamtime, for countless and revealing theoretical discussions, and for lots of fun during the exploration of numerous conference venues.

- Detlev Grützmacher, Gregor Mussler, Claudio Falub and Gabriel Dehlinger for the growth of excellent samples, for the close and uncomplicated cooperation beyond the scope of projects, and for the 'ultrashort delay' between the outgoing sample design and the incoming physical sample.

- Nguyen Q. Vinh, Jonathan P. Philips and Britta Redlich for the excellent operation of the FEL as well as the support and hospitality during our beam time visits at the FOM Rijnhuizen.

- Prof. Manfred Helm, Stefan Winnerl, Michael Zier and Andreas Kolitsch for simulating and performing the ion-implantation required for the BIB fabrication. Prof. Manfred Helm and Stefan Winnerl for their support in the course of both the work on BIB devices and the time-resolved experiments, and for their hospitality during our stay at the FZ Dresden-Rossendorf.

- Prof. Ben N. Murdin for introducing us to the free-electron-laser FELIX and for his support in the course of the time-resolved experiments.

- Prof. Carl R. Pidgeon for numerous fruitful discussions.

- Alma Halilovic, Ursula Kainz and Stefan Bräuer and Ernst Vorhauer for their commitment to the ungrateful and invaluable job of maintaining the technical facilities, which form the basis of this institute.

- Eugen Wintersberger for the Sysiphus-like administration of the computer system and for conceptualizing complex IT issues in a colorful and picturesque language.

- Friedrich Binder for the technical support.

- Susanne Lechner and Alexandra Stangl for the guidance through the jungle of administration and for the always warm welcome in their office.

- My parents, my sister and grandparents for their love and support. My grandfather, who unfortunately did not live to see this work, for his proud confidence in my work.

- My girlfriend Veronika for patiently balancing my life during the demanding process of writing this thesis, for showing me a more pink though equally accurate view on difficulties I encountered during my work, and for her unconditional love and limitless support.

Die VDM Verlagsservicegesellschaft sucht für wissenschaftliche Verlage abgeschlossene und herausragende

Dissertationen, Habilitationen, Diplomarbeiten, Master Theses, Magisterarbeiten usw.

für die kostenlose Publikation als Fachbuch.

Sie verfügen über eine Arbeit, die hohen inhaltlichen und formalen Ansprüchen genügt, und haben Interesse an einer honorarvergüteten Publikation?

Dann senden Sie bitte erste Informationen über sich und Ihre Arbeit per Email an *info@vdm-vsg.de*.

Sie erhalten kurzfristig unser Feedback!

VDM Verlagsservicegesellschaft mbH
Dudweiler Landstr. 99 Telefon +49 681 3720 174
D - 66123 Saarbrücken Fax   +49 681 3720 1749
www.vdm-vsg.de

Die VDM Verlagsservicegesellschaft mbH vertritt

Printed by Books on Demand GmbH, Norderstedt / Germany